The IMA Volumes in Mathematics and its Applications

Volume 121

Series Editor
Willard Miller, Jr.

Springer
New York
Berlin
Heidelberg
Barcelona
Hong Kong
London
Milan
Paris
Singapore
Tokyo

Institute for Mathematics and its Applications IMA

The **Institute for Mathematics and its Applications** was established by a grant from the National Science Foundation to the University of Minnesota in 1982. The IMA seeks to encourage the development and study of fresh mathematical concepts and questions of concern to the other sciences by bringing together mathematicians and scientists from diverse fields in an atmosphere that will stimulate discussion and collaboration.

The IMA Volumes are intended to involve the broader scientific community in this process.

Willard Miller, Jr., Professor and Director

* * * * * * * * * * *

IMA ANNUAL PROGRAMS

1982–1983	Statistical and Continuum Approaches to Phase Transition
1983–1984	Mathematical Models for the Economics of Decentralized Resource Allocation
1984–1985	Continuum Physics and Partial Differential Equations
1985–1986	Stochastic Differential Equations and Their Applications
1986–1987	Scientific Computation
1987–1988	Applied Combinatorics
1988–1989	Nonlinear Waves
1989–1990	Dynamical Systems and Their Applications
1990–1991	Phase Transitions and Free Boundaries
1991–1992	Applied Linear Algebra
1992–1993	Control Theory and its Applications
1993–1994	Emerging Applications of Probability
1994–1995	Waves and Scattering
1995–1996	Mathematical Methods in Material Science
1996–1997	Mathematics of High Performance Computing
1997–1998	Emerging Applications of Dynamical Systems
1998–1999	Mathematics in Biology
1999–2000	Reactive Flows and Transport Phenomena
2000–2001	Mathematics in Multimedia
2001–2002	Mathematics in the Geosciences
2002–2003	Optimization

Continued at the back

Philip K. Maini Hans G. Othmer
Editors

Mathematical Models for Biological Pattern Formation

With 166 Illustrations

Springer

Philip K. Maini
Mathematical Institute
University of Oxford
Oxford, OX1 3LB
UK
maini@maths.ox.ac.uk

Hans G. Othmer
School of Mathematics
University of Minnesota
Minneapolis, MN 55455
USA
othmer@math.umn.edu

Series Editor:
Willard Miller, Jr.
Institute for Mathematics and its Applications
University of Minnesota
Minneapolis, MN 55455, USA

Mathematics Subject Classification (2000): 92C10, 92C15, 92C17, 92E10

Library of Congress Cataloging-in-Publication Data
Mathematical models for biological pattern formation : frontiers in biological mathematics / [edited by] Philip K. Maini, Hans G. Othmer.
p. cm. — (The IMA volumes in mathematics and its applications ; v. 121)
Includes bibliographical references (p.).
ISBN 0-387-95103-2 (alk. paper)
1. Pattern formation (Biology)—Mathematical models. I. Maini, Philip K. II. Othmer, H.G. (Hans G.), 1943- III. Series.
QH491 .M29 2000
570′.5′118—dc21 00-044018

Printed on acid-free paper.

Production managed by A. Orrantia; manufacturing supervised by Joe Quatela.
Camera-ready copy prepared by the IMA.
Printed and bound by Sheridan Books, Inc., Ann Arbor, MI.
Printed in the United States of America.

9 8 7 6 5 4 3 2 1

ISBN 0-387-95103-2 SPIN 10774025

Springer-Verlag New York Berlin Heidelberg
A member of BertelsmannSpringer Science+Business Media GmbH

FOREWORD

This 121st IMA volume, entitled

MATHEMATICAL MODELS FOR BIOLOGICAL PATTERN FORMATION

is the first of a new series called *FRONTIERS IN APPLICATION OF MATHEMATICS*. The *FRONTIERS* volumes are motivated by IMA programs and workshops, but are specially planned and written to provide an entrée to and assessment of exciting new areas for the application of mathematical tools and analysis. The emphasis in *FRONTIERS* volumes is on surveys, exposition and outlook, to attract more mathematicians and other scientists to the study of these areas and to focus efforts on the most important issues, rather than papers on the most recent research results aimed at an audience of specialists.

The present volume of peer-reviewed papers grew out of the 1998-99 IMA program on "Mathematics in Biology," in particular the Fall 1998 emphasis on "Theoretical Problems in Developmental Biology and Immunology." During that period there were two workshops on Pattern Formation and Morphogenesis, organized by Professors Murray, Maini and Othmer. James Murray was one of the principal organizers for the entire year program.

I am very grateful to James Murray for providing an introduction, and to Philip Maini and Hans Othmer for their excellent work in planning and preparing this first *FRONTIERS* volume.

I also take this opportunity to thank the National Science Foundation, whose financial support of the IMA made the Mathematics in Biology program possible.

Willard Miller, Jr., Professor and Director

The editors are pleased to dedicate this volume to Professor James D. Murray, affectionately known as Jim to his friends. Jim has been a leader in the mathematical analysis of biological pattern formation for 25 years, and has influenced it dramatically by his unbending insistence that the problem is first and foremost a biological one, and therefore the biological details do really matter. The Centre for Mathematical Biology at Oxford University, which he founded in 1983, has been a magnet and haven for mathematicians who were interested in the many aspects of biological pattern formation, and its success is in no small part due to Jim's warmth and kindness to all, and his strong support of young researchers.

We wish Jim, and his soulmate Sheila, the best in the coming years.

Philip K. Maini
Hans G. Othmer

CONTENTS

BIOLOGICAL PATTERN FORMATION - A MARRIAGE OF THEORY AND EXPERIMENT

J.D. MURRAY*

Abstract. The interdisciplinary challenges to discover the underlying mechanisms in the generation of biological pattern and form are central issues in development. Here I briefly discuss a philosophy of such an integrative biology approach. I then describe, by way of example, the successful use of a very simple model —even linear — for the growth of brain tumours in an anatomically accurate brain. All of the model parameters are estimated from experiment and patient data. Even with such a basic model the results highlight the inadequacies of current medical intervention treatment of brain tumours. I conclude with some brief general views on the use of models in biology.

Although the biomedical world is in the throes of the genetic revolution the basic question which genes do not address is the development of spatio-temporal pattern and form, whether it is the growth of a tumour or the development of stripes on a fish. During the past 20 some years a large amount of research in mathematical biology, or biomathematics or whatever name is given to the application of mathematics to the biomedical sciences, has been devoted to trying to increase our understanding of the underlying biological processes involved in pattern formation processes. The relatively few people working in the field in the 1970's has blossomed into the several thousand who are now actively involved in modelling a vast and ever widening spectrum of biomedical problems. The collection of papers in this volume demonstrate not only how powerful such mathematical models can be, but how far the field has come in even just the past 10 years. Although we still do not know the complete detailed mechanism involved in any specific situation I am optimistic that we are approaching the situation when we shall. Several of the theoretical studies of pattern formation paradigms, such as the organisation of social amoebae like *Dictyostelium discoideum* and bacterial patterns [19, 21], have resulted in major advances in our understanding and guided the direction of illuminating experimental programmes. What is, I feel, indisputable, is that major progress has come about by genuine interaction between the theoreticians and the experimentalists. Gone are the days when papers in which functions describing blood cell density as being "imbedded in some appropriate Banach space" with statements such as "this will be of great interest to cardiologists" tagged at the end of a paper replete with theorems and lemmas with as much relevance to biological pattern formation as the length of the the latest pop singer's earing. The articles in this collection deal with real problems and, irrespective of the mathematical sophistication involved in the model analyses, relate directly to real biological problems and importantly increase our basic understanding of the biological processes.

*Department of Applied Mathematics, Box 352420, University of Washingon, Seattle, Washington 98195-2420.

Although it is generally accepted, it should perhaps be stressed again that mathematical descriptions of patterning phenomena are not explanations. One of the principal uses of any theory is in its biological predictions. From a theoretical point of view, the art of good modelling in biology relies not only on sufficient mathematical expertise (often not at all sophisticated), but also on: (i) a sound understanding and appreciation of the biological problem; (ii) a realistic mathematical representation of the important biological phenomena; and (iii) a biological interpretation of the mathematical analysis and results in terms of insights and predictions. Scientifically relevant mathematical or theoretical biology is unquestionably an interdisciplinary science *par excellence.*

An important point arising from theoretical models is that any pattern contains its own history. Consider a simple engineering analogy of our role in trying to understand a biological process [14]. It is one thing to suggest that a bridge requires a thousand tons of steel, that any less will result in too weak a structure, and any more will result in excessive rigidity. It is quite another matter to instruct the workers on how best to put the pieces together. In morphogenesis, for example, it is conceivable that the cells involved in tissue formation and deformation have enough expertise that given the right set of ingredients and initial instructions they could be persuaded to construct whatever element one wants. This is the hope of many who are searching for a full and predictive understanding. However, it seems very likely that the global effect of all this sophisticated cellular activity would be critically sensitive to the sequence of events occurring during development. As scientists we should concern ourselves with how to take advantage of the limited opportunities we have for communicating with the workforce so as to direct experiment towards an acceptable end-product. This is perhaps a little philosophical, but even a cursory look at many theories in the literature reveal a fixation on simplistic explanations. On the other hand, in situations which frequently arise especially in medical problems, the complexity is such that if we wish to be useful we often have to start with what is clearly an oversimplistic scenario and build into the models progressively more realism as we discover more about the problem. There are certainly no ground rules as to how complex or simple a model has to be to be useful.

None of the individual models that have been suggested for any biological patterning process, and not even all of them put together, could be considered a complete model. In the case of some of the widely studied problems (such as *Dictyiostelium discoideum*), each model has shed light on different aspects of the process and we can now say what the most important conceptual elements have to be in a complete model. These studies have served to highlight where our knowledge is deficient and to suggest directions in which fruitful experimentation might lead us. Indeed, a critical test of these theoretical constructs is in their impact on the experimental community.

Since the articles in this volume are primarily concerned with biological as opposed to medical spatial problems, it is perhaps appropriate to briefly describe a particularly simple model for the highly complex and poorly understood problem of the growth of human brain tumours (glioblastomas). The fast pace of medical discoveries, real and spurious, is a fruitful field for genuine integrative interdisciplinary research. Some of these discoveries bring new uses for extant theories. For example, the recent experimental work on the importance of anti-angiogenetic drug [8, 1] for the control of tumours first suggested by Judah Folkman in the 1970's [6, 7] has brought the developmental problem of the mechanisms that could be involved in angiogenesis to the fore [13, 15]: without angiogenesis the tumour cannot grow.

1. A simple mathematical model for virtual brain tumours (gliomas) – enhancing medical imaging. Gliomas are particularly nasty brain tumours that diffuse aggressively, thereby invading the surrounding normal tissue. That the spatial spread involves diffusion is now fairly generally accepted. Although other processes are probably involved, diffusion and cell mitosis play major (arguably the major) roles in the spread of cancer cells. Being a diffusion process there is a long tail where the cell density is extremely low. There is clearly a threshold level below which even the most sophisticated scans cannot detect in spite of the continuing development of medical imaging such as enhanced computerized tomography (CT) and magnetic resonance imaging (MRI). At least one inadequacy of current medical imaging is that even extensive surgical resection or local irradiation of gliomas, based on where the tumour "boundary" is as defined by the scans, is followed by tumour recurrence at or near the edge of the excised tumour [12].

In an attempt to try and get some understanding of the growth of such tumours, Dr. Elsworth Alvord MD (Pathology, Health Science, University of Washington), myself and several of my graduate students and post-doctorals over the past six years have looked at some very simple diffusion models to try and obtain some quantitative estimates of brain tumour growth, both with and without medical intervention [18, 5, 20, 2]. Perhaps the most damning demonstration of the inadequacies of current medical treatment has been given by the work [17] with Dr. Alvord and a former student, Dr. Kristin Swanson. We started with a basic diffusion model for the cancer cells involving exponential cell growth (justified by the data on such tumours). The major difference to previous work along these lines is that the diffusion was simulated within anatomically accurate heterogeneous brain tissue in three spatial dimensions. The work will be reported in detail elsewhere. Here I give only a brief sketch of the model and results, since it highlights the above point that even simple models can be clinically useful.

The availability of the BrainWeb [3] brain atlas database let us define the gross anatomical boundaries and to vary the degree of motility of glioma cells in grey or white matter in heterogeneous, anatomically accurate brain tissue. Glioma cells are reported to migrate more rapidly in white matter than in grey matter [9] so we allow the motility coefficient to differ depending on the local tissue composition.

Our mathematical model for glioma growth and invasion, including the differential motility of gliomas in grey and white matter, can be written as

$$\frac{\partial c}{\partial t} = \nabla \cdot \big(D(\mathbf{x})\nabla c\big) + \rho c \,, \tag{1}$$

where $c(\mathbf{x}, t)$ is the concentration of tumour cells at position $\mathbf{x}$ and time t. $D(\mathbf{x})$, a function of position $\mathbf{x}$ in the brain, is the diffusion coefficient defining the random motility of the glioma cells with $D(\mathbf{x}) = D_g, D_w$, constants for $\mathbf{x}$ in grey and white matter, respectively. ρ represents the net proliferation rate of the glioma cells. The diffusion coefficient in white matter is larger than that in grey, so $D_w > D_g$. The difference in the diffusion coefficients has been estimated to range from 2 to 100 fold [17], but we chose 5 as an arbitrary first approximation to illustrate the model's potential. To complete the model formulation, we required zero flux of cells across the brain boundaries and assumed that the tumour had grown to about 4,000 cells as a local mass before it began to diffuse and the model equation (1) applies.

The BrainWeb lets us simulate the growth of a virtual glioma in any of the 3 standard planes (coronal, sagittal and axial or horizontal) to demonstrate a pseudo-3-dimensional tumour. (The numerical simulation was a challenging problem.)

For every current medical imaging technique there is a threshold of detection below which gliomas cells are not detectable. Even microscopy has a limit beyond which individual cells cannot be detected.

Survival time. Previous models assumed that diagnosis is made when the volume of an enhanced CT-detectable tumour has reached a size equivalent to a sphere of an average 3 cm diameter, and that death occurs when the volume reaches an average 6 cm diameter. The difference between these two times can be defined as the survival time of the hypothetical or virtual patient. With earlier models, and even simpler brain structure, the comparison of calculated survival times [20] with extant data [11] was very good.

Crucial to all successful modelling, particularly those which give rise to simple models which have fewer parameters, is the ability to determine reasonable estimates of the critical parameters, here the growth rate ρ and the diffusion coefficient D. For high-grade gliomas (glioblastomas) previous estimates, based on extant data, have suggested a net proliferation rate of $\rho \approx 0.012/\text{day}$ [20, 2, 17, 4], corresponding to a volume-doubling time of 60 days, and a diffusion coefficient of $D \approx .0013\ \mathrm{cm}^2/\mathrm{day}$ [2, 17]. The actual ranges of these values are quite extreme but real values for any actual patient could be substituted if they could be measured.

Figure 1 shows three perpendicular cross-sections (coronal, sagittal and horizontal or axial) of the virtual human brain intersecting in a point marked by an asterisk in the superior frontal region where the virtual tumour originates. The grey and white matters of the brain domain appear grey and white, respectively, A contour plot of the tumour cell density is represented in color with red denoting a high density and blue a low density. In each image, a single thick black curve defines the edge of the tumour that the model suggests would be detectable on enhanced CT scan associated with a threshold of detection of 8000 cells/mm^3. The outermost light blue profile corresponds to an arbitrary threshold of detection 80 times more sensitive than enhanced CT (that is 100 cells/mm^3). The left column of images in Figure 1 represents the tumour at the time of detection, defined as an enhanced CT-detectable tumour with average diameter of 3 cm, while the right column represents the tumour at the time of death, defined by an enhanced CT-detectable tumour with average diameter of 6 cm. With our model it is possible to simulate the growth of a tumour starting at any point we wish.

What is abundantly clear from the figure is how far tumour cells have diffused beyond any current range of detection. It is also clear why surgical resection is so difficult and ineffectual since the tumour "boundary" is so diffuse. Even resecting a significant distance outside the detectable tumour fails to excise all the tumour cells. Previous studies of the motility of gliomas have demonstrated that diffusion is an accurate estimation for the method of spread of gliomas [17, 20]. A consequence of modelling cellular motility by Fickian or gradient-driven diffusion, is the lack of a definitive interface between malignant and normal tissue. This mathematical consequence is correlated with the actual biology of human gliomas. Consider using CT-images, or other visual detection procedures, to delineate the possible interface between cancerous and normal tissue. Radical excision of the tumour even well beyond these interfaces has been shown to fail in numerous studies as summarized by [16]. Clearly tumour cells invade peripheral to the CT or MRI defined boundaries of the tumour. Even standard histopathological analysis, one of our most sensitive means of detecting glioma cells, fails in locating all of the tumour cells.

Because of the diffuse nature of gliomas there is no clear boundary defining the interface of pathological and normal tissue, even though many attempts have been made to suggest that a boundary exists. Figures 1 shows the spatio-temporal invasion of virtual gliomas at the time of diagnosis and death. These simulations clearly reveal the subthreshold invasion of the tumour well beyond the detectable portion of the tumour. No matter the extent of resection, the mathematical model indicates that the gross tumour will ultimately recur and kill (see also [20]).

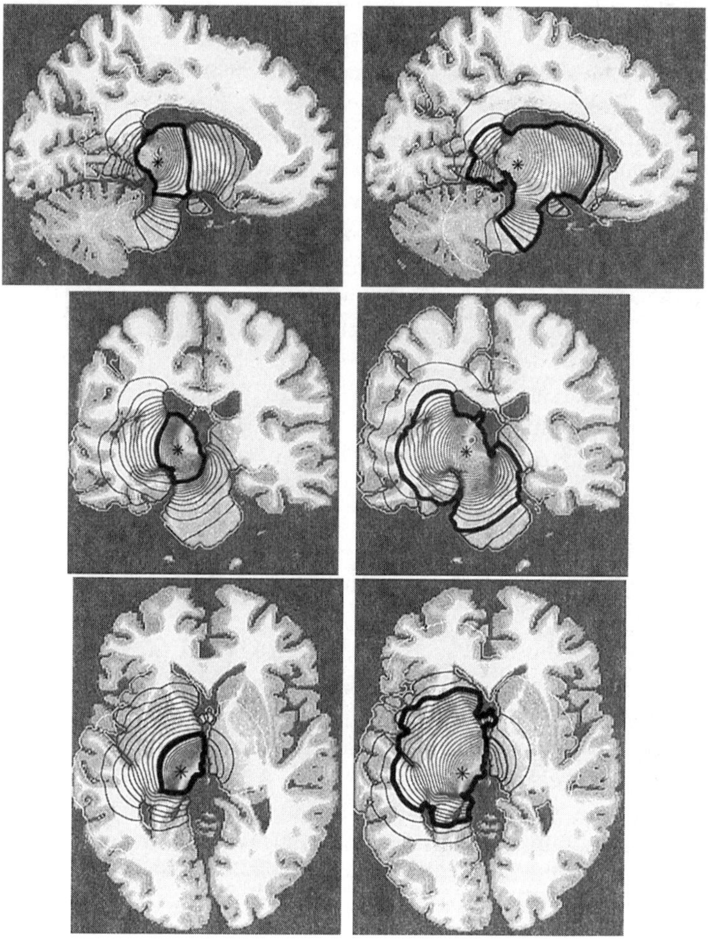

FIG. 1. *Sections of the virtual human brain in sagittal, coronal and horizontal planes that intersect at the site of the glioma originating in the superior frontal region denoted by an asterisk (*). Red denotes a high density of tumour cells while blue denotes a low density. A thick black contour defines the edge of the tumour detectable by enhanced computerized tomography (CT). Cell migration was allowed to occur in a truly three-dimensional solid representation of the brain.*

Unlike real patients with real gliomas, virtual patients with virtual gliomas can be analyzed by allowing any particular factor to vary while keeping all the other determining factors constant. Such isolation techniques, of course, require a mathematical model that has sufficient realism and involves the major variables and parameters. The recent availability of simulated MRI's, with proportions of grey and white matter accurately indicated, let us develop this model which is sufficiently complex to allow different diffusion rates in grey and white matter (for example, a 5-fold increase in diffusion or migration in white matter) as well as to prevent spread across certain parts of the brain.

The model is a simple one which focuses on only two key elements, namely diffusion and growth. Other variables can be introduced into the model as their relative importance is discovered. Previous studies [18, 20, 5] showed how to determine estimates for these parameters from patient scans. With these the present model can be depressingly predictive as to the where the tumour is likely to grow in real time. Of course many aspects, which can be included in more complex models, such as swelling and distortion of tissue should be included. The point of this brief discussion is to show how even a simple basic model can still be useful clinically. However, even without these other effects included what seems clear from these theoretical studies of virtual gliomas is that current imaging techniques are woefully inadequate for definitive clinical decisions as to what constitutes the optimal treatment for patients with gliomas.

2. General concluding remarks. Theoretical modelling has been proven to be useful in the study of a remarkably diverse spectrum of biological problems such as wound healing, quantifying disease control strategies, the effect of introducing genetically engineered organisms in the environment and suggesting experiments associated with limb development, to name just a few.

Pattern formation studies are sometimes criticized for their lack of inclusion of genes in the models. But then criticism can be levelled at any modelling abstraction of a complex system to a relativley simple one. It should be remembered that the generation of pattern and form, particularly in development, is usually a long way from the level of the genome. Of course genes play crucial roles in development, but they do not actually create patterns. Many of the evolving patterns could hardly have been anticipated solely by genetic information.

Why use mathematics to study something as intrinsically complicated and ill-understood as development, angiogenesis, wound healing, infectious disease dynamics, regulatory networks and so on? We suggest that mathematical modelling must be used if we ever hope to genuinely and realistically convert an understanding of the underlying mechanisms into a predictive science. Mathematics is required to bridge the gap between the level on which most of our knowledge is accumulating (cellular and below) and the macroscopic level of the patterns we see. A mathematical approach lets us explore the logic of pattern formation. Even if the mechanisms were well understood — and they certainly are far from it at this stage — mathematics would be required to explore the consequences of manipulating the various parameters associated with any particular scenario. In the case of such things as wound healing, tumour growth and it will be increasingly so in angiogenesis with the cancer connection, the number of options that are fast becoming available to wound and cancer managers will become overwhelming unless we can find a way to simulate particular treatment protocols before applying them in practice. The latter has already been of use in understanding the efficacy of various treatment scenarios with brain tumours [18, 20, 17] and new two step regimes for skin cancer [10].

There is no doubt that we are a long way from being able to reliably simulate actual developmental scenarios, notwithstanding the multitude of theories that abound. The active cellular control of key processes is poorly understood. Despite such limitations, we argue that exploring the logic of biological processes is worthwhile, in some current situations even essential in our present state of knowledge. It allows us to take an hypothetical mechanism and examine its consequences in the form of a mathematical model, make predictions and suggest experiments that would verify or invalidate the model; the latter is frequently biologically informative. In fact, the very process of constructing a mathematical model can be useful in its own right. Not only must one commit to a particular mechanism, one is also forced to consider what is truly essential to the process and what the key players are. We are thus involved in constructing frameworks on which we can hang our understanding. The equations, the mathematical analysis and the numerical simulations that follow serve to reveal quantitatively, as well as qualitatively, the consequences of that logical structure.

The best integrative biology studies have served to highlight where our knowledge is deficient and to suggest directions in which fruitful experimentation might lead us. A crucial aspect of this research is the interdisciplinary content and, as already mentioned, a crucial test of all theoretical models should be in their impact on the experimental community. The field of mathematical or theoretical biology or integrative biology has now achieved some level of maturity, and we believe that future dialogue between experimentalists and theoeticians will lead us more rapidly towards a fuller understanding, if not a complete one, of several biological processes involving pattern formation.

REFERENCES

[1] T. Boehm, J. Folkman, T. Browder, and M. O'Reilly. Antiangiogenesis therapy of experimental cancer does not induce acquired drug resistance. *Nature*, 404–407, 1997.

[2] P.K. Burgess, P.M. Kulesa, J.D. Murray, and E.C. Alvord, Jr. The interaction of growth rates and diffusion coefficients in a three-dimensional mathematical model of gliomas. *J Neuropathol and Exp Neurol*, **56**:704–713, 1997.

[3] D.L. Collins, A.P. Zijdenbos, V. Kollokian, J.G. Sled, N.J. Kabani, C.J. Holmes, and A.C. Evans. Design and construction of a realistic digital brain phantom. *IEEE Transactions on Medical Imaging*, **17**:463–468, 1998.

[4] V.P. Collins, R. K. Loeffler, and H. Tivey. Observations on growth rates of human tumors. *Am J Roentgenol Radium Ther Nucl Med*, **76**:988–1000, 1956.

[5] G.C. Cruywagen, D.E. Woodward, P. Tracqui, G.T. Bartoo, J.D. Murray, and E.C. Alvord, Jr. The modelling of diffusive tumours. *J Biological Systems*, **3**:937–945, 1995.

[6] J. Folkman. Anti-angiogenesis: New concept for therapy of solid tumors. *Annals of Surgery*, **75**:409–416, 1971.

[7] J. Folkman. Tumor angiogenesis: therapeutic implications. *New England Journal of Medicine*, **285**:1182–1186, 1972.

[8] J. Folkman. Angiogenesis in cancer, vascular, rheumatoid and other diseases. *Nature Medicine*, **1**:27–31, 1995.

[9] A. Giese, L. Kluwe, B. Laube, H. Meissner, M. Berens, and M. Westphal. Migration of human glioma cells on myelin. *Neurosurg*, **38**:755–764, 1996.

[10] T. Jackson, S.R. Lublin, N.O. Siemers, P.D. Senter, and J.D. Murray. Mathematical and experimental analysis of localization of anti-tumor antibody-enzyme conjugates. *British Journal of Cancer*, **80**:1747–1753, 1999.

[11] F.W. Kreth, P.C. Warnke, R. Scheremet, and C.B. Ostertag. Surgical resection and radiation therapy versus biopsy and radiation therapy in the treatment of glioblastoma multiforme. *J Neurosurg*, **78**:762–766, 1993.

[12] B.C. Liang and M. Weil. Locoregional approaches to therapy with gliomas as paradigm. *Curr. Opinion in Oncol.*, **10**:201–206, 1998.

[13] D. Manoussaki, S.R. Lubkin, R.B. Vernon, and J.D. Murray. A mechanical model for the formation of vascular networks *in vitro*. *Acta Biotheretica*, **44**:271–282, 1996.

[14] J.D. Murray, J. Cook, R. Tyson, and S.R. Lubkin. Spatial pattern formation in biology: I dermal wound healing. ii bacterial patterns. *Journal of the Franklin Institute*, **335**B:303–332, 1998.

[15] J.D. Murray, D. Manoussaki, S.R. Lubkin, and R.B. Vernon. A mechanical theory of *in vitro* vascular network formation. In C. Little, V. Mironov, and E. Helene Sage, editors, *Vascular Morphogenesis in vivo, in vitro, in mente*, pages 173–188. Birkhauser, Boston, 1998.

[16] J.M. Nazzaro and E.A. Neuwelt. The role of surgery in the management of supratentorial intermediate and high-grade astrocytomas in adults. *J. Neurosurg.*, **73**:331–344, 1990.

[17] K.R. Swanson. *Mathematical modeling of the growth and control of tumors*. PhD thesis, University of Washington, 1999.

[18] P. Tracqui, G.C. Cruywagen, D.E. Woodward, G.T. Bartoo, J.D. Murray, and Jr. E.C. Alvord. A mathematical model of glioma growth: the effect of chemotherapy on spatial-temporal growth. *Cell Poliferation*, **28**:17–31, 1995.

[19] R. Tyson, S.R. Lubkin, and J.D. Murray. A minimal mechanism for bacterial patterns. *Proc. Roy. Soc. Lond.*, pages 299–304, 1998.

[20] D.E. Woodward, J. Cook, P. Tracqui, G.C. Cruywagen, J.D. Murray, and Jr. E.C. Alvord. A mathematical model of glioma growth: the effect of extent of surgical resection. *Cell Prolif*, **29**:269–288, 1996.

[21] D.E. Woodward, R. Tyson, M.R. Myerscough, J.D. Murray, E.O. Budrene, and H.C. Berg. Spatio-temporal patterns generated by *Salmonella typhimurium*. *Biophys. J.*, **68**:2181–2189, 1995.

SPATIOTEMPORAL PATTERN FORMATION IN EARLY DEVELOPMENT: A REVIEW OF PRIMITIVE STREAK FORMATION AND SOMITOGENESIS

S. SCHNELL*, K.J. PAINTER†, P.K. MAINI* , AND H.G. OTHMER†

Abstract. The basic body plan of a number of vertebrates results from two processes that occur early in the development of the blastoderm: large scale rearrangements of tissue via a process called gastrulation, and axial subdivision of tissue in a process called somitogenesis. The first step of gastrulation in avians is formation of the primitive streak, which marks the first clear manifestation of the anterior-posterior axis. Cell movements that occur through the streak ultimately convert the single layered-blastoderm into a trilaminar blastoderm comprising prospective endodermal, mesodermal and ectodermal tissue. During streak formation a group of cells moves anteriorly as a coherent column from the posterior end of the blastoderm, and as it proceeds other cells stream over the lateral edges of the furrow left behind. The anterior end of the streak is a specialized structure called Hensen's node, which serves as an organizing center for later axis formation and determination of the left-right asymmetry of the body. Soon after the primitive streak forms, Hensen's node regresses towards the tail, leaving the notochord and a pair of segmental plates parallel to the primitive streak in its wake. The posterior end of the segmental plate moves down the cranio-caudal axis with the node, as more cells are added to it by cell division within the plate and by cells entering from the primitive streak. A pair of somites forms from the anterior ends of the two plates at regular intervals. Despite the fact that much is known about the basic biological processes, the mechanisms that underlie the formation of the primitive streak and somitogenesis are still unknown, and elucidating them is one of the major unsolved problems in developmental biology. Mathematical modelling has been a useful tool in this process, as it provides a framework in which to study the outcome of proposed interactions and can make experimentally testable predictions. In this paper we outline the biological background of these processes and review existing models of them.

Key words. Primitive streak formation, somitogenesis, theoretical models, mathematical models, *Hox* genes, *c-hairy-1*, Notch-Delta genes.

1. Introduction. Early vertebrate development is a complex process that involves cell division, cell-cell signaling, cell movement, and cell differentiation. Many adult vertebrates exhibit common structures, but the developmental processes that produce them may or may not be similar. For example, formation of a primitive streak is central to avian, reptilian and mammalian gastrulation, and while it is not present in amphibian blastulae, they contain an analogous structure, called the blastopore. On the other hand, somitogenesis is common to all vertebrates. This review focuses on experimental and theoretical aspects of primitive streak formation and somitogenesis in avian embryogenesis. The chick embryo is a widely-used model system for experimental studies and, as a result, there is a

*Centre for Mathematical Biology, Mathematical Institute, Oxford University, Oxford, OX1 3LB, UK.

†Department of Mathematics, University of Minnesota, Minneapolis, MN 55455, USA.

large amount of experimental data. We begin with a brief description of the early events: details of these events can be found in [35], [88], and [50].

The chick embryo develops from a small, disk-shaped blastodisc floating on top of the yolk. After the egg is fertilized cells divide repeatedly, forming a multicellular stratified structure called the blastoderm. The period from just prior to laying through several hours afterwards has been subdivided into 14 stages [31, 50]. Cell division is dominant during stages I - VI, and morphogenetic movements begin during stages VII-X, when cells of the central blastodisc, called the *area pellucida* (*cf.* Figure 1), separate from the yolk, producing a hollow region beneath the disc called the subgerminal cavity [75, 99]. Subsequently some cells from the central blastodisc move into the subgerminal cavity (either actively or passively), and simultaneously the disc expands radially over the yolk. The opaque marginal zone of the blastoderm, known as the *area opaca*, remains in contact with the yolk and may play an active role in the radial movement (Figure 1 A). The result is that during stages VII-X the central part of the disc changes from a layer 4-6 cells deep to a translucent layer one cell thick called the epiblast. The anterior-posterior axis of the embryo is also determined during these stages [50]. After stage X some cells within the marginal zone migrate posteriorly, and then leave the marginal zone at the posterior marginal zone (PMZ)(Figure 1 B). They spread across the subgerminal cavity beneath the epiblast as a loosely-connected sheet, incorporating islands of cells shed from the blastodisc earlier. By stage XIV this sheet connects with the anterior margin of the disc and forms the hypoblast, and at this stage the blastoderm is bi-layered with the epiblast and hypoblast separated by the blastocoel cavity. Fate maps for cell movements in these stages are available [39].

During hypoblast formation the embryonic shield or Koller's sickle develops at the posterior end of the epiblast (cf Figure 2(a)). This consists of a thickened epiblast [93] comprising primitive streak precursor cells that have migrated to this area by a series of 'polonnaise movements' [105]. The first visible sign of gastrulation is formation of the primitive streak, which arises from Koller's sickle at the posterior midline of the blastodisc [52] (Figure 1 C and D). The sickle narrows and the primitive streak moves anteriorly between the epiblast and the hypoblast. The tip of the ingressing streak moves $\sim 60\%$ of the way across the blastoderm before it stops, and later, regresses. At full primitive streak stage (Hamburger and Hamilton stage 4, [38]) the organizer of the avian embryo, Hensen's node, develops as a bulbous structure at the anterior tip of the streak. The period between the accumulation of cells at the posterior region and full primitive streak is approximately 12 hours. The structure of the blastoderm at this stage is illustrated in Figure 2(b). During the advance of the node, epiblast cells move through the streak and into the interior. Those that migrate through the node form anterior structures, those that migrate through the lateral parts of the primitive streak become endodermal and mesodermal cells, and

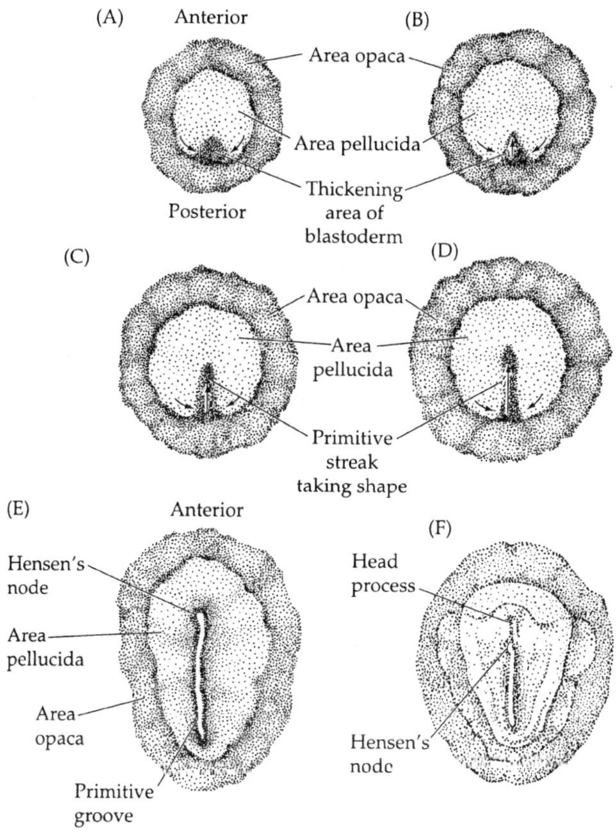

FIG. 1. *A schematic of the stages in early development of the chick embryo (A) 3-4 hours post-laying, (B) 5-6 hours, (C) 7-8 hours, (D) 10-12 hours, (E) 15-16 hours, (F) 19-22 hours, . (Reproduced with permission from [35])*

the remainder constitute the ectoderm. Simultaneously, the *area pellucida* changes from circular to pear-shaped, narrowing in the posterior portion. The head structure, notochord and somites are laid down during regression of the node, and when regression is complete the embryo is a flat trilaminar blastoderm comprising the ectodermal, mesodermal and endodermal layers. These will form various organs during subsequent morphogenesis, in addition to the structures formed during regression. The regressing node and anterior portion of the streak eventually form the tail bud [94]. Regression proceeds on a slower time scale than progression, taking approximately 24 hours for the node to regress after the streak reaches its maximum length of approximately 1.9 mm [94].

During regression of the primitive streak the neural folds begin to gather at the center of the embryo, and the segmental plates, which are

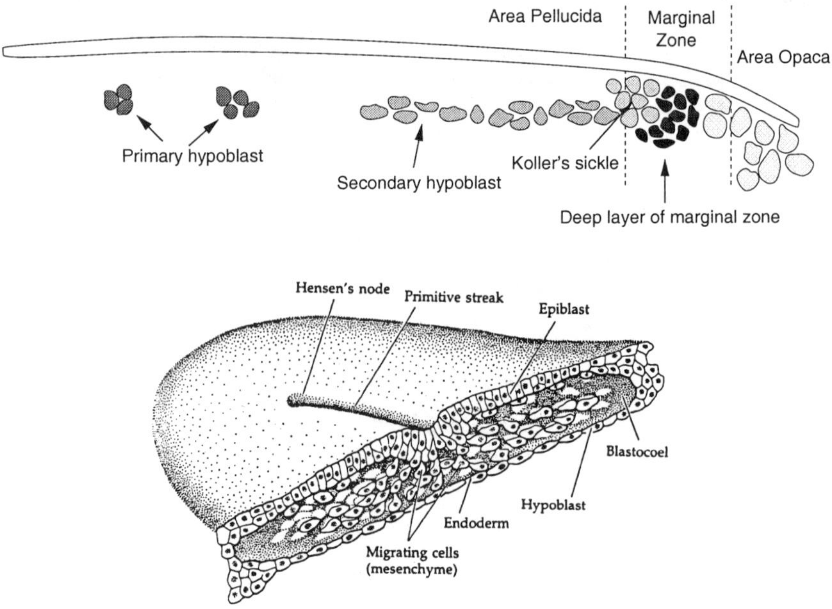

FIG. 2. *(a) A schematic cross-section of the blastoderm prior to primitive streak formation. (b) The blastoderm at the stage of maximal streak ingression (Reproduced with permission from [35])*

often referred to as paraxial mesoderm or presomitic mesoderm (PSM), separate into blocks of cells known as somites. They form as paired epithelial spheres arranged bilaterally along the anterior-posterior axis and emerge in strict cranio-caudal order [36]. Simultaneously, new cells are incorporated into the PSM from the regression of Hensen's node at the same rate as new somites are formed rostral to the PSM [16, 83]. Figure 3 is a schematic representation of these early processes. Somites are divided by a fissure into anterior and posterior halves that differ in their gene expression and differentiation [104, 36].

The formation and differentiation of somites is the result of three distinct morphological events progressing in a strict spatio-temporal order: (1) the prepatterning of the PSM; (2) somite and somitic boundary formation; and (3) the differentiation of a somite into anterior and posterior halves [36]. Several experimental observations confirm these events. Scanning electron microscopy observations [42] and transplantation experiments [49] show that the PSM displays a prepattern prior to segmentation. In addition, *Hox* and Notch-Delta pathway genes are involved in all these events [104, 25]. These molecular results suggest the existence of a conserved mechanism for segmentation in protostomes and deuterostomes [61].

The segmental pattern of somites in turn governs the segmental pattern of the peripheral nervous system and determines the shapes and ap-

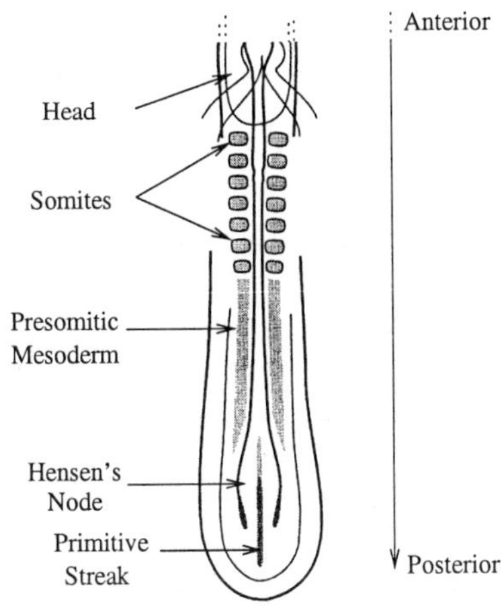

FIG. 3. *A schematic diagram illustrating the main structures involved in somitogenesis. Segmentation of the presomitic mesoderm occurs in an anterior-posterior sequence and the time taken for the formation of a somite is approximately 90 minutes in the chick. See text for details. (Redrawn from [10].)*

pendage characteristics of the vertebrae. Somites are also the source of cells for muscles, and influence the metameric distribution of blood vessels. Genetic or/and environmental factors disturbing somitogenesis produce malformations and abnormal development [117, 27, 36].

Although the sequence of events in early avian development is well documented, less is known about the mechanisms that give rise to primitive streak formation and somitogenesis. A number of theoretical models have been proposed to explain somitogenesis, and while these models are satisfactory in some respects, none can explain the complete set of observations. In the following subsections we present a brief exposition of the current experimental facts on primitive streak formation and somitogenesis. We then describe the theoretical models developed to explain some of these observations.

1.1. Formation of the primitive streak and the organizer. The ability of specific parts of the embryo to induce a primitive streak and node has been identified by a number of experiments. In particular, two regions have been tested, the PMZ and Koller's sickle. We should stress that references below to the PMZ may include Koller's sickle, except where stated explicitly.

I. *Posterior Marginal Zone (PMZ).*

- At stage X, transplants or rotation of the PMZ to lateral or anterior positions can form an ectopic primitive streak; at stage XI the inner region in contact with the PMZ also has the potential to form primitive streak, and at stage XII the PMZ has lost the ability to induce a primitive streak [53]. At both stages X and XI the size of the transplanted fraction is also critical in its capacity to initiate an ectopic axis [30].

- If a fragment of the PMZ is removed and replaced by lateral marginal zone (LMZ) tissue at stage X, a single primitive streak always originates in the normal position, but if the fragment of PMZ is replaced by beads which prevent healing of the wound, then two primitive streaks form [54].
- If donor PMZ tissue is inserted at 90° to the host PMZ at stage X, a single primitive streak develops at the site of the host PMZ. However, if the host PMZ is removed two small primitive streaks develop, one at the normal site and one at the transplant site. Khaner and Eyal Giladi [54] have also demonstrated that transplantation of a portion of the PMZ into the LMZ of a host embryo induces a second primitive streak to grow at 90° to the primitive streak growing from the PMZ.
- Any part of the blastoderm, provided it contains a portion of the PMZ and is sufficiently large, has the potential to develop a normal embryo. The streak is normally initiated along a radius [96]. When the blastodisc is cut in half, perpendicular to the anterior-posterior axis, the posterior half will form a streak initiated from the posterior margin. The anterior half can also form a streak, which is more likely to be initiated from the LMZ, but it may form from the anterior margin. When the cut is made parallel to the anterior-posterior axis, two streaks form, one on either side of the cut.
- Fate map experiments demonstrate that PMZ tissue has the capacity to induce an ectopic primitive streak without contributing cells to the streak [6]. This suggests that the PMZ may function as an avian equivalent of the Nieuwkoop center [66] - a region of the amphibian blastula that induces an organizer in adjacent cells without contributing to it. The experiments further demonstrate that: (i) PMZ does not give rise to hypoblast but remains stationary; (ii) transplants of quail PMZ (cut to exclude Koller's sickle) to the anterior side of a chick anterior region can induce a primitive streak from the anterior pole in a significant number of cases, and grafts to the posterior side of the anterior region results in a high frequency of streaks from the posterior end. In neither case, however, does the graft contribute cells to the streak. These experiments suggest that the PMZ determines the position of the streak.

II. *Koller's sickle.*

- It is known that Koller's sickle begins to form in the PMZ at stage X, and if cell movement in this area is blocked, no primitive streak is formed [95].

- Transplants of Koller's sickle to lateral portions of host embryos [13, 41, 14] can induce an ectopic primitive streak. In normal development, cells of Koller's sickle contribute to the primitive streak [41].
- Detailed fate mapping of midline cells [6] show that the epiblast above Koller's sickle and Koller's sickle itself both contribute cells to the node and primitive streak. The epiblast above and anterior to Koller's sickle, and cells in the anterior part of Koller's sickle, contribute cells to the node and anterior streak, whereas those cells immediately dorsal to the sickle and in the posterior part of the sickle contribute to the posterior part of the streak. Transplants of quail PMZ cut in a manner to include Koller's sickle (compare with previous item) were able to form a primitive streak when grafted to the anterior-most part of a chick anterior fragment with much greater frequency than when Koller's sickle was excluded. The quail cells were found to contribute to the streak when the graft included Koller's sickle.
- Grafts of PMZ including the sickle retain the competence to induce a primitive streak at later stages than grafts excluding the sickle [6]. The ability of Koller's sickle alone to induce an ectopic axis is lost by stage XIII, but a large fragment of the PMZ together with Koller's sickle can still induce an ectopic axis [52].

Stimulated in part by the wealth of data unearthed in other model developmental systems, many recent experiments have been directed at discovering the genes regulating development. For example, the *Hox* gene *goosecoid* is first found in a small population of cells corresponding to Koller's sickle [41]. Later this gene characterizes cells of the primitive streak, and expression is highest in cells of Hensen's node and the anterior portion of the streak. *Brachyury (Ch-T)* genes are expressed in forming mesoderm in response to inducing factors and at stage XII in a broad arc in the posterior epiblast. These gene expression patterns suggest that primitive streak formation can be regulated by gradients of organizer genes [5].

The signals involved in streak formation, particularly the transforming growth factors, have also been studied recently. A number of members of the transforming growth factor *beta* family (*TGF-β*) have been shown to induce primitive streak formation. For example, activin has been shown to induce development of axial structures [65, 118, 23], but it does not have the spatial and temporal distribution expected of an inducer. *cVg1* expressed in the PMZ of pre-primitive streak embryos has been shown to induce development of an ectopic primitive streak [91]. The activation of the Wnt proto-oncogene pathway potentiates the activity of activin and *cVg1*. In contrast, the bone morphogenetic protein-4 (BMP-4) inhibits primitive streak formation [102]. Furthermore, BMP antagonists such as *chordin* can induce both primitive streak formation and organizer genes.

These experiments suggest that areas of the LMZ can form a primitive streak if they are exposed to fragments of PMZ, but they are inhibited from doing so by neighboring PMZ. Thus cells in the PMZ are already differentiated from those in other parts of the marginal zone and the remainder of the blastoderm when ingression of the primitive streak begins.

Traditionally the blastoderm has been considered homogeneous prior to streak formation, but recent findings suggest earlier cell diversity and considerable cell movement in the early epiblast [98]. Canning and Stern [15] identified a subpopulation of cells testing positive for the epitope HNK-1, which is first expressed on the surface of cells of the PMZ and on those which later form primary hypoblast. Later it is found in the area of streak formation, distributed with a distinct anterior-posterior gradient. A primitive streak does not form when these cells are removed. This has led to the suggestion that HNK-1 cells are the source of streak-derived tissue [98]. The precise role of the epitope itself is not clear, but it may have a role in modulating cell adhesion (see [97] and references therein).

Given the critical role of the organizer in patterning the embryo (for example, formation of the axial structures and left-right asymmetry), it is surprising that in embryos where the node and anterior portion of the streak has been extirpated [37, 113, 112, 84], or replaced in reverse orientation [1], a new organizer can be regenerated and development proceeds normally (albeit delayed). In fact, a lateral isolate of the embryo, cut such that both the primitive streak and Hensen's node have been excluded, can reconstitute a primitive streak and organizer [114, 115].

Using labeling techniques, Joubin and Stern [43] have demonstrated that the organizer is not a static population of cells, as was traditionally believed, but is a transitory population of cells that have moved into the node, acquired organizer characteristics (i.e. express specific organizer genes), and then left the node. It appears that the central third of the primitive streak (axially), characterized by the overlapping expression of *cVg-1* and *Wnt-8c*, induces the cells anterior to it to acquire organizer characteristics. The organizer prevents neighboring tissue from acquiring organizer status by releasing an inhibitory signal. The issue is confused, however, by the observation of a resident population of cells within the epiblast which remain part of the node during its regression [89, 90, 83]. It has been suggested that this population constitutes stem cells which divide and produce notochord/somite progeny.

1.2. Somitogenesis. During somitogenesis, as in other segmentation processes, the body axis is divided along the anterior-posterior axis into similar repetitive structures formed from the embryonic layers. In insects, such as *Drosophila melanogaster*, segments are generated by the simultaneous division of the syncitial blastoderm. In other invertebrates such as annelids and crustaceans, and in vertebrates, the mechanism of metamerisation is different; the segments are formed at the cranial end of a multicellular embryo and segmentation propagates caudally [110].

During somitogenesis, continuous inductive interactions with Hensen's node, notochord, neural tube and endoderm are not necessary for somite formation [7, 11, 100]. For example, explants of PSM are able to form somites in the absence of all surrounding structures. Further experiments, in which the PSM is cut into several parts and these parts are rearranged, show that somites do not form. However, if the disrupted PSM is in contact with epithelial structures then somites do form, suggesting that some factor derived from the epithelium may influence somite formation [69].

Scanning electron microscope images show that the PSM is not a homogeneous tissue. Prior to segmentation, the PSM displays metameric arrangements of groups of cells, named somitomeres by Meier [62], which are evidently the predecessors of somites [42, 36]. The existence of this prepattern is confirmed in microsurgical experiments [70, 18], where isolated parts of the PSM form somites in strict cranio-caudal order some time after their isolation, differentiating into anterior and posterior halves in each somite. The existence of a prepattern is also strongly supported by the periodic pattern of *Hox* and Notch-Delta gene expression in the PSM [104, 57, 25]. Furthermore, the prepattern of anterior and posterior halves is also established before the formation of a somite [49]. Transplantation experiments reversing the anterior-posterior axis of the PSM demonstrate that the anterior-posterior polarity of the resulting pattern of somites is also reversed, so somite halves develop according to their original orientation [2]. In addition, there is a change in the mechanical properties of the cells in the PSM before they differentiate into a somite. There is an increase in cell compaction, and in cell-cell and cell-substratum adhesion, followed by epithelialization [49, 104] of the ball of cells as they form a somite. Several studies suggest that adhesion molecules such as cadherins play a major role in these processes [26, 85, 59]. It should be noted that cell labeling experiments indicate that cells of the PSM can contribute to more than one somite, suggesting that the prepattern of somitomeres does not preclude mixing between the prospective somites [101].

The total number of somites is regulated in an embryo. The *Amputated* mouse mutant, which is shorter than the wild-type mouse, has the same number of somites, but their somites are considerably smaller than those of the wild-type embryos [32]. However, the number of somites can be altered experimentally [49]. For example, heat shock applied to chick embryos can induce the formation of an extra somite [106, 82], or can result in up to four repeated somite anomalies, confined to one or to both rows, separated by relatively constant distances of six to seven normal somites [82]. The repeated anomalies suggest that heat shock affects an oscillatory process within the somite precursors [101].

There appears to be some degree of cell cycle synchrony between cells in the PSM which are destined to segment together to form a somite. The cell cycle synchrony is observed in the early somite two cell cycles after segmentation [101, 81]. To some extent, cells of the PSM seem to be arranged in order of developmental age, with cells at a given level having relatively synchronous cell cycles. The rostral end of the PSM has an increased mitotic index, which indicates that this region has a high proportion of cells in mitosis [82].

Recently, the study of the expression of the transcriptional factor *c-hairy-1* in the PSM of chick embryos has provided molecular evidence for the existence of a segmentation clock [72, 22]. During segmentation, the cells of the PSM go through 12 cycles of *c-hairy-1* expression before becoming part of a somite, while more cells are continuously incorporated into the posterior end of the PSM. This observation suggests that the segmentation clock controls the time duration of cells in the PSM before they will form part of a somite. During the time taken for one somite to form, the expression of *c-hairy-1* sweeps along the PSM in the posterior-anterior direction, narrowing as it propagates (see Figure 4). This wavefront-like expression finally stops and is maintained in a half somite-sized domain which gives rise to the caudal half of the forming somite. The *c-hairy-1* expression is independent of cell movements and does not result from the propagation of a signal in the plane of the PSM; it is an intrinsic cell autonomous property of this tissue [61, 79]. More recently, studies by McGrew *et al.* [60] and Forsberg *et al.* [33] have shown that *lunatic fringe* (*l-fng*) gene expression resembles the expression of *c-hairy-1* in PSM. In fact, they show that both expressions are coincident and are responding to the same segmentation clock [80]. In *Drosophila*, it is known that *l-fng* plays an important role in the formation of the wing margin by potentiating Notch activation by Delta and the inhibition of Notch activation by the alternative ligand Serrate [74, 116]. In *l-fng* mutant mice, the formation of somites is disrupted and if a somite forms its anterior-posterior patterning is disturbed [27, 117].

Finally, it is important to mention that the principle differentiation pattern of all the somites is very similar. However, during morphogenesis subsequent differentiation forms unique anatomic structures, depending on the position along the anterior-posterior axis. Experiments in chick embryos demonstrate that the positional specification of somites occurs early during somitogenesis [55, 20, 21, 19, 107, 17, 12]. When cervical somites are replaced with somites from the trunk region, rib-like structures develop in the cervical vertebral column of the embryo. When thoracic somites are replaced by cervical somites, embryos do not develop ribs [55]. There is now a large body of experimental work showing that positional specification of the PSM requires members of the *Hox* gene family [57]. *Hox* gene activation during development correlates with gene position in the *Hox* complex, a property referred to as colinearity. The spatial and temporal colinearity in the expression of these genes results in unique combinations of *Hox* genes in defined groups of somites and their derivatives along the anterior-posterior axis [34, 40]. This led to the suggestion that a *Hox* code specifies the identity of somites [48, 47]. The role of *Hox* genes in positional specification has been analyzed by interfering with or altering the

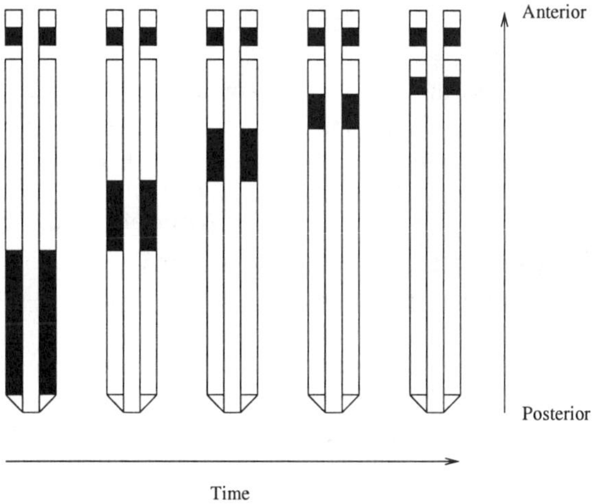

FIG. 4. *Schematic illustration of the wave of* c-hairy-1 *sweeping in the posterior-anterior direction (bottom to top) along the PSM with time (left to right). The shading denotes expression of this factor. It begins as a broad wave but narrows as it moves anteriorly until it finally correlates with the posterior half of the new forming somite. Then a new wave begins at the posterior margin of the PSM. Similar behavior is observed for* lunatic fringe.

expression of single *Hox* genes or by simultaneously perturbing the expression with retinoic acid, which is implicated in the specification of the axes during development [103].

2. Questions. Early organization of the avian blastoderm clearly involves a carefully controlled sequence of events. At present, very little is known concerning the mechanisms regulating this development and here we list some of the major unresolved questions. In the following section we describe some of the theories postulated to explain these processes.

2.1. Early development.

1. How is the posterior site of the embryo determined? Formation of the *area pellucida* involves a gravity induced directional shedding of cells (posterior to anterior) to form a one-cell thick layer [56]. How is this translated into the structural differences associated with the posterior region (e.g. Koller's sickle, secondary hypoblast formation)?

2. Development of the primary hypoblast involves an apparent dropping of cells in the *area pellucida* to form isolated islands in the subgerminal cavity [15]. What leads to the early diversification of such cells, and how do they separate from the *area pellucida*? One possibility is to link the diversification with the cell cycle, such that at the time of primary hypoblast formation a randomly scattered population in a specific phase of the cycle experiences a change in its cellular properties, for example adhesion. This change in adhesion may result in such cells being forced from the *area pel-*

lucida. To test such a hypothesis, it is necessary to construct a discrete cell model which incorporates cell adhesion [73].

3. What controls formation of the secondary hypoblast, and does the hypoblast influence streak formation? The role of the hypoblast in streak formation is controversial, and earlier experiments in which the hypoblast has been shown to induce streak formation [108, 3, 4] have been challenged by recent experiments [51]. However it is still not known whether the hypoblast is able to exert some influence over streak formation.

4. What initiates motion and guides the early migration of cells in the lateral regions toward the PMZ? Stern [97] observed migration of a subpopulation of the *area pellucida* to the posterior marginal zone prior to streak formation and speculated that a chemoattractant is produced at that site. Although collagen-gel assays support this theory, no chemoattractant has been identified.

5. What cues guide elongation and movement of the primitive streak? A simple anterior-posterior gradient of a diffusible morphogen cannot be used for positional information along that axis [53], for if it were the 90° transplants of the primitive streak would ingress toward the anterior pole rather than along a ray through the center of the disk.

6. What is the role of cell division in streak formation? Recent results by Wei and Mikawa [109] suggest that a subpopulation of cells in the posterior region may divide in a directional manner to form the primitive streak. It remains to be understood whether this division is essential for streak formation, or if it is simply an associated phenomenon.

7. What mechanisms can account for the fact that the primitive streak maintains its rod-like structure during ingression? Does the primitive streak ingress by convergent extension [46], whereby cells intercalate at the posterior marginal zone and push the primitive streak forward? Are there adhesive differences between cells in the primitive streak and those in the hypoblast and epiblast, or is the structure maintained by chemotactic attraction between cells in the primitive streak? Alternatively, is the streak maintained as a rod by the forces occurring throughout the blastoderm at these stages.

8. There appears to be a gradient within the marginal zone of potential to form a streak, with the posterior being the most capable and the anterior the least. At what stage is this potential determined, and by what mechanisms?

9. The primitive streak seems to inhibit other streaks from forming. What is the nature of this inhibition, and is it confined to act along the marginal zone?

10. The size and age of a blastodisc segment or donor implant are important in determining the site of streak formation. How do the key properties involved change with time?

11. How is the organizer defined, and how are the movements of cells through the organizer to form notochord, head process, paraxial mesoderm, etc., regulated? Recent experiments have revealed that the organizer is a transitory population continuously defined by cells in the middle part of the primitive streak [43]. Previous results, however, suggest that there exists a resident population of cells within the organizer that moves back with the node throughout regression. What is the relevance, if any, of this resident population?

12. What mechanisms control regression of the streak/organizer? Does regression of the streak simply occur through the disappearance of anterior cells into axial structures. Does the node regress by being pushed back by cells that are ingressing through it? Ablation of the node results in the regeneration of a new node, yet the new node must regenerate before regression proceeds. Does the static population of cells within the node control the movements of the node during regression?

13. How is the left-right asymmetry established? The earliest indication of left-right asymmetry in the avian embryo occurs with the asymmetric expression of *sonic hedgehog (shh)* in the avian node [58]. Studies in the mouse have revealed the presence of a nodal fluid flow from right to left as a result of unidirectionally rotating cilia on node cells [67, 68] and this has been linked with the establishment of the left-right axis. However, no such cilia have been located in the chick, and the cause of left-right asymmetry remains unknown.

2.2. Somitogenesis.

1. What regulates the number and size of somites?

2. What determines differentiation into anterior and posterior halves within a somite?

3. What are the differentiation and mechanical properties involved in the epithelialization of somites?

4. What determines the regional specification of somites - that is, certain somites form certain structures. What is the precise role of the *Hox* family in this process and how is it controlled?

5. What drives the segmentation clock? Is there a relation between the cell-cycle and the segmentation clock?

6. What is the precise role of the segmentation clock during somitogenesis?

7. How is the interplay between the segmentation clock and Notch-Delta and related components established?

8. What regulates the refinement of the *c-hairy-1* and *l-fng* cycles in the forming somite? How do these cycles interact with the segmentation clock?

9. How can the heat shock experiments be explained?

3. Models of streak formation and somitogenesis.

3.1. Formation of the primitive streak.

PI. *Model of Induction by Gravity:* Eyal-Giladi [29] proposed that substances needed for the initiation of primitive streak formation become nonuniformly distributed by gravity while the embryo is tilted, moving from the vegetal pole toward the region that is incorporated into the PMZ. Alternatively, Eyal-Giladi also suggests that these factors can be located under the embryo and shifted toward the posterior by the sliding of the yolk, and could later be found in the PMZ and Koller's sickle. Classic experiments in chick embryos have established that labile anterior-posterior polarity is determined 20 hours after fertilization. During this period there is a critical 2-hour time window where the outer albumen layers are rotated by the uterus while the yolk remains stationary but slightly tilted within a layer of low friction thin albumen [28]. In these experiments, the side of the embryo that is tilted upward during the critical window is defined as posterior. This model is unsatisfactory in some aspects. Little work has been done on this hypothesis due to the difficulty of obtaining uterine eggs. In addition, this model does not address the ingression and regression of the primitive streak.

PII. *Model of Induction by the PMZ:* In this model, proposed by Bachvarova [5], the PMZ is considered analogous to the Nieuwkoop Center of the frog embryo, which is the structure responsible for induction during the first stages of amphibian development. The PMZ of the chick embryo acts as an extra-embryonic signaling center promoting formation of the primitive streak in the adjoining posterior central disc epiblast. According to the model, factors such as *Vg1* and *Wnt8c* produced in the PMZ activate organizer genes such as *goosecoid* in Koller's sickle and *chordin* in the posterior central-disc epiblast. In turn, *chordin* suppresses BMP and this decrease promotes activation of organizer genes in the posterior midline. Lower concentrations of *Vg1* or *TGF-β* factors induce *Brachyury*-like genes in a broader crescent of posterior central disc epiblast, leading to mesoderm formation. BMP activity from lateral and anterior marginal zone induces epidermis in the adjoining central disc. Finally, the activation of the Wnt pathway in the late uterine and freshly laid egg plays an important role in the asymmetry observed in cells of Koller's sickle and the hypoblast.

This model incorporates several aspects of primitive streak formation. However, as in the previous model it does not address the ingression and regression of the primitive streak. Furthermore, as indicated by Bachvarova [5], many outstanding problems remain with this model. For example, it is not clear if factors such as *Vg1* are required in normal development. In addition, other factors such as *TGF-β* cannot be present and active in early embryos; and little is known about the Wnt pathway.

A mathematical formulation of this model could help understand the outcome of the complex of interactions proposed and make experimentally-testable predictions.

PIII. *Chemotaxis Model:* Chemotaxis (or haptotaxis) is a plausible mechanism for the observed oriented cell movements both prior to and during primitive streak formation, and this mechanism has been incorporated into a model designed for formation and subsequent maintenance of the streak (though not the determination of the initial site of outgrowth) [71].

The model assumes that there is a specialized subpopulation of cells residing at or close to the posterior marginal zone that both respond to and modulate the level of an attractant. This population serves to mark the site of the primitive streak and guide the movements of elongation and regression. Several cell populations have been identified [41, 99, 109] as having a role in primitive streak formation. The model does not, however, postulate how other cells ingress through the streak. In Figure 5 we show the pattern of movements predicted on a two-dimensional domain. To achieve movement of cells as a rod, rather than a general spreading of cells, it is necessary to choose conditions such that the chemoattractant initially has its highest concentration at the center of the domain (corresponding to the center of the *area pellucida*) and decreases to zero at the marginal zone. Plausible mechanisms for generating such conditions are given in [71].

The model makes a number of experimentally-testable predictions (Figure 6). Firstly, it predicts that any ectopically induced embryonic axis will develop along radial lines. Secondly, it predicts that disruption of the center of the *area pellucida* will have a significant effect on the morphology of the streak. It also predicts the natural development of an *organizer* region at the anterior portion of the streak as a region of higher cell density, and demonstrates a decrease in the rate of regression as the streak moves back, in agreement with experimental results [94]. However there is no experimental evidence for chemotactic motion in streak formation, and it is unclear whether the same mechanism that drives propagation of the streak is also responsible for regression. Thus this model simply demonstrates that chemotaxis can produce the observed behavior.

PIV. *Cell Division Model:* Wei and Mikawa [109] have proposed a model for formation of the streak based on directional cell division. In this model, a specific subpopulation of cells (localized at stage XII to the epiblast-midline region of the PMZ) undergoes oriented cell division along the anterior-posterior axis to form the Hamburger and Hamilton stage 3 primitive streak. The model is supported by cell marking experiments which demonstrate that the Hamburger and Hamilton stage 3 streak comprises only cells derived from this region, and not cells which have migrated in from lateral regions, as has previously been assumed. Furthermore, cells in the streak were shown to have metaphase chromosome plates (which indicate cleavage direction) perpendicular to the anterior-posterior axis. The calculation, based on the number of cells in the pre-streak region and Hamburger and Hamilton stage 3 streak, of a cell cycle time of approximately 4 hours is consistent with the mitotic index for cells of the chick gastrulae.

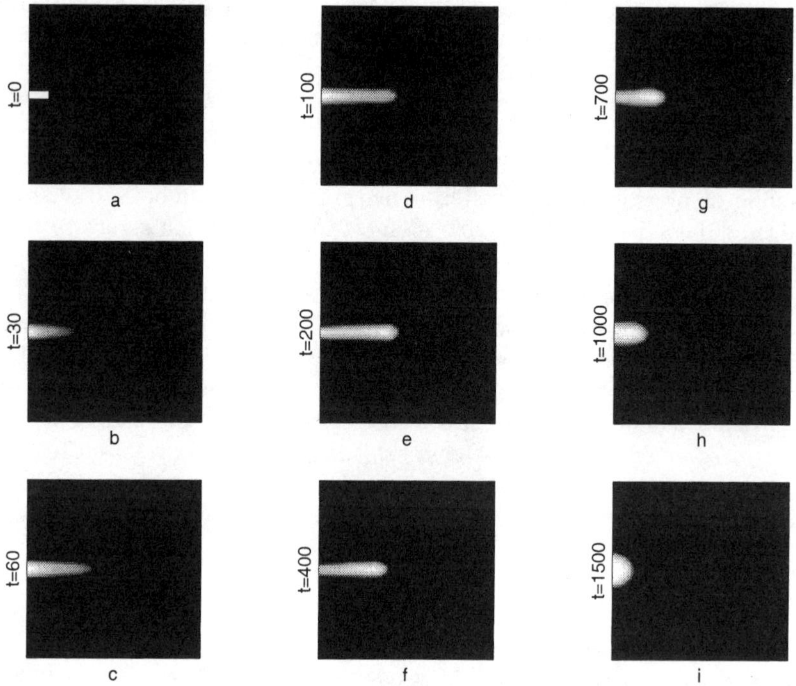

FIG. 5. *A time sequence showing the cell density for model PIII on a two-dimensional rectangular domain. White represents high cell density, black represents a zero cell density. The results show cell movement across the domain to form a rod which extends approximately half the way across the domain (e). Subsequent development shows a period of reverse movement, which occurs on a slower time scale.*

This model is consistent with the observation that the epiblast portion of the posterior marginal zone contributes to the primitive streak, and with the idea that a PMZ-derived signal induces primitive streak in the adjoining epiblast (see model II above). However, it is not yet clear if directional cell division would be able to induce the streak to form a long straight rod alone, nor is there any suggestion as to how regression of the streak is controlled.

PV. *Convergent-Extension Model:* Schoenwolf [88] has postulated that primitive streak formation may occur via a convergent-extension mechanism similar to that observed in developing amphibia [44, 45]. In this model, prospective primitive streak cells from either side of the midline would converge at the midline, intercalating with those on the opposite side and thereby producing an elongating primitive streak. This also raises the possibility that regression may occur through a reverse process.

This model is speculative, yet some evidence for it can be found in the general cell movements observed to take place in the epiblast during primi-

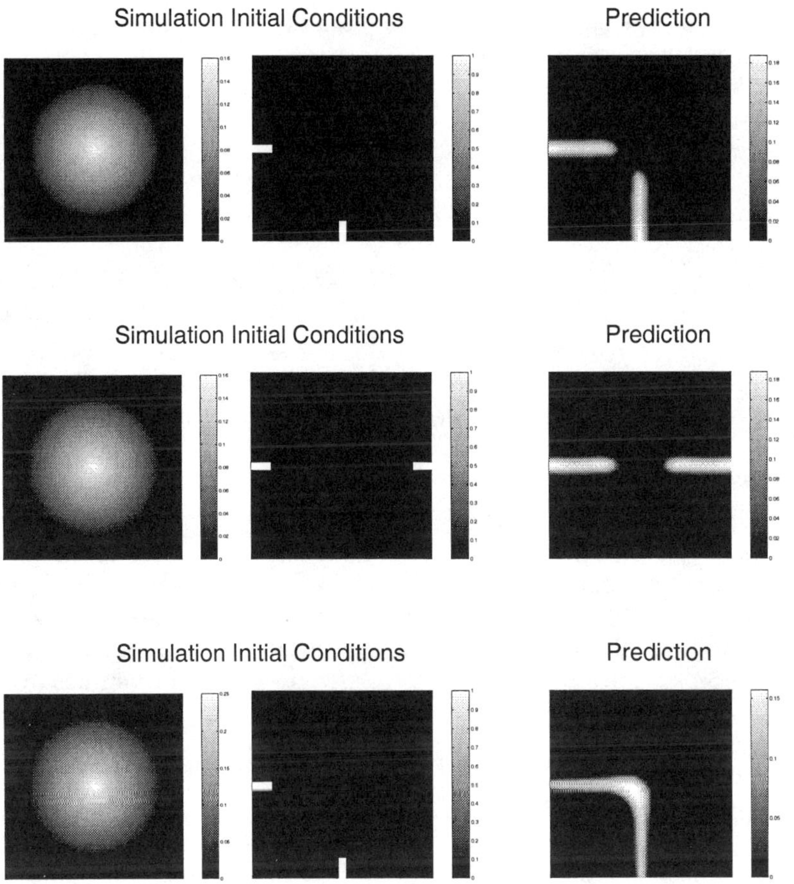

FIG. 6. *The time course for the development of an ectopic streak following 'transplantation' in model PIII. When a second population of "able" cells is placed at another point along the marginal zone (top: lateral, middle: anterior), an ectopic streak develops which moves towards the center of the domain. Fairly small changes in model parameters can result in the fusing of these streaks at the anterior ends. In the bottom figures, this has been effected by increasing the concentration gradient of the chemoattractant.*

tive streak formation [105]. Furthermore, the mechanism could provide an explanation for the change in morphology of the blastoderm from circular to pear-shaped during formation of the streak, as intercalation would result in a streak being driven in both anterior and posterior directions. However the author does not suggest what determines the posterior marginal zone as the site of streak formation, nor what the mechanisms are for guiding cell movement during the convergent-extension process so as to maintain the rod-like morphology of the streak.

3.2. Somitogenesis. During the last three decades, several models have been proposed to explain the formation of somites [24, 32, 8, 9, 42, 49, 63, 64, 76, 78, 77, 82, 81, 87]. Some of these incorporate the different aspects of somitogenesis previously mentioned, and are satisfactory in many respects. It is important to emphasize that these models cannot explain all the experimental facts presented in section 1.2, but they do lend insight to many of the observations. The models can be divided into four main categories:

SI. *Induction Models:* In these models, somite formation is explained in terms of inductive interactions with neighboring tissue [8]. These models are unsatisfactory in many aspects. No single tissue has been shown to induce somite formation. As we previously indicated, somites can be formed in the absence of Hensen's node, notochord, neural tube and endoderm [7, 11, 100], but the midline structures are necessary after experimental disruption of the PSM [69].

SII. *Prepattern Models:* These models postulate that there is a spatially-periodic prepattern present in the PSM before formation of the somites. Bellairs and Veini [11] proposed that somitogenic clusters are generated during PSM formation. Meier [62] suggested that prior to segmentation the PSM displays metameric arrangements of groups of cells, named somitomeres. The observation of the prepattern has been confirmed in microsurgical and transplantation experiments [70, 18, 2]. However, this sort of model does not address the key problems of how the prepattern is set up and how it is maintained and regulated.

SIII. *Positional Information Models:* These assume that a spatial pattern in chemical morphogen is set up, either via a gradient or a reaction-diffusion mechanism, and this prepattern determines cell differentiation. There are two main models:

1. The wave gradient model was proposed by Wilby and Ede [111] and Flint *et al.* [32]. This model proposes that regression of Hensen's node creates two strips of paraxial mesoderm, and that cells recruited into them start to synthesize a morphogen. The morphogen concentration increases in the cells until a threshold is reached, at which point an irreversible change from synthesis to destruction of morphogen occurs. The morphogen concentration in these cells falls, establishing a sink relative to cells that are still producing the morphogen. Neighboring cells maintain a morphogen concentration below the threshold, as morphogen diffuses from them into the sink, but cells further back in the paraxial mesoderm exceed the morphogen concentration threshold, and another trough of concentration is formed. Thus, a pattern of alternating peaks and troughs is created, which later gives rise to somites and fissures respectively. In this model, the size of the somites is determined by either the rate of incorporation of cells into the PSM or the speed of node regression. If the rate of node regression depends on the size of the embryo, then this model can account for the observation that there is regulation of somite number so, for example, in the *Amputated* mutant mouse embryo, which is only two thirds the normal size, the number of somites formed is still the same as in the normal case. This would be consistent with the assumption that the node regresses more slowly in smaller embryos. However, this model cannot easily explain the observations after the anterior-posterior axis of the PSM is reversed. It seems likely that the pattern of morphogen concentration would be severely disrupted during these experiments and the model would not then predict somite formation in the normal way.

2. Meinhardt [63, 64] proposed a reaction-diffusion type model, with two cell states A and P, which locally exclude each other, but stimulate each other over a long range. Cells switch from one state to the other until finally reaching a stable state. These can lead to a pattern of stable ... APAP ... stripes forming from anterior to posterior. If the transition from, say P to A, allows a change of segmental specification then each AP pair (or segment) will have a more posterior specification than its predecessor. Thus a segmental pattern can be generated in which segments have different regional characteristics.

To set up this pattern, Meinhardt proposed two alternative mechanisms, one involving a morphogen gradient in which threshold concentrations of the morphogen are required for successive P to A transitions, the other involving outgrowth in which new segments are added as the domain grows. Meinhardt's model is in agreement with two observations of Palmeirim *et al.* [72]: one full cycle of *c-hairy-1* oscillation corresponds to the formation of one somite, and *c-hairy-1* expression seems to be reminiscent of the spatiotemporal dynamics of one of the autocatalytic substances, because its wavefront expression stops and is maintained in the posterior half of the somites. It is also the only model, to our knowledge, that addresses the regional differences of somites and the anterior-posterior pattern of somites.

This model cannot easily explain the results of the experiment in which an isolated part of the PSM forms normal somites and the experiments that involve reversing the anterior-posterior axis of the PSM. In the former, one would expect any diffusion-based structures to be disrupted by the experiment, while in the latter, the model would predict that somites would form first in the anterior part of the reversed PSM and somitogenesis would proceed as normal, but in reality they develop according to their original location. One would have to assume that rostral-caudal determination occurs very early and is fixed before isolation or rotation of the PSM. This possible explanation requires more detailed investigation. It is not clear that this scenario is consistent with that envisioned for the *c-hairy-1* dynamics. Furthermore this model does not explain the cell-autonomous nature of somite formation which is strongly suggested by the experiments of Palmeirim *et al.* [72], McGrew *et al.* [60] and Forsberg *et al.* [33]. In these experiments, a portion of one side of the PSM is removed but the *c-hairy-1* waves propagate in synchrony in both sides of the PSM, including the isolated portion.

As it stands, the model does not appear to explain the heat shock effects which seem to require a link between cell fate and cell cycle. Such a link is not apparent in this model.

IV. *Clock or Oscillator Models:* There are a number of models along these lines:

1. Cooke and Zeeman [24] were the first to propose he existence of a cellular oscillator, which they assumed interacts with a progressing wave of cell determination travelling along the anterior-posterior axis of the PSM. This model, known as the clock and wavefront model, is able to explain the control of somite number [92], but is contradicted by the results of the experiments which reverse the anterior-posterior polarity of the PSM, because, as in **PIII**, this model would predict that segmentation should continue in the anterior-posterior direction without disturbance. To be consistent with the observation of the repetitive anomalies observed after the single heat shock experiments [82, 81, 101] this model would have to additionally assume that the cellular oscillator was closely linked to the cell cycle. The model does not address the formation of the anterior and posterior halves of a somite.

2. Stern *et al.* [101] proposed that the cell cycle plays the role of the oscillator. This cell cycle model relies on an intracellular oscillator that controls cell division and interacts with a kinematic wave which produces a signal that recruits other cells in the vicinity shortly before segmentation [49, 101, 81]. It explains the periodic anomalies of the heat shock experiments, the cell cycle synchrony observed in the PSM, as well as the isolation and transplantation experiments. This model addresses pattern formation at the cellular level and therefore does not address molecular issues such as the oscillations of *c-hairy-1* and its pattern in the PSM. A direct link between this model and the *c-hairy-1* oscillations is not obvious, because in chick embryos, the period of the cell cycle in the PSM is 9 hours while the period of the oscillations is only 90 minutes [72]. Furthermore, heat shock experiments in zebrafish embryos show that the periodic unit of somite defect (four normally formed somite + one abnormally formed), which correspond to 2.5 hours, does not match the overall cell cycle length (4 hours). This suggests that the proposed relationship between segmentation clock and cell cycle in vertebrates should be re-evaluated [86].

3. In a similar model to the one above, Polezhaev [76, 78, 77] proposed that a wave of cell determination moves along the PSM causing cell differentiation in a particular phase of the cell cycle, resulting in these cells secreting an inhibitor which impedes the differentiation of other cells. This model can explain the results of the heat shock experiments, is consistent with the observations of the isolation and transplantation experiments, and the epithelialization observed just before overt segmentation [49, 104].

However, as in the previous model, this model does not address events at the molecular level, nor does it address the formation of the anterior and posterior halves. To explain the regulation of somite number [49] one would have to assume that the cell determination wave moved at different rates (as in **PIII** 1).

4. Recently, Schnell and Maini [87] have proposed a clock and induction model in which, as a group of cells destined to form a somite traverses the PSM, cells undergo a series of *l-fng* expression pulses, followed by a longer final pulse which will remain at the posterior half of the newly forming somite. *l-fng* expression synthesizes a protein associated with the cell membrane, which increases its membrane levels in a ratchet-like fashion proportional to the segmentation clock oscillations experienced. The formation of a somite is then assumed to be triggered at a threshold level of l-fng protein. Elements of the Notch-Delta pathway associated with *l-fng* would allow the formation of a somite boundary and anterior-posterior pattern, through an induction mechanism. This model is consistent with the rhythmical expression of *c-hairy-1* and *l-fng* and the expression of the Notch-Delta pathway genes in PSM. The model can explain the isolation and transplantation experiments, and the heat shock defects. However, it cannot explain the cell cycle synchronization or epithelialization.

4. Discussion. Building the early embryo involves an architectural challenge that higher organisms have addressed through two processes that occur in early development: large scale rearrangements of tissue via a process called gastrulation, and the axial subdivision of tissue in a process called somitogenesis. Remarkably, somitogenesis has many elements in common with limb development. In fact both of these phenomena can be considered as examples of segmentation. For example, in limb development the anterior-posterior specification of digit elements (see, Dillon, this volume) is determined by the *Hox* genes, and differentiation and boundary formation is determined by the Notch-Delta pathway, as in somitogenesis.

In this paper, we have reviewed the theoretical and mathematical models developed to explain primitive streak formation and somitogenesis. Most of these models have been designed to explain particular aspects of these processes and are successful in doing so. In our critique of previous models, we have compared the models only with the experimental results that are widely accepted and which address the gross mechanisms of primitive streak formation and somitogenesis. As the models stand at present, none of them can easily explain all of these experimental observations. It should be noted that the majority of these models were developed before the discovery of the molecular evidence for the control of primitive streak formation and somitogenesis and are based on cell and tissue level observations.

A challenging future problem for theoretical and mathematical modelling will involve linking the pattern formation mechanisms at the cellular level with the molecular control of cell properties. In Section 2 we listed 22 key questions connected with the problems of primitive streak formation and somitogenesis. The reader will note that the models presented in Section 3 addressed only a small fraction of these.

Acknowledgments. This research (SS) has been funded by José Gregorio Hernández and ORS Awards, CONICIT(Venezuela) and a Lord Miles Senior Scholarship in Science. K. Painter and H. G. Othmer are supported in part by grant GM29123 from the National Institutes of Health.

We thank Daragh Mcinerney for helping with Figure 3 and Paul Kulesa for helpful comments on the manuscript.

REFERENCES

[1] M. Abercrombie, *The effects of antero-posterior reversal of lengths of the primitive streak in the*, Phil. Trans. Roy. Soc. Lond. B., 234 (1950), pp. 317–338.

[2] H. Aoyama and K. Asamoto, *Determination of somite cells: Independence of cell differentiation and morphogenesis*, Development, 104 (1988), pp. 15–28.

[3] Y. Azar and H. Eyal-Giladi, *Marginal zone cells, the primitive streak inducing component of the primary hypoblast in the chick.*, J. Embryol. Exp. Morphol., 52 (1979), pp. 79–88.

[4] ———, *Interaction of epiblast and hypoblast in the formation of the primitive streak and the embryonic axis in the chick, as revealed by hypoblast rotation experiments.*, J. Embryol. Exp. Morphol., 61 (1981), pp. 133–144.

[5] R. Bachvarova, *Establishment of anterior-posterior polarity in avian embryos*, Curr. Opin. Gen. Dev., 9 (1999), pp. 411–416.

[6] R. Bachvarova, I. Skromme, and C.D. Stern, *Induction of primitive streak and hensen's node by the posterior marginal zone in the early chick embryo*, Development, 125 (1998), pp. 3521–3534.

[7] R. Bellairs, *The development of somites in the chick embryo*, J. Embryol. Exp. Morph., 11 (1963), pp. 697–714.

[8] ———, *The segmentation of somites in the chick embryo*, Bull. Zool., 47 (1980), pp. 245–252.

[9] R. Bellairs, D.A. Ede, and J.W. Lash, eds., *Somites in Developing Embryos*, Plenum Press, New York, NY, USA; London, UK, 1986.

[10] R. Bellairs and M. Osmond, *The Atlas of Chick Development*, Academic Press, London, 1998.

[11] R. Bellairs and M. Veini, *An experimental analysis of somite segmentation in the chick embryo.*, J. Embryol. Exp. Morph., 55 (1980), pp. 93–108.

[12] J. Butler, E. Cosmos, and P. Cauwenbergs, *Positional signals: Evidence for a possible role in muscle fibre-type patterning of the embryonic avian limb*, Development, 102 (1988), pp. 763–772.

[13] M. Callebaut and E.V. Nueten, *Rauber's (koller's) sickle: the early gastrulation organizer of the avian blastoderm.*, Eur. J. Morph., 32 (1994), pp. 35–48.

[14] M. Callebaut, E. van Nueten, F. Harrisson, L. van Nassauw, A. Schrevens, and H. Bortier, *Avian gastrulation and neurulation are not impaired by the removal of the marginal zone at the unincubated blastoderm stage*, Eur. J. Morphol., 35 (1997), pp. 69–77.

[15] D. Canning and C. Stern, *Changes in the expression of the carbohydrate epitope hnk-1 associated with mesoderm induction in the chick embryo.*, Development, 104 (1988), pp. 643–655.

[16] M. Catala, M.-A. Teillet, and N.M.L. Douarin, *Organization and development of the tail bud analyzed with the quail-chick chimera system*, Mech.

Dev., 51 (1995), pp. 51–65.

[17] K. Chada, J. Magram, and F. Constantini, *An embryonic pattern of expression of a human fetal globin gene in transgenic mice*, Nature, 319 (1986), pp. 685–689.

[18] E.A.G. Chernoff and S.R. Hilfer, *Calcium dependence and contraction in somite formation*, Tiss. Cell., 14 (1982), pp. 435–449.

[19] A. Chevallier, *Role of the somitic mesoderm in the development of the thorax in bird embryos. II. Origin of thoracic and appendicular musculature*, J. Embryol. Exp. Morphol., 49 (1979), pp. 73–88.

[20] A. Chevallier, M. Kieny, and A. Mauger, *Limb-somite relationship: origin of the limb musculature*, Journal of Embryology and Experimental Morphology, 41 (1977), pp. 245–258.

[21] ———, *Limb-somite relationship: Effects of removal of somitic mesoderm on the wing musculature*, J. Embryol. Exp. Morph., 43 (1978), pp. 263–278.

[22] J. Cooke, *A gene that resuscitates a theory — somitogenesis and a molecular oscillator*, Trends in Genetics (Personal edition), 14 (1998), pp. 85–88.

[23] J. Cooke, S. Takada, and A. McMahon, *Experimental control of axial pattern in the chick blastoderm by local expression of wnt and activin: the role of hnk-1 positive cells.*, Dev. Biol., 164 (1994), pp. 513–527.

[24] J. Cooke and E.C. Zeeman, *A clock and wavefront model for control of the number of repeated structures during animal morphogenesis*, Journal of Theoretical Biology, 58 (1976), pp. 455–476.

[25] I. del Barco Barrantes, A.J. Elia, K. Wünsch, M.H. de Angelis, T.W. Mak, J. Rossant, R.A. Conlon, A. Gossler, and J.L. de la Pompa, *Interaction between notch signalling and lunatic fringe during somite boundary formation in the mouse*, Curr. Biol., 9 (1999), pp. 470–480.

[26] J. Duband, S. Dufour, K. Hatta, M. Takeichi, G.M. Edelman, and J.P. Thiery, *Adhesion molecules during somitogenesis in the avian embryo*, J. Cell Biol., 104 (1987), pp. 1361–1374.

[27] Y. Evrard, Y. Lun, A. Aulehla, L. Gan, and R.L. Johnson, lunatic fringe *is an essential mediator of somite segmentation and patterning*, Nature, 394 (1998), pp. 377–381.

[28] H. Eyal-Giladi, *Gradual establishment of cell commitments during the early states of chick development*, Cell Differ., 14 (1984), pp. 245–255.

[29] ———, *Establishment of the axis in chordates: facts and speculations*, Development, 124 (1997), pp. 2285–2296.

[30] H. Eyal-Giladi and O. Khaner, *The chick's marginal zone and primitive streak formation*, Dev. Biol., 134 (1989), pp. 215–221.

[31] H. Eyal-Giladi and S. Kochav, *From cleavage to primitive streak formation: a complementary normal table and a new look at the first stages of the development of the chick. i. general morphology*, Dev. Biol., 49 (1976), pp. 321–337.

[32] O.P. Flint, D.A. Ede, O.K. Wilby, and J. Proctor, *Control of somite number in normal and* Amputated *mutant mouse embryos: an experimental and a theoretical analysis*, Journal of Embryology and Experimental Morphology, 45 (1978), pp. 189–202.

[33] H. Forsberg, F. Crozet, and N.A. Brown, *Waves of mouse* lunatic fringe *expression, in four-hour cycles at two-hour intervals, precede somite boundary formation.*, Curr. Biol., 8 (1998), pp. 1027–1030.

[34] S.J. Gaunt, *Mouse homeobox gene transcripts occupy different but overlapping domains in embryonic germ layers and organs: A comparison of Hox-3.1 and Hox-1.5*, Development (Cambridge), 103 (1988), pp. 135–144.

[35] S.F. Gilbert, *Developmental Biology*, Sinauer Associates, fifth ed., 1997.

[36] A. Gossler and M. Hrabe de Angelis, *Somitogenesis*, in Curr. Topics in Dev. Biol., vol. 38, Academic Press, 1998, pp. 225–287. Jackson Laboratory, Bar Harbor, Maine 04609, USA.

[37] C. Grabowski, *The effects of the excision of hensen's node on the early development of the chick embryo*, J. Exp. Zool., 133 (1956), pp. 301–344.

[38] V. Hamburger and H. Hamilton, *A series of normal stages in the development of the chick embryo.*

[39] Y. Hatada and C.D. Stern, *A fate map of the epiblast of the early chick embryo*, Development, 120 (1994), pp. 2879–2889.

[40] P.W.H. Holland and B.L.M. Hogan, *Expression of homeo box genes during development: A review*, Genes. Dev., 2 (1988), pp. 773–782.

[41] J. Izpisua-Belmonte, E.D. Robertis, K. Storey, and C. Stern, *The homeobox gene goosecoid and the origin of organizer cells in the early chick blastoderm.*, Cell., 74 (1993), pp. 645–659.

[42] A. Jacobson and S. Meier, *Somites in Developing Embryos*, vol. 118 of NATO ASI series. Series A, Life sciences, New York: Plenum Press., 1986, ch. Somitomeres: The primordial body segments, pp. 1–16.

[43] K. Joubin and C. Stern, *Molecular interactions continuously define the organizer during cell movements of gastrulation*, Cell, 98 (1999), pp. 559–571.

[44] R. Keller, *Vital dye mapping of the gastrula and neurula of* xenopus laevis. *i. prospective areas and morphogenetic movements of the superficial layer*, Dev. Biol., 42 (1975), pp. 222–241.

[45] ———, *The cellular basis of epiboly: An sem study of deep cell rearrangement during gastrulation in* xenopus laevis, J. Embryol. Exp. Morphol., 60 (1980), pp. 201–234.

[46] R. Keller, J. Shih, and P. Wilson, *Cell Motility, Control, and Function of Convergence and Extension During Gastrulation in* Xenopus, in *Gastrulation*: Movements, Patterns, and Molecules, W. C. R. Keller and F. Griffen, eds., Plenum Press, New York, NY, USA; London, UK, 1991.

[47] M. Kessel, *Respecification of vertebral identities by retinoic acid*, Development (Cambridge), 115 (1992), pp. 487–501.

[48] M. Kessel and P. Gruss, *Homeotic transformations of murine prevertebrae and concomitant alteration of* Hox *codes induced by retinoic acid*, Cell, 67 (1991), pp. 89–104.

[49] R. J. Keynes and C.D. Stern, *Mechanisms of vertebrate segmentation*, Development (Cambridge), 103 (1988), pp. 413–429.

[50] O. Khaner, *Axis determination in the avian embryo*, Curr. Topics in Dev. Biol., 28 (1993), pp. 155–180.

[51] ———, *The rotated hypoblast of the chicken embryo does not initiate an ectopic axis in the epiblast*, Proc. Nat. Acad. Sci. USA, 92 (1995), pp. 10733–10737.

[52] ———, *The ability to initiate an axis in the avian blastula is concentrated mainly at a posterior site.*, Dev Biol, 194 (1998), pp. 257–266. Department of Cell and Animal Biology, Hebrew University, Jerusalem, Israel.

[53] O. Khaner and H. Eyal-Giladi, *The embryo forming potency of the posterior marginal zone in stage x through xii of the chick.*, Dev. Biol., 115 (1986), pp. 275–281.

[54] ———, *The chick's marginal zone and primitive streak formation. I Coordinative effect of induction and inhibition*, Developmental Biology, 134 (1989), pp. 206–214.

[55] M. Kieny, A. Mauger, and P. Sengel, *Early regionalization of somite mesoderm as studied by the development of axil skeleton of the chick embryo*, Dev. Biol., 28 (1972), pp. 142–161.

[56] S. Kochav and H. Eyal-Giladi, *Bilateral symmetry in chick embryo, determination by gravity*, Science, 171 (1971), pp. 1027–1029.

[57] R. Krumlauf, Hox *genes in vertebrate development*, Cell, 78 (1994), pp. 191–201.

[58] M. Levin, *Left-right asymmetry and the chick embryo*, Sem. Cell. Dev. Biol., (1998).

[59] K. Linask, C. Ludwig, M. D. Hang, X. Liu, and K. K. G. L. Radice, *N-Cadherin/Catenin-mediated morphoregulation of somite formation*, Dev. Biol., 202 (1998), pp. 85–102.

[60] M. McGrew, J. Dale, S. Fraboulet, and O. Pourquié, *The* lunatic fringe *gene is a target of the molecular clock linked to somite segmentation in avian embryos*, Curr. Biol., 8 (1998), pp. 979–982.

[61] M.J. McGrew and O. Pourquie, *Somitogenesis: segmenting a vertebrate*, Current opinion in genetics and development, 8 (1998), pp. 487–493.

[62] S. Meier, *Development of the chick embryo mesoblast: Formation of the embryonic axis and establishment of the metameric pattern*, Dev. Biol., 73 (1979), pp. 24–45.

[63] H. Meinhardt, *Models of Biological Pattern Formation*, Academic Press, New York, USA, 1982.

[64] ———, *Somites in Developing Embryos*, vol. 118 of NATO ASI series. Series A, Life sciences, New York: Plenum Press., 1986, ch. Models of segmentation, pp. 179–189.

[65] E. Mitrani and Y. Shimoni, *Induction by soluble factors of organized axial structures in chick epiblasts*, Science, 247 (1990), pp. 1092–1094.

[66] P. Nieuwkoop, *Origin and establishment of embryonic polar axes in amphibian development*, Curr. Topics Dev. Biol., 11 (1977), pp. 115–132.

[67] S. Nonaka, Y. Tanaka, Y. Okada, S. Takeda, A. Harada, Y. Kanai, M. Kido, and N. Hirokawa, *Randomization of left-right asymmetry due to loss of nodal cilia generating leftward flow of extraembryonic fluid lacking kif3b motor protein*, Cell, 95 (1998), pp. 829–837.

[68] Y. Okada, S. Nonaka, Y. Tanaka, Y. Saijoh, H. Hamada, and N. Hirokawa, *Abnormal nodal flow precedes situs inversus in* iv *and* inv *mice*, Molecular Cell, 4 (1999), pp. 459–468.

[69] D.-J. Packard, R. Zheng, and D.C. Turner, *Somite pattern regulation in the avian segmental plate mesoderm*, Development, 117 (1993), pp. 779–791.

[70] D.S.-J. Packard and A.G. Jacobson, *The influence of axial structures on chick somite formation*, Dev. Biol., 53 (1976), pp. 36–48.

[71] K.J. Painter, P.K. Maini, and H.G. Othmer, *A chemotactic model for the advance and retreat of the primitive streak in avian development*, Bull. Math. Biol., 62 (2000). To appear.

[72] I. Palmeirim, D. Henrique, D. Ish-Horowicz, and O. Pourquie, *Avian* hairy *gene expression identifies a molecular clock linked to vertbrate segmentation and somitogenesis*, Cell, 91 (1997), pp. 639–648.

[73] E. Palsson and H.G. Othmer, *A model for individual and collective cell movement in* dictyostelium discoideum, Proc. Nat. Acad. Sci, (2000). Submitted.

[74] V. Panin, V. Papayannopoulos, R. Wilson, and K.D. Irvine, *Fringe modulates notch-ligand interactions*, Nature, 387 (1997), pp. 908–912.

[75] P. Penner and I. Brick, *Acetylcholinesterase and polyingression in the epiblast of the primitive streak chick embryo.*, Roux's Arch. Dev. Biol., 193 (1984), pp. 234–241.

[76] A.A. Polezhaev, *A mathematical model of the mechanism of vertebrate somitic segmentation*, Journal of Theoretical Biology, (1992).

[77] ———, *Mathematical model of segmentation in somitogenesis in vertebrates*, Biophysics, 40 (1995), pp. 583–589.

[78] ———, *Mathematical modelling of the mechanism of vertebrate somitic segmentation*, J. Biol. Sys., 3 (1995), pp. 1041–1051.

[79] O. Pourquié, *Clocks regulating developmental processes*, Curr. Opin. Neurobiol., 8 (1998), pp. 665–670.

[80] O. Pourquié, *Notch around the clock*, Curr. Opin. Gen. Dev., 9 (1999), pp. 559–565.

[81] D.R.N. Primmett, W.E. Norris, G.J. Carlson, R.J. Keynes, and C.D. Stern, *Periodic segmental anomalies induced by heat shock in the chick embryo are associated with the cell cycle*, Development (Cambridge), 105 (1989), pp. 119–130.

[82] D.R.N. Primmett, C.D. Stern, and R.J. Keynes, *Heat shock causes repeated segmental anomalies in the chick embryo*, Development (Cambridge), 104 (1988), pp. 331–339.

[83] D. Psychoyos and C.D. Stern, *Fates and migratory routes of primitive streak cells in the chick embryo*, Development (Cambridge), 122 (1996), p. 1523.

[84] ———, *Restoration of the organizer after radical ablation of hensen's node and the anterior primitive streak in the chick embryo*, Development, 122 (1996), pp. 3263–3273.

[85] G.L. Radice, H. Rayburn, H. Matsunami, K.A. Knudsen, M. Takeichi, and R.O. Hynes, *Developmental defects in mouse embryos lacking n-Cadherin*, Dev. Biol., 181 (1997), pp. 64–78.

[86] M.N. Roy, V.E. Prince, and R.K. Ho, *Heat shock produces periodic somitic disturbances in the zebrafish embryo*, Mech. Dev., 85 (1999), pp. 27–34.

[87] S. Schnell and P.K. Maini, *Clock and induction model for somitogenesis*, Dev. Dyn., 217 (2000), To appear.

[88] G.C. Schoenwolf, *Cell movements in the epiblast during gastrulation and neuralation in avian embryoes*, in Gastrulation, R. Keller, ed., Plenum Press, New York, NY, USA; London, UK, 1991, pp. 1–28.

[89] M. Selleck and C. Stern, *Fate mapping and cell lineage analysis of hensen's node in the chick embryo*, Development, 112 (1991), pp. 615–626.

[90] ———, *Formation and Differentiation of Early Embryonic Mesoderm*, Plenum Press, New York, 1992, ch. Evidence for stem cells in the mesoderm of Hensen's node and their role in embryonic pattern formation, pp. 23–31.

[91] S.B. Shah, I. Skromne, C.R. Hume, D.S. Kessler, K.J. Lee, C.D. Stern, and J. Dodd, *Misexpression of chick Vg1 in the marginal zone induces primitive streak formation*, Development (Cambridge), 124 (1997), pp. 5127–5138. Department of Physiology and Cellular Biophysics, College of Physicians and Surgeons of Columbia University, New York, NY 10032, USA.

[92] J.M.W. Slack, *From Egg to Embryo. Regional specification in early development*, Cambridge: Cambridge University Press, 1991.

[93] N. Spratt, *Location of organ-specific regions and their relationship to the development of the primitive streak in the early chick blastoderm*, J. Exp. Zool., 89 (1942), pp. 69–101.

[94] ———, *Regression and shortening of the primitive streak in the explanted chick blastoderm.*, J. Exp. Zool., 104 (1947), pp. 69–100.

[95] ———, *Some problems and principles of development*, Am. Zool., 6 (1966), pp. 215–254.

[96] N. Spratt and H. Haas, *Integrative mechanisms in development of the early chick blastoderm. I Regulative potentiality of separate parts*, J. Exp. Zool., 145 (1960), pp. 97–137.

[97] C. Stern, *Gastrulation: Movements, Patterns, and Molecules*, Plenum, New York, 1991, ch. Mesodorm formation in the chick embryo revisited., pp. 29–41.

[98] C. Stern and D. Canning, *Origin of cells giving rise to mesoderm and endoderm in the chick embryo.*, Nature, 343 (1990), pp. 273–275.

[99] C.D. Stern, *The marginal zone and its contribution to the hypoblast and primitive streak of the chick embryo*, Development (Cambridge), 109 (1990), p. 667.

[100] C.D. Stern and R. Bellairs, *Mitotic activity during somite segmentation in the early chick embryo*, Anat. Embryol. (Berl.), 169 (1984), pp. 97–102.

[101] C.D. Stern, S.E. Fraser, R.J. Keynes, and D.R.N. Primmett, *A cell lineage analysis of segmentation in the chick embryo*, Development (Cambridge), 104 Supplement (1988), pp. 231–244.

[102] A. Streit, K. Lee, I. Woo, C. Roberts, T. Jessell, and C. Stern, *Chordin regulates primitive streak development and the stability of induced neural cells, but is not sufficient for neural induction in the chick embryo*, Development, 125 (1998), pp. 507–519.

[103] D. Summerbell and M. Maden, *Retinoic acid, a developmental signalling molecule*, Trends in neurosciences (Regular ed.), 13 (1990), pp. 142–147.

[104] P.P.L. Tam and P.A. Trainor, *Specification and segmentation of the paraxial mesoderm*, Anat. Embryol. (Berl.), 189 (1994), pp. 275–305.

[105] L. Vakaet, *Chimeras in Developmental Biology*, Academic Press, London, 1984, ch. Early development of birds.

[106] M. Veini and R. Bellairs, *Somites in Developing Embryos*, vol. 118 of NATO ASI series. Series A, Life sciences, New York: Plenum Press., 1986, ch. Heat shock effects in chick embryos, pp. 135–145.

[107] F. Wachtlet, B. Christ, and H.J. Jacob, *Grafting experiments on determination and migratory behaviour of presomitic, somitic and somatopleural cells in avian embryos*, Anat. Embryol. (Berl.), 164 (1982), pp. 369–378.

[108] C. Waddington, *Induction by the endodrem in birds.*, Roux's Arch. Dev. Biol., 128 (1933), pp. 502–521.

[109] Y. Wei and T. Mikawa, *Formation of the avian primitive streak from spatially restricted blastoderm: evidence for polarized cell division in the elongating streak*, Development, 127 (2000), pp. 87–96.

[110] R.G. Weisblat, C.J. Wedeen, and R.G. Kostriken, *Evolution of developmental mechanisms: Spatial and temporal modes of rostrocaudal patterning*, Curr. Top. Dev. Biol., 29 (1994), pp. 101–134.

[111] O.K. Wilby and D.A. Ede, *A model for generating the pattern of cartilage skeletal elements in the embryonic chick limb*, Journal of Theoretical Biology, 52 (1975), pp. 199–217.

[112] S. Yuan, D. Darnell, and G. Schoenwolf, *Identification of inducing, responding and suppressing regions in an experimental model of notochord formation in avian embryos*, Dev. Biol., 172 (1995), pp. 567–584.

[113] ———, *Mesodermal patterning during avian gastrulation and neurulation: experimental induction of notochord from non-notochordal precursor cells*, Dev. Genets., 17 (1995), pp. 38–54.

[114] S. Yuan and G. Schoenwolf, *De novo induction of the organizer and formation of the primitive streak in an experimental model of notochord reconstitution in avian embryos*, Development, 125 (1998), pp. 201–213.

[115] ———, *Reconstitution of the organizer is both sufficient and required to reestablish a fully patterned body plan in avian embryos*, Development, 126 (1999), pp. 2461–22473.

[116] Y.P. Yuan, J. Schultz, M. Mlodzik, and P. Bork, *Secreted fringe-like signalling molecules may be glycosyl-transferases*, Cell, 88 (1997), pp. 9–11.

[117] N. Zhang and T. Gridley, *Defect in somite formation in* lunatic fringe-*deficient mice*, Nature, 394 (1998), pp. 374–377.

[118] T. Ziv, Y. Shimoni, and E. Mitrani, *Activin can generate ectopic axial structures in chick blastoderm explants*, Development, 115 (1992), pp. 689–694.

MATHEMATICAL MODELING OF VERTEBRATE LIMB DEVELOPMENT*

ROBERT H. DILLON†

Abstract. Vertebrate limb development is a model system in developmental biology for the study of tissue growth, pattern formation and differentiation. This paper gives an overview of the development process and experimental results as well as a description of several modeling approaches. In addition, a new model is described. This model incorporates both outgrowth due to growth as well as the production and transport of signaling molecules produced in specialized regions of the limb. Results are shown from several example simulations. These demonstrate the model's ability to predict key phenomena described in the experimental literature for normal and experimentally manipulated embryos.

1. Introduction. The embryonic vertebrate limb is an ideal model system for the study of growth, differentiation and pattern formation. The development of the avian, mammal and amphibian limb has been the subject of extensive experimental and theoretical study for many decades. Earlier experimental work was motivated in part by the accessibility of the embryonic chick limb to microsurgical manipulation. The limb is also accessible to molecular manipulations and in the past fifteen years there has been much progress made in understanding the molecular mechanisms governing limb development. In this paper we give an overview of the developing limb bud and describe several modeling approaches including a new model that combines the processes of growth, morphogenesis, and cell signaling from organizing centers. Finally, we show the results of example simulations that indicate the ability of the model to predict key phenomena found in normal development as well as in experimentally manipulated embryos.

2. Overview of limb development. Although the mechanisms governing limb development are thought to be similar in all tetrapods, much of the experimental work has been done using chickens and mice as model systems. We shall discuss limb development in the framework of limb development in chicks. The chick develops from fertilized egg to hatchling in about three weeks via a sequence of developmental events which include cleavage, gastrulation and the formation of the embryonic axis. At the end of the third day the limb bud begins to emerge from the embryonic body and rapidly elongates. The humerus begins to appear during the fourth day and by the end of the seventh, the cartilage prepattern of the limb's skeleton is complete.

*This work was supported in part by NSF grant DMS-9805501.

†Department of Pure and Applied Mathematics, Washington State University, Pullman, WA 99164; dillon@math.wsu.edu. This work was supported in part by NSF grant DMS-9805501.

Development in chick can be described using the Hamburger-Hamilton (H-H) system of normal stages [15]. During development the dorsal mesoderm of the chick segments into blocks known as somites in an anterior-posterior sequence. Each H-H stage represents the appearance of three somites. Up to stage 23 each stage represents about four hours; thereafter, each represents about six hours.

The three orthogonal axes of the limb bud, which are designated the anterior-posterior (AP), dorsal-ventral (DV), and proximal-distal (PD) axes are illustrated in Figure 1a. These axes play a key role in the description and modeling of limb bud development. The wing bud's AP polarity

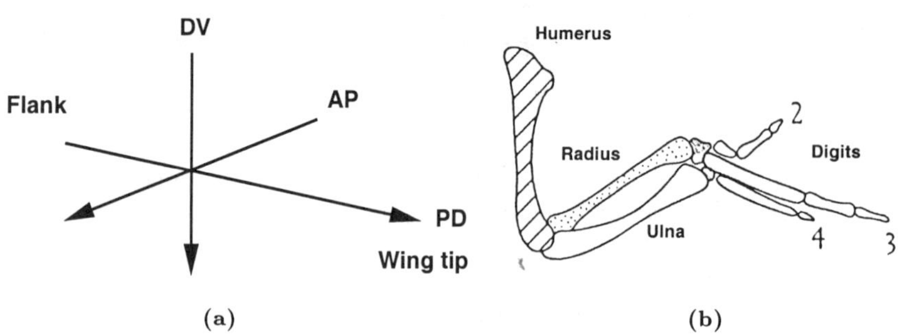

FIG. 1. *(a) The orientation of axes used to describe the limb. (b) A schematic of the adult wing skeleton in chick. From [9].*

is determined prior to stage 8, the time of wing site determination. Its DV polarity is established during stage 11. The specialized region known as the apical ectodermal ridge (AER), located along the distal margin of the limb bud and associated with outgrowth along the PD axis, is induced by the underlying mesoderm during stages 14–16 [62].

In its early stages, the limb bud consists of a central mass of mesodermal cells jacketed by a thin layer of ectoderm. The interior consists primarily of two morphologically similar cell lines: precartilage cells and premyogenic cells. The former are derived from the flank region of the embryo known as the lateral plate and will become the cartilage, bone, connective tissue, and blood vessels of the fully formed embryo. The latter, which are derived from the adjacent somites, will develop into the muscle masses.

When the wing bud emerges from the embryo body it is flat and approximately elliptical in cross section with its major and minor axes aligned with the bud's AP and DV axes. By stage 21 the dorsal side of the wing bud is rounded and the ventral side flattened. During wing bud outgrowth, the distance from somite to wing tip, known as the PD length, increases rapidly. Between stages 18, when the AER appears, and stage 25, when distinct cellular condensations within the humeral region are first observed, the PD length increases from approximately 0.23 mm to 1.74 mm [27]. As the bud elongates, the posterior half grows more rapidly than the anterior [22]. The AP length grows from approximately 0.8mm in width at stage 21 to 2.0 mm at stage 28 [20].

The formation of the limb's bone structure occurs in two stages. During the first, a cartilage prepattern of the bone structure is established. In the second, the cartilage is replaced with bone through the process of osteogenesis. The first signs of cartilage differentiation can be observed at stage 22 with the uptake of radioactively-labeled sulphate (^{35}S-sulfate) into mucopolysaccharides [51]. The Y-shape of the prospective humerus, radius, and ulna is first seen in autoradiographs at the end of stage 23 [51]. At about stages 24–25, distinct cellular condensations within the humeral region are observed [16]. The cartilage elements can be detected in a proximal-distal and posterior-anterior sequence with alcian green staining. The humerus and ulna can be seen at stage 24; the radius, at stage 25; posterior wrist parts, at stage 26; anterior wrist parts, at stage 28. The first digit appears at stage 26 and, when the tip of digit 2 appears at stage 34, the full cartilage pattern is complete [27].

2.1. Specialized regions. Several regions of the developing limb bud have an important role in determining the final cartilage pattern. A thickened ridge of ectoderm along the distal edge of the limb bud known as the apical ectodermal ridge (AER) has an essential role in normal limb development. If the AER is removed, the proximal limb develops normally. However, the distal limb is truncated, with the level of truncation dependent upon the developmental stage at which the AER was removed [49].

The distal subridge or "progress zone", which consists of the limb bud mesoderm extending proximally 200–400 μm from the AER, is a growth center [49] of higher mitotic index than the proximal limb tissue [55]. The subridge fails to develop if the AER is removed and it has been hypothesized that the AER maintains the tissue of the subridge in an "embryonic state" preventing differentiation and encouraging a higher rate of cell division [51]. Premyogenic cells are of somitic origin and migrate into the limb bud between stages 15 and 18 [50]. As the limb bud elongates, the premyogenic cells normally migrate distally. Since few premuscle cells are found in the subridge, the leading edge of premuscle cell colonization is found at the proximal boundary of the subridge [34].

Normal wings develop a digit pattern of 234 (see Figure 1b). Saunders and Gasseling [48] showed that grafting a block of tissue from the posterior margin of the limb bud into the anterior margin of a host could induce anterior mirror image duplication of the distal limb bud, often with a digital sequence of 432234 [48]. The region of limb bud capable of eliciting this mirror image duplication is known as the zone of polarizing activity (ZPA). The potency of ZPA tissue to stimulate duplication in anterior grafts is inhomogeneously distributed within the ZPA itself and is stage dependent [26, 57, 19, 20]. ZPA activity is strongest in stages 19–28, though activity can be detected as early as stage 15 [18]. The nature of the duplication obtained in ZPA anterior transplants is also dependent upon the location of the transplant in the host [59]. To be effective, the donor tissue must be transplanted into a zone of undifferentiated mesenchyme at the tip of the limb rimmed by the AER, either adjacent to the AER or immediately under the AER [54].

A intriguing development in understanding the phenomenon of mirror image limb duplication was the discovery that implants under the anterior AER of material soaked with retinoic acid (RA) could produce mirror image limb duplication closely resembling ZPA-induced duplication [60, 56]. In both cases, the extent of duplication depends on the dose. Depending upon the RA concentration contained in the implanted bead, extra digits may develop. With an optimal dosage concentration of RA a complete mirror image duplication of 432234 often results. At the optimal level and at lesser levels of RA concentration digit patterns including those of 2234, 3234, 43234, and 4334 may be found.

2.2. Molecular mechanisms. The classical experimental work on limb development relied extensively on microsurgical intervention. Research in the past two decades has been focused more on the molecular and cellular mechanisms of limb development. As a result many details concerning the molecular and cellular basis of limb development are now known (see [35] for a recent review). These discoveries include the identification of key signaling molecules produced in the AER and ZPA as well as transcription factors that are thought to be key determinants of pattern formation in the limb.

It has long been known that the AER is an essential signaling center for limb development. As mentioned above, AER removal usually leads to distal truncations. This fate can be avoided by attaching beads soaked with growth factors to the limb bud after AER removal. In particular, several members of the fibroblast growth factor family (FGF) including FGF-2, FGF-4 can substitute for the AER and result in nearly normal limb development. *Fgf-4* is normally expressed in the AER [36, 24].

A member of the hedgehog family known as *Sonic hedgehog* (*Shh*), encodes a protein that may be the primary ZPA signal. Misexpression of *Shh* in the limb bud's distal anterior can lead to mirror-image duplication of the limb's digits in a manner similar to that found in anterior ZPA transplants and RA-soaked bead implants. The *Shh* protein is secreted by cells in the ZPA. *Shh* expression begins at stage 17 and matches the location of the ZPA as mapped by Honig and Summerbell [20].

The spatial and temporal distribution of many other proteins or RNA has also been described (see [61]). Perhaps the most important is the expression patterns of the vertebrate HOX genes which are thought to be primary genes involved in the spatial patterning of the embryo and of the limb as well. The spatial patterns of expression of *Hoxa* and *Hoxd* at various stages of development have been mapped in detail. Five *Hoxd* genes, labeled 9–13 from 3' to 5' on the chromosome, are expressed in a nested pattern that is centered at the ZPA (see Figure 2). *Hoxd-9* and *Hoxd-10* are

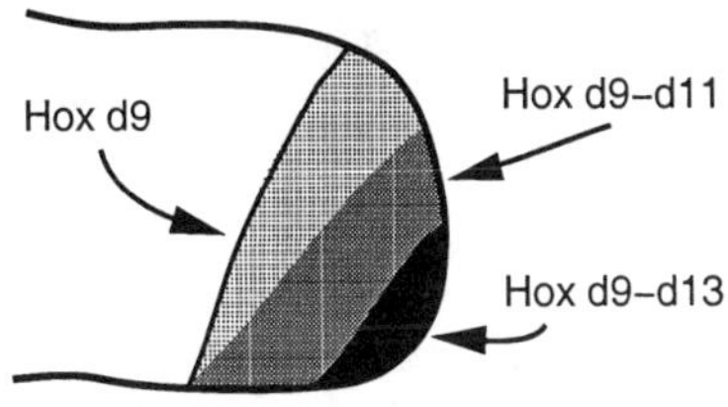

FIG. 2. *A schematic of the spatial pattern of* Hoxd. *After Robertson and Tickle [47].*

expressed throughout the limb at stage 16. *Hoxd-11* appears at stage 18 and *Hoxd-12* and *Hoxd-13* shortly thereafter [32]. The 5' members of the *Hoxa* family, 9–13, are also expressed in a nested pattern centered at the AER (see Figure 3). There is evidence that *Shh* can induce *Hoxd*. However, the

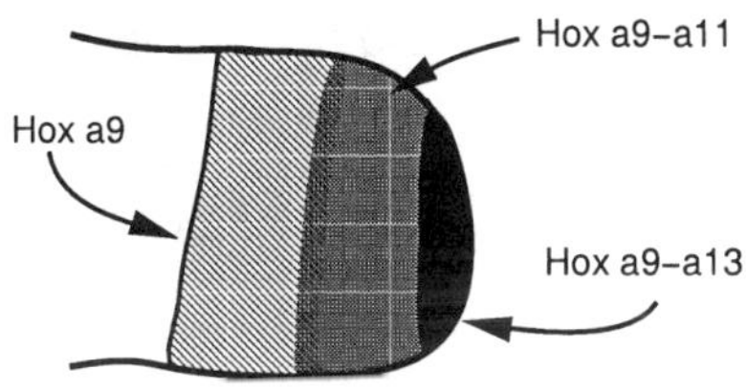

FIG. 3. *A schematic of the spatial pattern of* Hoxa. *After Robertson and Tickle [47].*

details of the relationship are largely unknown. The relationship between *Hox* expression and cartilage pattern formation is also unclear.

Sonic Hedgehog is one of the Hedgehog (HH) family of genes found in vertebrates. The members of this family are homologs to the hedgehog gene *Hh* in Drosophila. One of the features of the HH family of proteins is that they can act as long-range or short-range patterning signals in both vertebrates and Drosophila. In vertebrates, floor-plate induction depends on a short-range signal that requires contact between the neural plate and cells that express *Shh*. In contrast, motor neuron induction by *Shh* can occur over distances of several hundred microns (for a review of Hedgehog see [28]). The *Sonic Hedgehog* protein can be autocleaved into an amino-terminal product that accounts for all of the signaling activity and a carboxy-terminal domain that contains the determinants for autoprocessing of the Hedgehog protein precursor. The carboxy-terminal domain diffuses freely but the amino-terminal domain is usually found in tight association with the cell surface. Because of this cell-association, it is not clear how long range *Shh* signaling is effected. The long range signal may be due to secondary signaling molecules induced by *Shh* or low level concentrations of the amino-terminal domain that do manage to diffuse over a longer range [46, 21, 74].

There is experimental evidence for coupling between the AER and the ZPA through the interaction of *Shh* and *FGF-4* either directly or through intermediaries such as *Bmp-2* [24]. In one experiment, a replication competent virus that expresses *Sonic Hedgehog* cDNA was injected into the distal anterior margin of stage 18–20 limb buds. This lead to ectopic expression of *Shh* and *Hoxd-11*. In proximal-medial injections, *Shh* was expressed but not *Hoxd-11*. In another experiment, a *Sonic Hedgehog* virus was injected into the anterior margin of the limb bud after removal of the anterior half of the AER in stage 20/21 limbs. Since the AER still covered the posterior margin, nearly normal outgrowth and pattern formation were seen. However, the injections failed to induce anterior mesodermal proliferation, *Hox* expression or *Bmp-2* expression. Thus a signal is required from the AER along with *Shh*. Moreover, if a bead soaked in FGF-4 protein was stapled to the anterior distal limb following anterior AER removal, anterior Sonic Hedgehog virus injection induced expression patterns of *Hoxd-11*, *Hoxd-13* and *Bmp-2* at levels typically seen in the presence of a complete AER [24].

Dorsal-ventral patterning is also important in the limb and may be controlled by *Wnt7a* [23]. Three dimensional pattern formation is thus controlled by a complex network of signals originating in the AER, the ZPA and the non-AER ectoderm. A model for these interactions is shown below in Section 3.

3. Models.

Positional information and the gradient model. The concept of positional information has had a strong influence in the field of limb development. The basic idea of positional information is that the spatial pattern of differentiation is established in two steps. In the first, a cell parameter is fixed reflecting the position of the cell in the developing organism. In the second, the cell differentiates according to its developmental history and value of the cell parameter [66, 58, 68, 71]. In this theory, gradients of biochemicals could determine a cell's positional identity. These diffusible signaling molecules are known as morphogens, a term coined by Alan Turing in the 1950s [64].

The discovery of the mirror-image duplication property of ZPA tissue in anterior transplants led to the hypothesis that the ZPA produces a morphogen which diffuses throughout the limb bud and is degraded in it thereby establishing an exponential gradient along the AP axis which could serve to specify positional information with respect to the AP axis for the cells [66]. Cells might take on discrete positional values associated with fixed thresholds of the gradient field.

Several arguments have been raised against the positional information model in its gradient model form. For example, the model incorrectly predicts duplications or eliminations of the humerus when polarizing regions are placed at different positions along the anteroposterior axis [72, 71]. An additional mechanism may be needed to separate the digits. The phenomenon of ectopic digit formation takes place after the ZPA has lost much of its effectiveness and, would require reassignment of positional values to obtain the extra digits [17]. The gradient model suggests that each of the digits is fundamentally unique. The ectopic digit research suggests that the differences between digit shape in the length and the number of elements may be largely determined by the initial size of the condensation and the persistence of the distal AER [17].

Progress zone model. The progress zone model is a second type of positional information model which offers a mechanism for specifying position along the PD axis. In the progress zone model, cells differentiate according to the length of time spent in the progress zone, as measured by the number of cell divisions undergone by the time they leave [58]. Cells leaving early form proximal structures while cells leaving late form distal structures. Several experimental results support this model. As was discussed earlier, when the ridge is removed, the progress zone disappears and distal structures do not develop. If limb tips of different ages are exchanged, the tips continue to develop according to their original expected fate. This leads to deletion or duplication of limb structures as would be predicted by the progress zone model [67]. However, removal of sections of limb bud tissue orthogonal to the PD axis can lead to normal development, showing that the early limb bud may have some regulative properties not predicted by the progress zone model [55].

Turing's model. Turing suggested a mechanism by which an initially-homogeneous distribution of morphogens could give rise to a spatial pattern through the interaction of reaction and diffusion [64]. In Turing's model, cells differentiate according to the local morphogen concentration level. For example, in limb development precartilage cells might differentiate into cartilage if the steady state morphogen concentration is above a fixed threshold level, but differentiate into connective tissue if the steady state concentration is below this threshold (see [69] and [31] for a review). A detailed description and analysis of the standard Turing system is given in [7]. Reaction diffusion mechanisms of this type have been extensively studied and applied to several developing systems [31]. A major drawback to reaction-diffusion models is the problem of identifying morphogens in a real developing system. A reaction diffusion system involving fibronectin and the growth factor TGF-β has been proposed by Newman et al. [33] Recently, Turing-type structures have been found in the chlorite-iodide-malonic acid reaction [5, 40, 25, 12]. Aside from the difficulty of identifying morphogens and the reactions in a biological context, there are several properties of Turing systems that limit their applicability. For example, since the spatial patterns in a Turing system typically arise from an instability, the parameters must be tightly controlled to obtain the onset of the instability at the desired point in parameter space. Because the instabilities result from the interaction of reaction and diffusion, the patterns that arise are sensitive to the overall scale of the system. As a result, it is difficult to obtain the degree of scale-invariance that is observed in various biological systems. However, modifications of Turing's model can circumvent this difficulty [39, 41]. Generalized Turing systems with mixed boundary conditions, inhomogeneous domains and spatially varying diffusivities have also been investigated [3, 8, 6, 7]. More generally, models based on reaction diffusion systems of activator-inhibitor type have been proposed by Meinhardt (see for example, [29, 30]).

Mechanochemical models. In the Turing model, a chemical prepattern precedes differentiation. In the mechanochemical model developed by Oster, Murray, and coworkers [37, 38, 31], pattern formation and structural change occur simultaneously. In this model, it is assumed that regions of high cell density form cartilage. A particular motivation is the work of Toole [63] suggesting that the higher levels of hyaluronic acid (HA) in the distal subridge prevent cells from aggregating. When the cells leave the progress zone, the enzyme hyaluronidase (HAase) is produced. The degradation of HA by HAase reduces intracellular HA concentration and allows cells to aggregate. The process of aggregation could deform the limb into its characteristic paddle shape, leading to bifurcation in the aggregations. The model includes equations representing changes in cell and ECM density, the transport and production of HA and HAse, and the balance of cell-ECM forces [27]. The strongest criticism has been that of Wolpert and Hornbruch, who argue that cells are differentiated prior to the time of overt condensation [70, 73] contrary to the underlying assumption of the mechanochemical model that pattern and aggregation occur simultaneously.

In addition, the assumption that a significant increase in cell density precedes differentiation may be incorrect. Fell, who first described the phenomena [13] of "precartilage condensation", and many observers since, noticed a significant increase in cell density in the precartilage aggregations. More recent work suggests that the apparent increase in cell density associated with the precartilage aggregations may be an artifact of the fixation method [52, 1, 53]. It is still not certain if cell density increases or not, yet the belief persists that a significant increase occurs. For example, a 1992 review article on condensations [14] asserts that cell densities can increase by 75% between stages 22 and 26 and 200% in the chick wing bud between stages 19 and 25.

Growth and morphogenesis. The processes governing outgrowth and the associated transformations in limb bud geometry were modeled by Ede and Law [11]. They described a two-dimensional model of the limb on a grid in which an individual cells occupied an individual grid site. Cell division was represented by the placement of a new cell in an unoccupied site. The processes of growth and morphogenesis depended on explicit rules modeling differential growth rates and adhesivity, as well as cell movement.

Molecular models. As we noted in the Introduction, pattern formation in vertebrate limbs is controlled by a complex network of signals originating in the AER, the ZPA and the non-AER ectoderm. A model for these interactions as proposed by Duprex et al [10] is shown in Figure 4(a). A simplified model shown in Figure 4(b) assumes a direct interaction

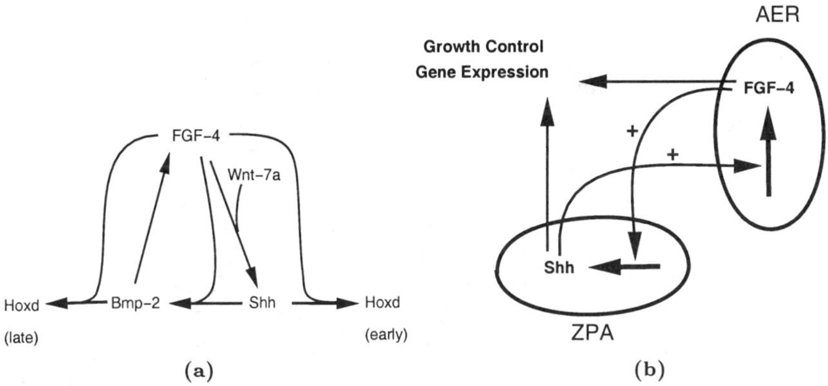

FIG. 4. *(a) A model for the interactions of* Fgf-4, Shh, *and* Bmp-2, *and* Wnt-7a. *Adapted from [10]. (b) A schematic of the reduced kinetic interactions between* Fgf-4 *and* Shh. *From [9].*

between *Shh* and *Fgf-4* with no intermediaries and ignores the DV signaling of *Wnt-7a*. In this model, the signaling molecules encoded by *Fgf-4* and *Shh* control growth through concentration levels of the growth factor FGF-4. In addition, downstream control of *Hox* and other genes is assumed to be under the control of the two signaling molecules.

4. A model for outgrowth and spatial patterning. In this section we describe a new type of model for limb development that combines the processes of growth, morphogenesis and cell signaling from specialized regions. A more detailed description can be found in [9]. The initial version is two-dimensional and models the limb bud outgrowth and patterning processes in the PD-AP plane. A schematic of the model is shown in Figure 5. The model consists of a fluid-mechanical component that describes limb bud outgrowth, a moving boundary that represents the mechanical properties of the limb bud ectoderm, and a reaction-diffusion-advection component that determines the spatio-temporal distribution of the signaling molecules that are produced in the AER or ZPA. The initial limb bud shape is an idealization of a stage 19 chick limb bud. At later stages, the force generated by the growth process produce transformations in the shape and size of the limb.

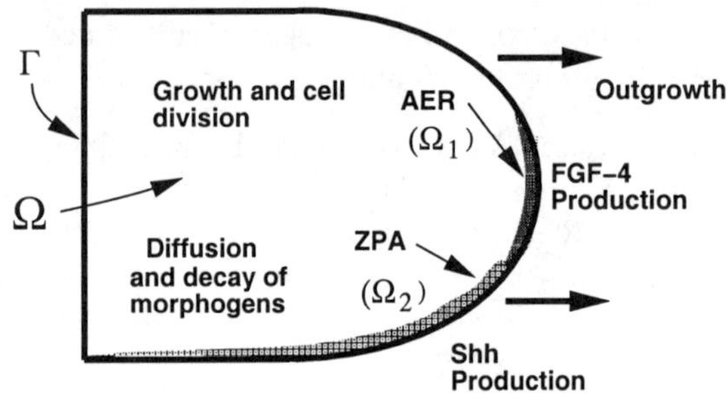

FIG. 5. *A schematic of the growing limb and the processes involved in the limb. The interior of the limb is denoted Ω, the AER region is denoted Ω_1, the ZPA region is denoted Ω_2, and the boundary of the limb is denoted Γ. The anterior edge of the limb bud is at the top; the posterior edge at the bottom. From [9]*

The mesoderm of the embryonic limb bud is a complex mixture of cells. In a long-term culture, embryonic tissue masses exhibit liquid-like characteristics in response to stress [43, 45, 44]. The short-term response is more like that of an elastic solid. Thus, the embryonic tissue is a complex viscoelastic material. Since we are primarily interested in the slow motion due to growth we neglect the elastic component and model the tissue as a viscous fluid. The process of cell growth and cell division requires transport of nutrients and fluid via diffusion and convection across the limb bud ectoderm and through the extracellular matrix, and at a later stage, through the capillary system. We idealize this complex process as a distributed source $S(\mathbf{c}, \mathbf{x}, t)$ of volume within the limb bud. The local source strength may depend upon the local concentration of growth factors $\mathbf{c}$, the location $\mathbf{x}$ of the tissue within the limb bud, and the age of the limb.

We assume that the fluid density is constant and that in the absence of growth, the fluid is incompressible. However, with growth the continuity equation for the local fluid velocity $\mathbf{u}$ takes the form

$$\nabla \cdot \mathbf{u} = S(\mathbf{c}, \mathbf{x}, t). \tag{4.1}$$

We assume that the fluid motion is governed by the Navier-Stokes equations, which provide the simplest description of a viscous fluid. These are given by [2]

$$\rho \frac{\partial \mathbf{u}}{\partial t} + \rho(\mathbf{u} \cdot \nabla)\mathbf{u} = -\nabla p + \mu\left(\nabla^2 \mathbf{u} + \frac{1}{3}\nabla S\right) + \rho \mathbf{F}. \tag{4.2}$$

Here ρ is the fluid density, $\mathbf{u}$ is the fluid velocity vector, p is the pressure, and μ is the fluid viscosity. The term $\mathbf{F}$ is the force density (force per unit area in two dimensions) that limb bud ectoderm exerts on the fluid surrounding it. As will be seen below, $\mathbf{F}$ is nonzero only in a thin layer surrounding the limb bud boundary.

Eqns. (4.1) and (4.2) describe the dynamics of the mesodermal tissue in the interior of the limb. In addition, the growth model includes a moving boundary Γ that represents the limb bud boundary. The configuration Γ at time t is given by the function $\mathbf{X}(s,t)$, where s is a Lagrangian label for a point on the boundary. The boundary moves at the local fluid velocity

$$\frac{\partial \mathbf{X}}{\partial t} = \mathbf{u}(\mathbf{X}(s,t),t). \tag{4.3}$$

The limb boundary is treated as an elastic material and the force per unit length $\mathbf{f}(s,t)$ at each point on the boundary is a function of the instantaneous configuration. In a three dimensional model the limb bud boundary could be modeled entirely by tangential elastic spring forces. In two space dimensions, we include elastic links between the anterior and posterior edges to represent the circumferential forces in the three-dimensional ectoderm. These anterior-posterior links prevent the limb bud from ballooning outward as the limb grows. The boundary is taken to be neutrally buoyant and thus the limb bud boundary forces are transmitted directly to the fluid via the force density $\mathbf{F}$, which is given by

$$\mathbf{F}(\mathbf{x},t) = \int_\Gamma \mathbf{f}(s,t)\,\delta(\mathbf{x}-\mathbf{X}(s,t))\,ds. \tag{4.4}$$

In this equation the integration is over the points of the boundary Γ and δ is the two-dimensional Dirac delta function. The limb bud grows out from the flank of the embryo, and for simplicity we regard the flank as an immovable boundary. This is accomplished by tethering the points on the proximal boundary in Figure 5 to fixed points in space with stiff elastic spring forces. The formulation of the fluid-mechanical system for this model is based on the immersed boundary method which was originally introduced by Peskin to model the blood flow in the heart [42]. A detailed description of the numerical implementation for the limb model is shown in [9].

In our initial study, we use the reduced biochemistry model outlined in Figure 4b in which FGF is produced exclusively within the AER and SHH within the ZPA. In the reduced model each species enhances the production of the other. Both species are assumed to diffuse freely throughout the limb bud mesoderm and to degrade everywhere within the tissue.

In mathematical terms, we represent the evolution of the morphogens $\mathbf{c} = (c_1, c_2)$ in the limb bud interior Ω by a system of advection-reaction-diffusion equations of the form

$$\frac{\partial \mathbf{c}}{\partial t} + \nabla \cdot (\mathbf{u}\mathbf{c}) = D\nabla^2 \mathbf{c} + \mathbf{R}(\mathbf{c}) \tag{4.5}$$

The first species c_1 represents the AER signaling molecule and second c_2 represents the ZPA signal. The diffusion matrix D is a diagonal matrix whose entries are the diffusion coefficients of the two proteins. We have assumed here for simplicity that the diffusion coefficients are constants. The morphogens are convected at the local velocity of the limb bud mesoderm $\mathbf{u}$.

As we indicated previously, the AER species is only produced in the AER (Ω_1) and the ZPA species is only produced in the ZPA (Ω_2). Thus $\mathbf{R} = (R_1, R_2)$ has the form

$$R_k = \begin{cases} r_k(\mathbf{c}) - \kappa_k c_k & \mathbf{x} \in \Omega_k \\ -\kappa_k c_k & \text{otherwise} \end{cases} \tag{4.6}$$

As shown in the equations, the AER is regarded as a specialized region having the competency to produce FGF. However, the actual production rate depends upon the local concentration levels of SHH which is produced in the ZPA and carried to the AER via diffusive transport. The ZPA is similarly modeled as a region of tissue within the limb identified as having the competency to produce SHH. The actual SHH production rates depend upon the local concentration levels of FGF. We assume Michaelis-Menten kinetics in Eqn. (4.6) of the form

$$r_1(\mathbf{c}) = V_1 \frac{c_2}{K_1 + c_2}, \qquad r_2(\mathbf{c}) = V_2 \frac{c_1}{K_2 + c_1} \tag{4.7}$$

with rate constants v_k and K_k. The source term S in Eqn. (4.1) has the form

$$S = s_1 c_1 + s_2 \tag{4.8}$$

with constants s_1 and s_2. Thus the local growth rate is modeled as a constant plus a term proportional to the local concentration of c_1.

On the boundary limb bud boundary Γ we specify homogeneous Neumann or zero-flux boundary conditions

$$\mathbf{n} \cdot D\nabla \mathbf{c} = 0, \tag{4.9}$$

for the morphogens.

5. Numerical simulations. A detailed study of the model system in the case of normal development is shown in [9]. The model has been extended to include the possibility of studying the effects of microsurgical interventions such as bead implants, microfilters, ZPA transplants, etc. In order to illustrate the model's capabilities we show the results of two numerical simulations. In Figure 6 we show a simulation suggestive of normal development. In order to generate the initial concentration level of FGF and SHH shown here, we begin with a steady state solution to Eqn. (4.5) with zero fluid velocity obtained numerically. Panel (a) shows the contours of the ZPA and AER species as well as the initial configuration of the limb. Panel (b) shows the limb bud and concentration contours at the end of the

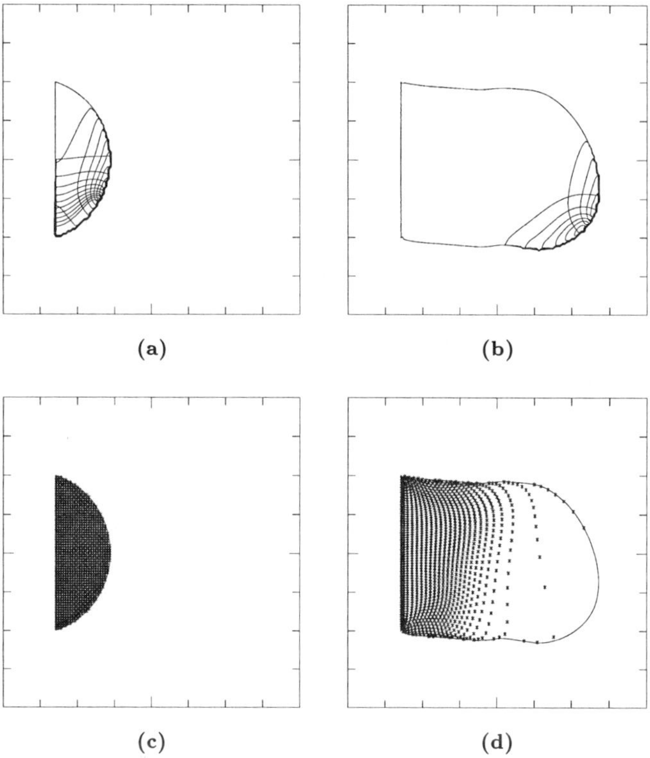

FIG. 6. *Numerical simulation of "normal" development. Panels (a) and (b) show the initial and final FGF and SHH contours. Panels (c) and (d) show the initial and final locations of fluid markers. Adapted from [9].*

simulation. In Panel (c), we show the limb bud in its initial configuration with fluid markers at each grid point within the limb. Panel (d) shows the limb and fluid markers at the end of the simulation. The fluid markers move with the local fluid velocity within the limb may be regarded as proxies for the limb bud cells and their cell progeny. Since the local growth rate depends linearly on the local concentration of the AER species, the local growth rates vary throughout the limb and are particularly elevated in the distal posterior region of the limb bud where the AER species has its highest concentration. The effects of the spatially varying local growth rate is clearly evident in the spacing of the fluid markers at the end of the simulation.

In the simulation shown in Figure 7 the limb bud does not have an AER region. As mentioned earlier, beads soaked in FGF and implanted or attached to the limb can substitute for the removal of the AER. In the simulation shown here, we have inserted a bead into the limb. The presence of the bead has an effect on both the chemistry and on the fluid flow within the limb.

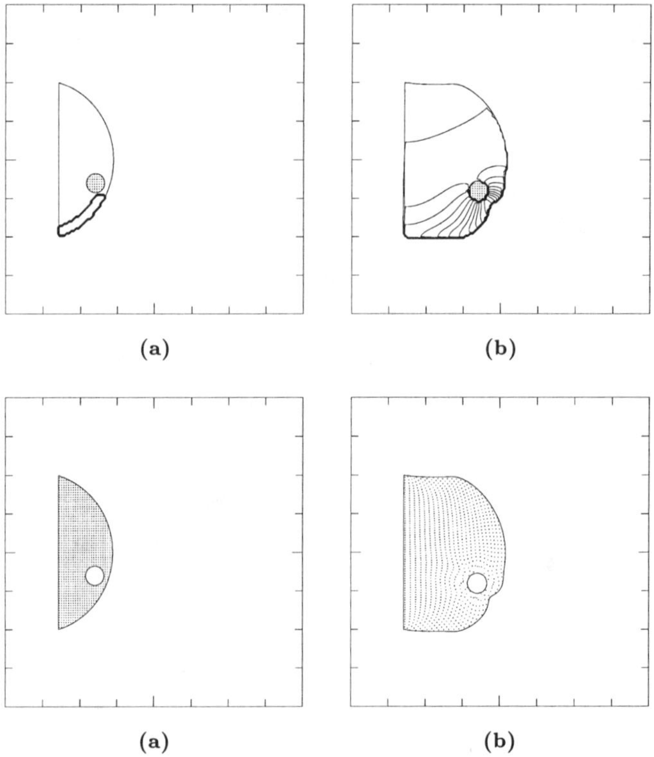

FIG. 7. *Numerical simulation with implanted bead and excised AER.*

Mechanically, the bead consists of a ring of immersed boundary points. Each point on the ring is connected to its two neighbors by stiff elastic springs and to a central immersed boundary point via stiff elastic springs that function as spokes. Additional rigidity is built into the bead by connecting each point on the rim to two additional nonneighboring points on the rim. Because of its rigidity, the bead is highly resistant to changes in geometry and thus the bead moves essentially as a rigid body. Since the immersed boundary points of the bead rim move at the local fluid velocities via Eqn. (4.3), the presence of the bead has a strong effect on the fluid flow near the bead.

We assume that the bead is impermeable to the ZPA species and impose Neumann boundary conditions at the rim of the bead for SHH. Additionally, we regard the bead as saturated with FGF and model this as a Dirichlet boundary condition for FGF at the rim of the bead. That is, we impose the boundary condition $c_1 = 1$, in dimensionless concentration units, at the rim of the bead. Thus the FGF concentration is fixed at the bead and diffuses outward. If the bead is placed near enough to the presumptive ZPA region, production of SHH is maintained as shown in this simulation. The FGF diffusivity and degradation rate terms used here (and in the first simulation as well) are such that the FGF concentrations fall off rapidly away from the bead. Although the FGF concentration profiles are not shown here, we can see in the distribution of fluid markers in panel D, that growth rates are elevated somewhat near the bead. The simulation in Figure 6 represents about 30 hours while that in Figure 7 represent about 43 hours. There is insufficient growth factor to maintain the same rate of outgrowth. We also see a deformation in limb bud shape.

6. Discussion. This model is designed to address several questions such as:

- What is the spatial distribution of the signaling molecules and how does their distribution depend on parameters such as the production rates and diffusion coefficients?
- Does the spatio-temporal history of cells correspond with results obtained from fate maps?
- Can one define a threshold-based combinatorial scheme of interpretation of the instantaneous concentration landscape or the history of the landscape that will lead to the observed spatial pattern of gene expression?
- Can the model explain some of the transplant results that are not explicable by the existing one-dimensional models?

The model does provide a detailed picture of the spatial distribution of the signaling molecules at each point in time. Of course, the details of these distributions depend on the mechanisms governing the production, transport and degradation of these signals. The model can be readily modified when more information about these processes becomes available.

We have compared our model simulations with fate maps shown in Bowen et al [4] and Vargesson et al [65]. We cannot hope to produce precise fate maps with our 2D model, but the fate maps we produce are qualitatively similar to the fate maps shown in these studies. As a result, the fluid markers as shown in Figure 6 can be regarded as proxies for individual cells and cells lines. A static model would be capable of predicting the morphogen concentration fields within the limb, but our model can show the location of individual cells or cell lines as the limb grows and ascertain the concentration levels of the morphogens at the cell sites over the course of limb outgrowth.

The spatio-temporal distributions of the *Hox* gene products and other genes as well, as discussed in the Introduction, depend either directly or indirectly on substances produced in the AER and ZPA. Transduction of extracellular signals into gene expression usually involves complex signal transduction mechanisms and presently the molecular basis of *Hox* gene activation has not been elucidated. As suggested in Yang et al [74] cells might respond to the morphogen signals in several ways. For example, a cell's response might by determined by the peak concentration levels of the signals. Alternatively, the response might depend on the morphogen levels experienced by the cell over a period of time. The new model described here provides detailed information on the location of individual cells and cell lines as well as a detailed history of the morphogen concentration levels at the site of the moving cells. We have used this information (cf. [9]) with a simple model of gene expression in which the concentration of the gene product depends on the entire history of a morphogen concentration. With this simple model of gene expression, it is possible to account for several of the *Hox* expression patterns at some stages of limb development. In future work we plan to explore the questions of gene expression in more detail.

One of the goals of the model is to develop a numerical simulation tool for studying limb development in normal and manipulated embryos. A current research direction is to simulate numerically several of the key limb experiments including bead implants. Implanted beads have been used for a variety of purposes including the delivery of retinoic acid in mirror image duplications. Also under development is the extension of the model to three-space dimensions and the development of a more realistic rheology.

REFERENCES

[1] A.L. Aulthouse and M. Solursh. The detection of a precartilage blastema-specific marker. *Dev. Biol.*, **120**(2):377–384, 1987.

[2] G.K. Batchelor. *An Introduction to Fluid Mechanics.* Cambridge Univ. Press, 1973.

[3] D.L. Benson, J.A. Sherratt, and P.K. Maini. Diffusion driven instability in an inhomogeneous domain. *Bull. Math. Biol.*, **55**:365–384, 1992.

[4] J. Bowen, J.R. Hinchliffe, T.J. Horder, and A.M.F. Reeve. The fate map of the chick forelimb-bud and its bearing on hypothesized developmental control mechanisms. *Anat. Embryol.*, **179**:269–283, 1989.

[5] V. Castets, E. Dulos, and P. De Kepper. Experimental evidence of a sustained standing Turing-type nonequilibrium chemical pattern. *Phys. Rev. Letts.*, **64**(24):2953–2956, 1990.

[6] R. Dillon. *A mathematical model of vertebrate limb development with modulated reaction and diffusion.* PhD thesis, University of Utah, 1993.

[7] R. Dillon, P.K. Maini, and H.G. Othmer. Pattern formation in generalized Turing systems I. Steady-state patterns in systems with mixed boundary conditions. *J. Math. Biol.*, **32**:345–393, 1994.

[8] R. Dillon and H.G. Othmer. Control of gap junction permeability can control pattern formation in limb development. In H.G. Othmer, P.K. Maini, and J.D. Murray, editors, *Experimental and Theoretical Advances in Biological Pattern Formation*, London, 1993. Plenum.

[9] R. Dillon and H.G. Othmer. A mathematical model for outgrowth and spatial patterning of the vertebrate limb bud. *J. Theor. Biol.*, **197**:295–330, 1999.

[10] D.M. Duprez, K. Kostakopoulos, P.H. Francis-West, C. Tickle, and P.M. Brickell. Activation of fgf-4 abd hoxD gene expression by BMP-2 expressing cells in the developing chick limb. *Development*, **122**:1821–1828, 1996.

[11] D.A. Ede and J.T. Law. Computer simulation of vertebrate limb morphogenesis. *Nature*, **221**:244–248, 1969.

[12] I.R. Epstein, I. Lengyel, S. Kádár, M. Kagan, and M. Yokoyama. New systems for pattern formation studies. *Physica A*, **188**:26–33, 1992.

[13] H.B. FELL. The histogenesis of cartilage and bone in the long bones of the embryonic fowl. *J. Morphol*, **40**:417–451, 1925.

[14] B.K. HALL AND T. MIYAKE. The membranous skeleton: the role of cell condensations in vertebrate skeletogenesis. *Anat Embryol*, **186**:107–124, 1992.

[15] V. HAMBURGER AND H.L. HAMILTON. A series of normal stages in the development of the chick embryo. *J. Morphol.*, **88**:49–92, 1951.

[16] T.F. HAYAMIZU, S.K. SESSIONS, N. WANEK, AND S.V. BRYANT. Effects of localized application of transforming growth factor $B1$ on developing chick limbs. *Dev. Biol.*, **145**:164–173, 1991.

[17] J.R. HINCHLIFFE AND T.J. HORDER. Lessons from extradigits. In J.F. Fallon, editor, *Limb Development and Regeneration*, pages 339–349. 4th Limb Development International Conference, California, J. Wiley, 1993.

[18] J.R. HINCHLIFFE AND D.R. JOHNSON. *The development of the vertebrate limb.* Oxford University Press, 1980.

[19] J.R. HINCHLIFFE AND A. SANSOM. The distribution of the polarizing zone in the leg bud of the chick embryo. *J. Embryol. Exp. Morphol.*, **86**:169–176, 1985.

[20] L.S. HONIG AND D. SUMMERBELL. Maps of strength of positional signaling activity in the developing chick wing bud. *J. Embryol. Exp. Morphol.*, **87**:163–174, 1985.

[21] R.V. PEARSE II AND C.J. TABIN. The molecular zpa. *J. Exp Zool.*, **282**:677–690, 1998.

[22] L.C. JAVOIS. Pattern specification in the developing chick limb. In G.M. Malacinski and S.V. Bryant, editors, *Pattern Formation: a primer in developmental biology*, pages 557–579. Macmillan, 1984.

[23] R.L. JOHNSON AND C.J. TABIN. Molecular models for vertebrate limb development. *Cell*, **90**(6):979–990, 1997.

[24] E. LAUFER, C.E. NELSON, R.L. JOHNSON, B.A. MORGAN, AND C. TABIN. *Sonic hedgehog* and *Fgf-4* act through a signaling cascade and feedback loop to integrate growth and patterning in the developing limb bud. *Cell*, **79**:993–1003, 1994.

[25] I. LENGYEL AND I.R. EPSTEIN. A chemical approach to designing Turing patterns in reaction-diffusion systems. *Proc. Natl. Acad. Sci.*, **89**:3977–79, 1992.

[26] J.A. MACCABE, JR J.W. SAUNDERS, AND M. PICKETT. The control of the anteroposterior and dorsoventral axes in embryonic chick limbs constructed of dissociated and reaggregated limb-bud mesoderm. *Dev. Biol.*, **31**:323–335, 1973.

[27] P.K. MAINI AND M. SOLURSH. Cellular mechanisms of pattern formation in the developing limb. *International Review of Cytology*, **129**:91–133, 1991.

[28] G. MARTIN. Pass the butter. *Science*, **274**:203–204, 1996.

[29] H. MEINHARDT. *Models of Biological Pattern Formation.* Academic Press, London, 1982.

[30] H. MEINHARDT. A boundary model for pattern formation in vertebrate limbs. *J. Embry. Exp. Morphol.*, **76**:115–137, 1983.

[31] J.D. MURRAY. *Mathematical Biology.* Springer-Verlag, 1989.

[32] C.E. NELSON, B.A. MORGAN, A.C. BURKE, E. LAUFER, E. DIMAMBRO, L.C. MURTAUGH, E. GONASLES, L. TESSAROLLO, L.F. PARADA, AND C. TABIN. Analysis of *Hox* gene expression in the limb bud. *Development*, 1996.

[33] S.A. NEWMAN, H.L. FRISH, AND J.K. PERCUS. On the stationary state analysis of reaction-diffusion mechanisms for biological pattern formation. *J. Theor. Biol.*, **134**:183–197, 1988.

[34] S.A. NEWMAN, M.P. PATOU, AND M. KIENY. The distal boundary of myogenic primordia in chimeric avian limb buds and its relation to an accessible population of cartilage progenitor cells. *Dev. Biol.*, **84**:440–448, 1981.

[35] J.K. NG, K. TAMURA, D.BÜSCHER, AND J.C. IZPISÚA-BELMONTE. Molecular and cellular basis of pattern formation during vertebrate limb development. *Curr. Top. Devel. Biol.*, **41**:37–66, 1999.

[36] L. Niswander, C. Tickle, A. Vogel, I. Booth, and G. Martin. FGF-4 replaces the apical ectodermal ridge and directs outgrowth and patterning of the limb. *Cell*, **75**(3):579–587, 1993.

[37] G.F. Oster, J.D. Murray, and A.K. Harris. Mechanical aspects of mesenchymal morphogenesis. *J. Embryol. Exp. Morph.*, **78**:83–125, 1983.

[38] G.F. Oster, J.D. Murray, and P.K. Maini. A model for chondrogenic condensations in the developing limb: The role of extracellular matrix and cell tractions. *J. Embryol. Exp. Morphol.*, **89**:93–112, 1985.

[39] H.G. Othmer. Synchronized and differentiated modes of cellular dynamics. In H. Haken, editor, *Dynamics of Synergetic Systems.* Springer-Verlag, 1980.

[40] Q. Ouyang and H.L. Swinney. Transition from a uniform state to hexagonal and striped patterns. *Nature*, **352**:610–612, 1991.

[41] E. Pate and H.G. Othmer. Applications of a model for scale-invariant pattern formation in developing systems. *Differentiation*, **28**:1–8, 1984.

[42] C.S. Peskin. Numerical analysis of blood flow in the heart. *J. Comp. Phys.*, **25**:220–252, 1977.

[43] H.M. Philips and M.S. Steinberg. Equilibrium measurements of embryonic chick cell adhesiveness. i. shape equilibrium in centrifugal fields. *Proc. Natl. Acad. Sci. USA*, **64**:121–127, 1969.

[44] H.M. Philips and M.S. Steinberg. Embryonic tissues as elasticoviscous liquids. I. Rapid and slow shape changes in centrifuged cell aggregates. *J. Cell Science*, **30**:1–20, 1978.

[45] H.M. Philips, M.S. Steinberg, and B.H. Lipton. Embryonic tissues as elasticoviscous liquids. II. Direct evidence for cell slippage in centrifuged aggregates. *Dev.Biol.*, **59**:124–134, 1977.

[46] J.A. Porter, S.C. Ekker, Woo-Jin Park, D.P. von Kessler, Y. Ma, A.S. Woods, R.J. Cotter, E.V. Koonin, and P.A. Beachy. Hedgehop patterning activity: Role of a lipophilic modification mediated by the carboxy-terminal autoprocessing domain. *Cell*, 1996.

[47] K.E. Robertson and C. Tickle. Recent molecular advances in understanding vertebrate limb development. *British journal of plastic surgery*, **50**(2):109–115, February 1997.

[48] J.W. Saunders and M.T. Gasseling. Ectodermal-mesenchymal interactions in the origin of limb symmetry. In R. Fleischmajer and R.E. Billingham, editors, *Epithelial-Mesenchymal Interactions*, pages 78–97, Baltimore, 1968. Williams and Wilkins.

[49] J.W. Saunders Jr. The proximo-distal sequence of origin of the parts of the chick wing and the role of the ectoderm. *Journal of Experimental Zoology*, **108**:363–403, 1948.

[50] A. Schmid, M. Dehlingerkremer, I. Schulz, and H. Gogelein. Voltage-dependent insp3 insensitive calcium channels in membranes of pancreatic endoplasmic reticulum vesicles. *Nature*, **346**:374–376, 1990.

[51] R.L. Searls. An autoradiographic study of the uptake of S-35 sulfate during the differentiation of limb bud cartilage. *Dev. Biol.*, **11**:155–168, 1965.

[52] C.T. Singley and M. Solursh. The spatial distribution of hyaluronic-acid and mesenchymal condensation in the embryonic chick wing. *Dev. Biol.*, **84**(1):102–120, 1981.

[53] M. Solursh. The role of extracellular matrix molecules in early limb development. *Seminars in Dev. Biol.*, **1**:45–53, 1990.

[54] D. Summerbell. A quantitative analysis of the effect of excision of the AER from the chick limb-bud. *J.Embryol.Exp.Morphol.*, **32**:651–660, 1974.

[55] D. Summerbell. Regulation of deficiencies along the proximal distal axis of the chick wing-bud: a quantitative analysis. *J.Embryol.Exp.Morphol.*, **41**:137–159, 1977.

[56] D. Summerbell. The effect of local application of retinoic-acid to the anterior margin of the developing chick limb. *J. Embryol. Exp. Morphol.*, **78**:269–290, 1983.

[57] D. Summerbell and L.S. Honig. The control of pattern across the antero-posterior axis of the chick limb bud by a unique signalling region. *Amer. Zool.*, **22**:105–116, 1982.

[58] D. Summerbell, J.H. Lewis, and L. Wolpert. Positional information in chick limb morphogenesis. *Nature*, **244**:492–496, 1973.

[59] C. Tickle. The polarizing region and limb development. In M. H. Johnson, editor, *Development in Animals*, pages 101–136. Elsevier/North-Holland Biomedical Press, 1981.

[60] C. Tickle, B. Alberts, L. Wolpert, and J. Lee. Local application of retinoic-acid to the limb bud mimics the action of the polarizing region. *Nature*, **296**(5857):564–566, 1982.

[61] C. Tickle and G. Eichele. Vertebrate limb development. In *Annual Review of Cell Biol.* Annual Reviews, Inc., 1994.

[62] W.L. Todt and J.F. Fallon. Development of the apical ectodermal ridge in the chick wing bud. *J.Embryol.Exp.Morphol.*, **80**:21–41, 1984.

[63] B. Toole. Hyaluronate turnover during chondrogenesis in the developing chick limb and axial skeleton. *Dev. Biol.*, **29**:321–329, 1972.

[64] A. M. Turing. The chemical basis of morphogenesis. *Phil. Trans. R. Soc. Lond. B*, **237**:37–79, 1952.

[65] N. Vargesson, J.D. Clarke, K. Vincent, C. Coles, L. Wolpert, and C. Tickle. Cell fate in the chick limb bud and relationship to gene expression. *Development*, **124**(10):1909–1918, May 1997.

[66] L. Wolpert. Positional information and the spatial pattern of cellular differentiation. *J. Theor. Biol.*, **25**:1–47, 1969.

[67] L. Wolpert. The development of pattern: mechanisms based on positional information. In *Membranes, dissipative structures and evolution.* Wiley, 1975.

[68] L. Wolpert. Pattern formation in limb morphogenesis. In H. W. Suer, editor, *(Advances in Zoologie, Vol. 26). Progress in Developmental Biology. Symposium*, pages 141–152, New York; Stuttgart, West Germany, 1981. Gustav Fischer Verlag.

[69] L. Wolpert. Positional information and prepattern in the development of pattern *Proceedings, Nato Conference on Cell-Cell signalling, Belgium*, 1989.

[70] L. Wolpert. Positional information and prepattern in the development of pattern. In A. Goldbeter, editor, *Cell to Cell Signalling: From Experiments to Theoretical Models*, pages 133–144, Cambridge, 1989. Academic Press.

[71] L. Wolpert. Positional information revisited. *Dev. Biol.*, **107**:3–12, 1989. (Supplement).

[72] L. Wolpert and A. Hornbruch. Positional signalling and the development of the humerus in the chick limb bud. *Development*, **100**:333–338, 1987.

[73] L. Wolpert and A. Hornbruch. Double anterior chick limb buds and models for cartilage rudiment specification. *Development.*, **109**:961–966, 1990.

[74] Y. Yang, G. Drossopoulou, P.-T. Chuang, D. Duprez, E. Marti, D. Bumcrot, N. Vargesson, J. Clarke, L. Niswander, A. McMahon, and C. Tickle. Relationship between dose, distance and time in *sonic hedgehog*-mediated regulation of anteroposterior polarity in the chick limb. *Development*, **124**:4393–4404, 1997.

MODELS FOR PIGMENT PATTERN FORMATION IN THE SKIN OF FISHES

K.J. PAINTER*

Abstract. The colours and patterns of the skin provides a fascinating system used for the study of pattern formation in experimental and theoretical research alike. In this article, a brief review of recent work on the pigmentation of the skin is presented. A mathematical model is shown to be able to capture many features associated with the evolving colour patterns on juveniles belonging to the genus of marine angelfish, *Pomacanthus*. Different forms of growth lead to very different patterning phenomena. The development of computational tools which can accurately reflect the geometry and growth of the real system will allow studies of the relationship between growth and patterning in species such as *Pomacanthus* or zebrafish.

1. Introduction. Across the animal kingdom, a large number of species rely on the colours and markings of their skin for purposes such as concealment and warning. The sophistication and control has reached astonishing levels in some animals. For example, species of bottom dwelling flatfish have been shown to adapt rapidly to background shades [54] and can assume a checkerboard type pattern when placed on the appropriate surface [46].

The ability to change skin colour and pattern is essential for survival in many species. In species such as the flatfish or the chameleon, the change occurs rapidly (on the order of seconds): such changes are termed *physiological* colour changes and are controlled by hormonal or nerve signals. The more slowly evolving changes are termed *morphological* colour changes, and these occur in response to continuous exposure to stimuli, as can be seen by prolonged exposure of skin to sunlight. A particularly striking example occurs in members of the marine angelfish genus *Pomacanthus* [15, 16]. Several species of *Pomacanthus* display a similar juvenile pattern consisting of several curved vertical white bars on a dark blue background. As the fish grows in size, new white bars are added between the older stripes, first emerging faint and narrow but widening as the fish continues to grow. This process repeats once or twice before the pattern evolves to the adult form, which can vary drastically between species. The aggressive adults attack fish with similar color patterns, and therefore juveniles adopt a distinctly different pattern to allow them to safely swim in an adults territory [17].

The relationships between age, growth and patterning have attracted interest from experimentalists and theoreticians alike. A simple one-dimensional model based on a Turing mechanism [62] was proposed to account for basic aspects of pattern evolution in the above angelfish [30], and further extensions have been incorporated to account for patterning

*Department of Mathematics, University of Utah, Salt Lake City, UT 84112, USA; Department of Mathematics, University of Minnesota, Minneapolis, MN 55455, USA.

details [63, 51]. An experimental study of a link between growth and patterning in members of the genus *Danio* (which includes the much studied zebrafish) has also been undertaken [36].

Pigment cells contain natural cell markers, namely, the pigment itself, and thus studies of pigmentation have provided experimentalists with a valuable system for understanding pattern formation in the developing embryo. In addition, studying these mechanisms may have important consequences for the clinical sciences. A number of diseases are attributed to defective pigmentation, including the condition vitiligo which affects approximately 1% of the population. This disease is a result of destruction of melanocytes and causes patches of white skin. Understanding the mechanisms by which pigment cells migrate and proliferate to pattern the skin may lead to a more successful course of treatment.

In this paper we present a brief review of the recent experimental and modelling research on the development of patterns and colours. We proceed to present a number of mathematical models for pigmentation in species of fish.

2. Formation of pigment patterns. A great body of research exists on the physiology of pigment cells and the formation of patterns within the skin. We refer to the following for greater detail: [11, 56, 13, 19, 18].

2.1. Pigments and chromatophores. Colour and pattern is the result of pigment cells (also called chromatophores) in the dermal and epidermal skin layers. These cells are large, branched and highly motile and contain membrane organelles called chromosomes holding the pigment granules. The organelles are rapidly aggregated or dispersed in response to nerve or hormonal signals. A number of chromatophore types exist, corresponding to the different pigments and colour. These include,

- *Melanophores* are a common chromatophore type residing in dermal and epidermal skin layers. In mammals and birds, they reside in the dermis where they are responsible for oranges and blacks by the transfer of pigment from melanophore dendrites into hairs, feathers and epithelial cells. In fish, melanophores give rise to black pigmentation, and are found in the dermal layers where they can respond to nerval and hormonal signals.
- *Iridophores* are smaller, round cells found in the dermal layers of fish. These cells contain platelets which reflect light resulting in a white/silvery colour.
- *Xanthophores* and *erythophores* result in bright yellows, oranges and reds.

2.2. Origin of pigment cells and migration to the skin. The identification of the neural crest, a transitory subpopulation of cells which develops above the neural tube in vertebrates, as the sole source of pigment cells was first established in amphibians [57] and then in other species. In fish, it is less clear as many species do not develop a distinct neural crest. It has, however, been clearly demonstrated in species such as the lamprey [47].

Pigment cell precursors migrate from the neural crest in a wave-like manner to uniformly seed the skin. It is thought that the various chromatophore types originate from a common neural crest precursor [3] and commitment to a specific type is not established until localization in the skin. The mechanisms controlling timing and migration of pigment cells from the neural crest are largely unknown. Cells do not acquire their characteristic pigment until after migration has ended, yet a number of markers have been developed which allow identification prior to pigment accumulation. A number of candidates have been proposed to control timing and migration of pigment cell precursors, including (i) a change in composition of the extra cellular matrix, (ii) the appearance of chemoattractant/repellent molecules, (iii) a change of cell adhesion properties. It is also possible that neural crest cell precursors may be forced onto a specific pathway due to the unattractive nature of other regions.

Studies of mouse mutations affecting coat pigmentation have provided an important tool for understanding factors involved in melanocyte development. Two essential receptor-ligand interactions have been revealed: the receptor tyrosine kinase, c-kit, with its ligand Steel factor (SLF, also called stem cell factor), e.g. [22, 65] and the endothelin receptor B together with endothelin 3 [5]. Soluble steel factor appears to have a role in regulating melanocyte precursor dispersal from the neural crest, whereas membrane-bound Steel factor is required for survival of the precursors within the dermis [64]. In addition to these observations, exogeneous SLF has been shown to be important for proliferation and differentiation of the melanocytes [32]. Intriguingly, SLF may have a role in promoting chemotactic activity in melanocytes [21, 64, 32]. Cells expressing functional c-kit receptors may be selectively attracted onto the lateral pathway by SLF, which diffuses from its site of production in the dermatomal epithelium.

In addition to SLF, a number of melanocyte mitogens (chemicals inducing cell division) have been identified, including leukotrines, endothelin-1 and certain fibroblast growth factors [40, 55, 24, 66]. Several of these have additionally been shown to induce melanocyte chemotaxis and chemokinetic movement [25].

2.3. Formation of colours, patterns and colour change. When one area of the skin is dominated by a specific colour, this can be attributed to an accumulation of pigment cells of the type producing that colour. White stripes (such as those on growing *Pomacanthus*) are often due to an abundance of iridophores, the red spots of certain carp are local aggregations of erythophore cells and black stripes in the angelfish are due to melanphores.

Despite few available pigments, a brief glance through an encyclopedia of fishes [9] reveals a staggering number of different colours. The macroscopically perceived colour can be attributed to the microscopic organization of pigment cells in the skin. In amphibians and reptiles, chromatophores are arranged in ordered layers in the dermis to form the *dermal chromatophore unit* [2]. These structures can create colours like green which cannot be formed by the available pigments alone. Similar arrangements exist in fishes. For example, blues of the damselfish are derived by a layer of iridophores backed by a layer of melanophores [27, 20].

Increases and decreases in the number of dermal chromatophores leads to morphological colour changes. The changes result from proliferation of chromatophores and/or the degradation of terminally differentiated pigment cells [54]. By counting the number of chromatophores over a long period of adaptation to light or dark backgrounds in *Oryzias*, it has been shown that dark adaptation occurs via increased numbers of melanophores and decreased numbers of leucophores, and *vice versa* for the reverse process [59].

The interactions and mechanisms which lead to pattern formation remain largely unknown, however establishment of patterns in larval salamanders has received attention. In the larval salamander *Ambyostoma tigrinum tigrinum*, melanophores scatter uniformly over the flank of the embryo, while xanthophores remain in aggregates in premigratory positions. As the xanthophores migrate, the melanophores recede short distances to form alternating bars [12, 48, 52]. At this time, a horizontal stripe over the lateral surface of the myotomes develops which is free of the otherwise abundant melanophores. A series of experiments [53] suggest that several factors are involved in development of this stripe, including interactions between xanthophores and melanophores and extracellular factors. The melanophore free region appears to develop through active retreat of melanophores from the forming lateral line. The horizontal stripes in other salamanders are thought to develop by response of pigment cells to cues in the ECM [61, 12].

2.4. Reaction-diffusion models in pigmentation. Turing [62] demonstrated that a simple system comprising of two reacting and diffusing chemicals can, under appropriate conditions, lead to stationary nonhomogeneous patterns. This was proposed by Turing as a mechanism for morphogenesis. The application of Turing patterns to pigmentation was made by Murray [41–44]. In essence, Murray proposed that a reaction-diffusion mechanism provides a morphogen prepattern which dictates cell differentiation. An attractive part of this theory is that the majority of mammalian patterns can be generated by such a mechanism. Similar models have been proposed by Bard [4] and Young [67].

The first modelling attempt to fish pigmentation patterns via reaction-diffusion theory was proposed by Kondo and Asai [30]. After observing the relationship between fish size and the number of stripes in the marine angelfish, *Pomacanthus semicirculatus*, a Turing model was proposed which predicts the doubling of the number of peaks of chemical concentration as the domain length doubles. A comparison of stripe rearrangement in adult *P. imperator* and numerical simulations of the model showed close agreement. Attempts to address additional aspects of pigmentation in *Pomacanthus* have since been proposed. Varea *et al.* [63] considered domains whereby one side is shaped to reflect the curved geometry of the fish skin. Through considering "enhanced" boundary conditions such that model parameters are elevated along certain boundaries, they have demonstrated several additional aspects of *Pomacanthus* patterning, such as the complicated patterns seen along boundary edges in *Pomacanthus imperator* and the orientation of stripes with respect to the boundary. Painter *et al.* [51] have augmented the model to include cell movement. This model is able to replicate the slow insertion of new stripes between older stripes. Additionally, features such as the curvature of juvenile stripes and the transition from stripes to spots as *P. semicirculatus* matures to adult have been shown in this model.

An activator inhibitor model has also been applied to the shell patterns of mollusks. The book by Meinhardt [39] demonstrates how many of the shell patterns seen in nature can by replicated in a reaction-diffusion model.

3. The cell movement model. In this section we consider a mathematical model for pigment pattern formation in the skin of fishes. In this model, the following aspects are taken into account:

- When a specific area of the skin is dominated by one colour, it is due to an aggregation of chromatophores of the type producing that colour.
- Morphological colour changes of the skin are due to an increased population of chromatophores. We should mention, however, that such colour changes can also be brought about by increased deposition of pigment in epithelial cells.
- Patterning in the skin is controlled by distribution of one or more chemical factors. A number of candidates have been discovered, yet no definitive identity exists for a so-called morphogen. Kirschbaum [29] demonstrated via transplantation experiments that differentiation of melanophores in the zebrafish occurred through response to local cues in the dermis.
- Increases in the total number of pigment cells are via a combination of proliferation from chromatophores or from undifferentiated stem cells. The stem cells are undifferentiated pigment cell precursors. Further differentiation of stem cells into a specific pigment cell type is likely to result from extracellular factors in the dermis/extracellular matrix.
- Initial pigment cell distribution varies from species to species. In the zebrafish, the larval pigment pattern consists of four lateral lines of melanophores: dorsal, septal, abdominal and ventral. Pigment cells in larval *Pomacanthus arcuatus* initially form a uniform

gray pigmentation when the fish is between 5 and 7 mm in length. The juvenile pigment pattern consisting of five vertical white bars develops from this uniform distribution [28].

- Several lines of evidence point to chemotactic guidance of migrating melanocyte precursors from the neural crest to the dermis [11, 6, 60, 21]. Potential chemoattractants include steel factor [32] and endothelin-1 amongst others [25]. In reality, cell movement may be induced by many mechanisms, including random motility, haptotaxis and cell adhesion.

Very little is currently known concerning the various interactions between the cell types, between the cells and the chemicals, and between the chemicals with each other, and it is not feasible at this moment to develop a highly detailed model for pigmentation patterning. We shall restrict attention to simple models of patterning which replicate the biological data.

In this section we consider a mechanism for generating the striped patterns in *Pomacanthus* based on a model for cell movement in response to gradients of chemical prepatterns derived from the reaction-diffusion system. The dominant chromatophore types involved in juvenile stripe development are likely to be melanophores, which we denote $M(\mathbf{x}, t)$, and iridophores, $I(\mathbf{x}, t)$. During the transition to adult stages, it is probable that other chromatophore types become important, and the effects of these may be a factor in the observed transformation. For example, yellow horizontal stripes of *P. imperator* are likely to be derived from xanthophores. In this model, we make the simplification of assuming that just one cell type is chemotactic, and that it is chemotactic to a single chemical species. Here we assume it is the iridophore cells, although we could equally postulate that it is the melanophores without qualitative differences in model behaviour. On a fixed domain, Ω, the equations are given by

$$
\begin{aligned}
\frac{\partial M}{\partial t} &= -\nabla \cdot J_M + f_M(M, u, v), \\
\frac{\partial I}{\partial t} &= -\nabla \cdot J_I + f_I(I, u, v), \\
\frac{\partial u}{\partial t} &= D_u \nabla^2 u + f(n, u, v), \\
\frac{\partial v}{\partial t} &= D_v \nabla^2 v + g(n, u, v).
\end{aligned}
\tag{3.1}
$$

J_M and J_I are flux terms for the melanophores and iridophores, respectively, and this contains contributions from random cell movement from chemotaxis. We simplify the model by assuming that only the two cell types reside in the dermis, and that the total cell density remains constant. Therefore, on a constant sized domain, we can take $f_M = f_I = 0$ (proliferation and degradation of pigment cells balance) and $\nabla \cdot (J_M + J_I) = 0$. Thus, melanophores themselves do not respond to the chemical gradients, however they experience a chemotactic flux type component through displacement by the moving iridophore cells. Zero flux is assumed at the boundary.

3.1. Patterning for different growth types. During embryonic and juvenile stages, animals undergo considerable tissue growth and deformation. The effects of such growth on patterning is beautifully illustrated by the evolving pigmentation patterns seen on species of amphibia, reptiles and fish and, clearly, models proposed to explain features of embryonic development must consider the potential effect of domain growth on the patterning process. For example, by considering the growth of the developing alligator jaw, it has been demonstrated that a simple reaction-diffusion type mechanism may account for the sequence of tooth primordia [31]. The need to consider the effects of growth on patterning has been illustrated for the reaction-diffusion mechanism on one and two-dimensional domains [51, 50]. In one dimension, patterning such that the number of concentration peaks developing through the reaction-diffusion mechanism evolves through the mode doubling sequence $1 - 2 - 4 - 8 - \ldots$ only occurs for certain parameters. In two dimensions, faster rates of growth can lead to highly convoluted stripes or no spatial pattern at all, rather than a regular pattern of stripes.

In the models considered to date domain growth has been incorporated in a simple manner, and a rigorous incorporation of tissue growth and deformation has yet to be undertaken. Even under a simple treatment it is possible to demonstrate that different types of domain growth may lead to very different patterning. This is demonstrated in the following example for a reaction-diffusion equation. Suppose evolution of a chemical, denoted by $c(\mathbf{x}, t)$, evolves on the constant domain according to

$$\frac{\partial c}{\partial t} = D\nabla^2 c + f(c).$$

If we assume that Ω changes as a function of time, then it is straightforward to show through application of Reynolds Transport equation [10, 8]

$$\frac{\partial c}{\partial t} + \nabla \cdot \mathbf{u}c = D\nabla^2 c + f(c), \tag{3.2}$$

where $\mathbf{u} = \partial \mathbf{x}/\partial t$ defines the fluid flow. We consider two simple types of growth for a growing one-dimensional domain, $[0, L(t)]$.

3.1.1. Uniform growth. Under uniform growth, we assume that for $x_i(0) \in (0, L)$, $x_i(t) = x_i(0)L(t)/L(0)$. Clearly, we have $u = xL'/L$, (where L' is the derivative with respect to t), and Equation (3.2) is given by,

$$\frac{\partial c}{\partial t} + \frac{cL'}{L} + \frac{xL'}{L}\frac{\partial c}{\partial x} = D\frac{\partial^2 c}{\partial x^2} + f(c).$$

We can convert the above equation on a one-dimensional growing domain onto a domain of constant size by the transformation, $(x,t) \to (y,\tau) = (x/L(t),t)$. It is easy to demonstrate that this transformation leads to

$$\frac{\partial c}{\partial t} = \frac{D}{L^2}\frac{\partial^2 c}{\partial y^2} + f(c) - \frac{L'}{L}c.$$

The uniform domain growth model leads to an equation for evolution of pattern in the reaction-diffusion system which can be solved with a simple numerical scheme. This may represent a good approximation for growth of the skin, were we to assume that nutrients required for cell proliferation were supplied uniformly to the skin from beneath. It is, however, a simplification of domain growth in living systems. Experimental data collected on growth of zebrafish and related species through larval and juvenile stages to adult [36] indicate that different regions of the body are growing at different rates.

3.1.2. Boundary growth. Under boundary growth, we assume that for $x_i(0) \in (0,L)$, $x_i(t) = x_i(0)$. This represents growth at the boundary and we have $u = 0$. We follow the above procedure and use the same transformation onto a domain of constant size. This gives the following equation defining pattern evolution,

$$\frac{\partial c}{\partial t} = \frac{D}{L^2}\frac{\partial^2 c}{\partial y^2} + \frac{yL'}{L}\frac{\partial c}{\partial y} + f(c)$$

Boundary growth occurs during extension of the developing axons in the nervous system. The developing neuron consists of a nerve cell body (or soma), a long thin axon and, at the axon tip, the growth cone from which filopodia extend to sense the environment for guidance signals. Growth of the axon occurs at the level of the growth cone [23]. Boundary growth may also be important with respect to skin growth if the nutrients were supplied from specific body regions.

3.1.3. Simulations under uniform and boundary growth. We compare the different types of growth above by numerical simulation. Kinetics for the reaction-diffusion system are based on a two-species system proposed by Lengyel and Epstein [33] to account for spatial patterns generated in the CIMA chemical reaction [7, 49]. The model is,

$$\begin{aligned}\frac{\partial u}{\partial t} &= D_u\frac{\partial^2 u}{\partial x^2} + k_1 - u - \frac{4uv}{1+u^2}\\ \frac{\partial v}{\partial t} &= k_3 D_v\frac{\partial^2 v}{\partial x^2} + k_2 k_3\left(u - \frac{uv}{1+u^2}\right)\end{aligned} \tag{3.3}$$

where D_u, D_v, k_1, k_2, k_3 are all positive constants. It is straightforward to determine the parameter space for Turing patterning.

In Figure 1 we compare results of numerical simulations for the following four cases:

- Exponential uniform and boundary growth $L(t) = L_0 \exp(rt)$, Figure 1(a) and (b).
- Linear uniform and boundary growth $L(t) = L_0(1 + rt)$, Figure 1(c) and (d).

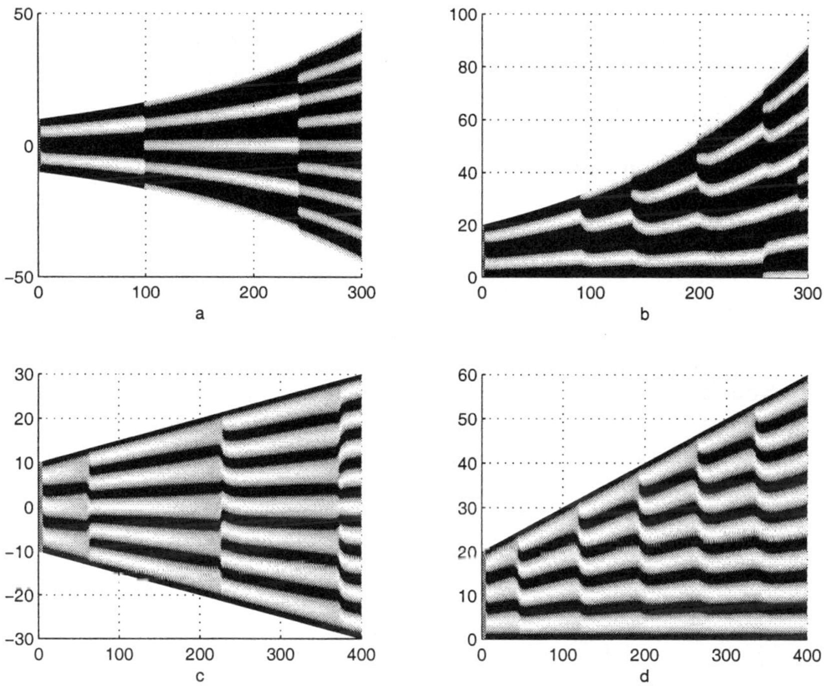

FIG. 1. *Chemical (u) concentrations on a domain growing in time (horizontal axes represents time, vertical axes represent space), for the four cases in the text. (a) Exponential uniform growth, (b) Exponential boundary growth, (c) Linear uniform growth, (d) Linear boundary growth. Parameters* $k_1 = 30.0$, $k_2 = 2.1$, $k_3 = 8.0$, $D_u = 1.0$, $r = 0.001$, $L_0 = 20.0$. *Each time unit in the figure represents 5 units of simulation time. Numerical simulations are solved with an adapted Euler method, with zero flux boundary conditions and random perturbations about the homogeneous steady state for initial conditions.*

Under uniform exponential domain growth, Figure 1(a), we see the mode doubling sequence of patterning which has been described previously [1, 30, 51], and proposed to provide a mechanistic basis for the doubling of the number of stripes as juvenile *Pomacanthus* doubles in length. With exactly the same parameters and conditions, yet using the boundary growth model, the regular sequence observed for uniform growth is lost. Initially new peaks/troughs develop near the growing boundary, however as the domain gets larger these are inserted in an irregular manner, Figure 1(b).

We have used exponential boundary growth to make a direct comparison with the results of simulations under exponential uniform growth. However, such growth is biological unrealistic, as it implies that the rate at which new tissue is added at the boundary occurs at a faster rate as time increases. A more plausible boundary growth would be linear, representing a constant rate of increase at the boundary. In Figure 1(c) and (d), we compare the uniform and boundary growth respectively for a linear function $L(t)$. With uniform linear growth, the regular sequence of mode doubling observed under exponential growth is lost. A regular sequence of patterning, however, now emerges for the linear growth under boundary growth. Peak splitting always occurs in the peak adjacent to the growing boundary, and those further from the growth retain their spatial location throughout growth. Thus the patterning of peaks evolves in the sequence $3-4-5-6-7-8-\ldots$

This latter behaviour is reminiscent of the mechanism by which new stripes emerge along the body of the adult zebrafish, Figure 2. Stripes 1 and 2 appear simultaneously from the larval pattern. Additional stripes appear in a specific sequence: 3 appears ventrally of 1, 4 appears dorsally of 1, 5 appears ventrally of 3, and 6 appears dorsally of 4. Stripes continue to be added, retaining this sequence, as space dictates. This sequence may suggest that growth in the zebrafish may be more closely approximated by boundary growth along the dorsal and ventral edges.

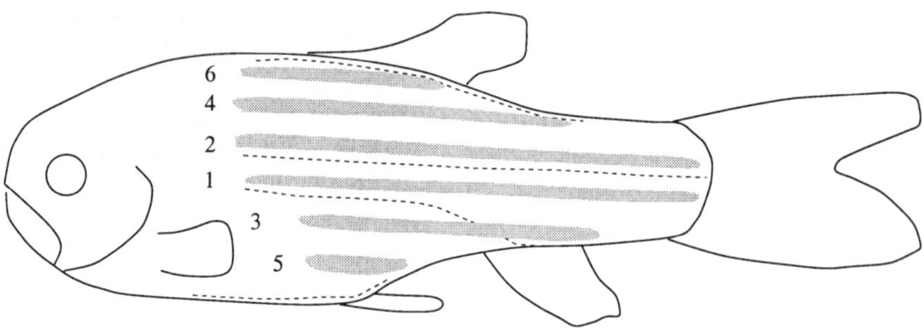

FIG. 2. *Order of adult stripe development in the zebrafish. Stripes 1 and 2 appear almost simultaneously from larval lines (dashed). Subsequent stripes appear in the order indicated.*

In summary, via the two types of growth we can force the patterning into different types of sequence. In boundary growth, the number of peaks of the reaction-diffusion sequence progress through the order $1-2-3-4-5-6-\ldots$, whereas exponential uniform growth gives a peak doubling sequence $1-2-4-8-16-\ldots$

3.2. Chemotactic-cell model under uniform growth. We consider numerical simulation of the full chemotactic-cell model with a uniformly growing domain incorporated. A detailed investigation into the various behaviours this model can show has been presented elsewhere [51]: Here we briefly explain how this model replicates the patterning phenomena of juvenile *Pomacanthus* development. The growth here is classified as *logistic*, stipulating that initially the fish grows in a manner approximating exponential growth, but eventually the growth rate slows and the fish approaches a maximum size.

This section considers two model formulations: (i) The zero cell feedback model, and (ii) the cell feedback model. In the former we have no effect on the chemicals by the cells: Chemical concentration patterns evolve independently and cells move in response to the gradients. In the second model we consider a form of chemical regulation by the cells by control of the rate of chemical synthesis. For the two-dimensional growing domain, $[0, L_1(t)] \times [0, L_2(t)]$, scaled onto a domain of constant size, we have

$$(3.4)\quad \begin{aligned} \frac{\partial I}{\partial t} &= \frac{1}{L_1^2}\frac{\partial}{\partial x}\left(D_I\frac{\partial I}{\partial x} - \frac{\chi_0}{(K+u)^2}I\frac{\partial u}{\partial x}\right) \\ &\quad + \frac{1}{L_2^2}\frac{\partial}{\partial y}\left(D_I\frac{\partial I}{\partial y} - \frac{\chi_0}{(K+u)^2}I\frac{\partial u}{\partial y}\right) \\ \frac{\partial u}{\partial t} &= D_u\left(\frac{1}{L_1^2}\frac{\partial^2 u}{\partial x^2}+\frac{1}{L_2^2}\frac{\partial^2 u}{\partial y^2}\right)+K(I)-u-\frac{4uv}{1+u^2}-(r_1+r_2)u, \\ \frac{\partial v}{\partial t} &= k_3 D_v\left(\frac{1}{L_1^2}\frac{\partial^2 v}{\partial x^2}+\frac{1}{L_2^2}\frac{\partial^2 v}{\partial y^2}\right)+k_2k_3\left(u-\frac{uv}{1+u^2}\right)-(r_1+r_2)v, \end{aligned}$$

where $L_1(t) = L_2(t) = L_0\exp(rt)/(a+\exp(rt))$. See [51] for a derivation of these equations. We consider zero flux boundary conditions, and initial conditions weighted such that a stripe pattern will initially form.

3.2.1. Case 1: No cell feedback, $K(I) = constant = k_1$. Simulations in this case have been described previously [51], and a succession of frames at different times are plotted in Figure 3 for the chemoattractant, u, (a)–(e), and the cell density, (f)–(j). Chemical concentrations evolve through a stripe doubling sequence and, with no feedback from cells to chemicals, cells will simply move in response to the morphogen gradients. This creates stripes appearing faint and narrow at first, but growing with time. As the domain approaches the maximum size, the regular sequence of stripes breaks into a pattern of spots. Domain growth appears to hold the stripe pattern during initial stages, but as the the domain approaches its maximum size, the pattern relaxes to its favored Turing wavelength. This has been compared to the sequence of stripes forming in juvenile *Pomacanthus semicirculatus* and the subsequent transition to the spotted adult pattern. During the transition from juvenile to adult pattern, a "mixed pattern" is observed whereby the pattern is reorganizing into spots, yet the stripes of the juvenile can still be seen. Such "transition patterns" are also observed during the transition from juvenile to adult pattern in *semicirculatus*. These transition stages are compared in Figure 4.

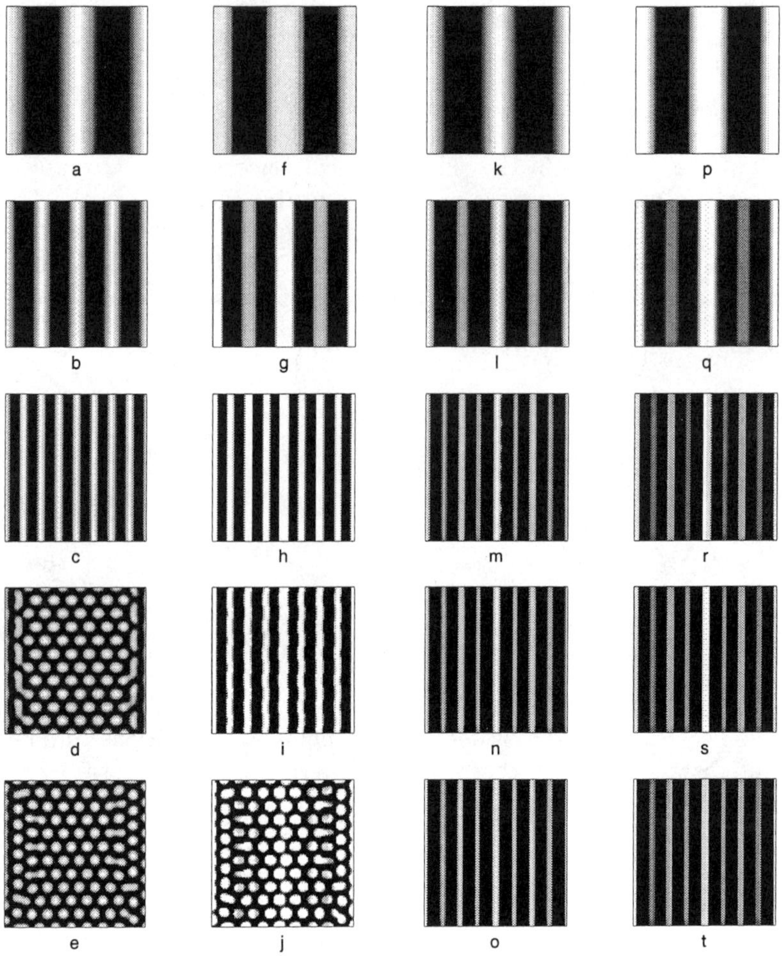

FIG. 3. *Numerical simulations for the cell movement model on a growing two-dimensional domain. For convenience of representation, the growing domain has been scaled onto one of constant size. (a)–(j) The zero feedback model shown for chemical u, (a)–(e), and cell density, (f)–(j), at* $t = 200$, *(a) and (f),* $t = 1600$, *(b) and (g),* $t = 3600$, *(c) and (h),* $t = 4200$, *(d) and (i), and* $t = 5200$, *(e) and (j). Corresponding plots for the feedback model are shown in (k)–(t). We use* $k_1 = 10.0$, $k_2 = 2.2$, $k_3 = 8.0$, $D_u = 0.01$, $D_v = 0.25$, $D_I = \chi_0 = 5.0 \times 10^{-5}$. $r = 0.001$, $a = 6.0$, $L_0 = 1.6$, $K = 1.0$. *Numerical simulations use an ADI method which has been adapted to include chemotactic cell movement.*

To understand the effects of differing chemotactic strengths, we vary the parameter χ_0 and examine the resulting cell density patterns. One dimensional results for three separate simulations are shown in Figure 5. For

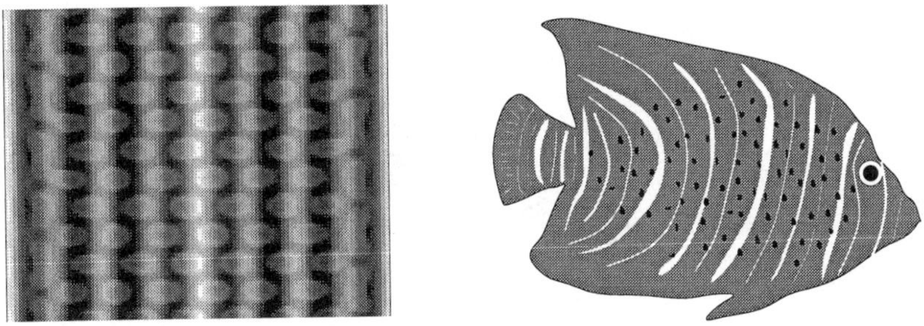

a b

FIG. 4. *(a) Frame from the numerical simulation of Figure 3, (f)–(j), showing the transition of cell density during the change from a striped to spotted pattern. The slow movement of cells results in intermediate patterns consisting of both stripes and spots. Such patterns are observed between frames (i) and (j) of Figure 3. (b)* Pomacanthus semicirculatus *also shows such transition patterns during the establishment of the adult coloring. Other details as for Figure 3.*

the "normal" chemotactic sensitivities, we observe the slow insertion of new stripes, consistent with the patterning on *Pomacanthus.* Similar patterning is observed when varying χ_0 by plus or minus an order of magnitude. Increases of χ_0 by plus or minus two orders of magnitude, however, result in very different patterning. In Figure 5(b), χ_0 has been increased by two orders in magnitude. Although initial peaks develop as previously, no new stripes form. Due to the strong chemotactic effect, all available iridophores are pulled into the initial stripes, leaving few to create secondary stripes. A decrease by two orders of magnitude, (c), results in the disappearance of pattern. When the chemotactic effect is very weak, emerging cell aggregates have small amplitudes, such that they are likely to be indiscernible at the macroscopic level.

3.2.2. Case 2: Cell feedback, $K(I) = k_4 I/(1.0 + I)$. We consider the effect of a cell feedback which takes the form of regulation of chemical synthesis. Simulations are plotted in Figure 3, (k)–(o), for the chemical concentrations and in (p)–(t) for the cell density. The parameters, growth rates and the time at which the panels are plotted are the same as those in (a)–(j). Here we set $k_4 = 2k_1$, which reflects the same rate of chemical production at the cell density homogeneous steady state.

The addition of feedback results in notable differences to the zero feedback case. First, we observe increased stability of the striped pattern. With feedback, no breakdown of stripes into spots results as the domain approaches maximum size: The stripes remain in an 8 stripe pattern, seen in the zero feedback case prior to break-up. Intuitively, this may be explained since the cells provide an additional reinforcement of the stripes: to rearrange the pattern into one of spots means also moving the cells. The

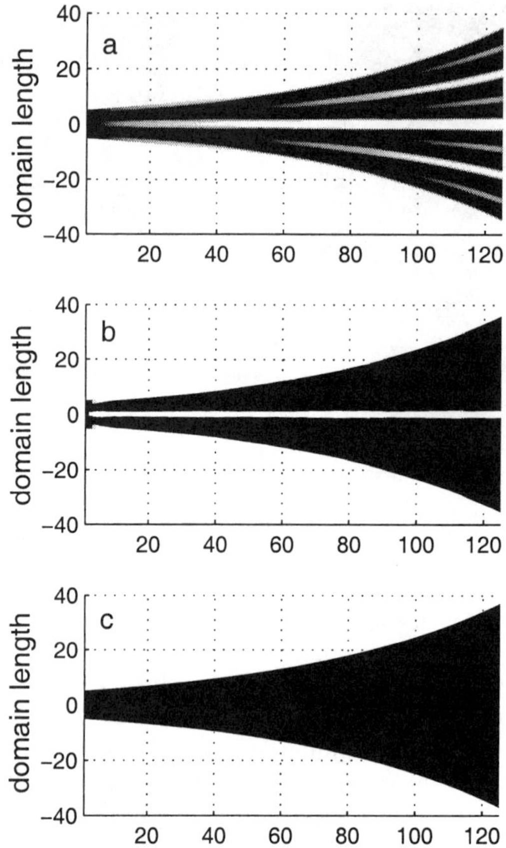

FIG. 5. *One dimensional space (vertical) - time (horizontal) plots for the cell density under varying chemotaxis strengths. In (a) we show the "standard" patterns with χ_0 set to 0.005 and other parameters as below. In (b), we use $\chi_0 = 0.5$. Here, the chemotaxis is strong and all available cells are pulled into the initial stripes. In (c), $\chi_0 = 0.00005$. Here, chemotaxis is weak such that the iridophore cell aggregations are indiscernible. Morphogen kinetics as used in Figure 3, $k_1 = 10.0$, $k_2 = 6.0$, $k_3 = 1.5$, $D_u = 0.01$, $k_3 D_v = 1.0$, $D_n = 0.001$ and $K = 10.0$. We use an exponentially growing domain, with growth rate $r = 0.01$ and an initial domain size of 1.6. White = cell density > 1.1, black = cell density <1.0.*

question of whether striped or spotted patterns arise in reaction-diffusion systems has been explored by several authors [14, 34, 35, 45, 37], however at present it is not completely clear which pattern will develop for a general reaction-diffusion model.

A second difference between the two cases lies in the form of chemical patterns. With no feedback, the chemicals evolve as of a standard reaction-diffusion system. Consequently, on doubling of the number of chemical stripes, the new stripes that emerge have the same amplitude and width as pre-existing stripes, Figure 6(a), see also Figure 1(a). This inability to produce peaks of distinct "ages" was one of the criticisms of the model proposed by Kondo and Asai for angelfish stripes [38, 51]. With chemical synthesis by cells, the underlying concentration peaks have clearly distinct amplitudes and widths, Figure 6(b).

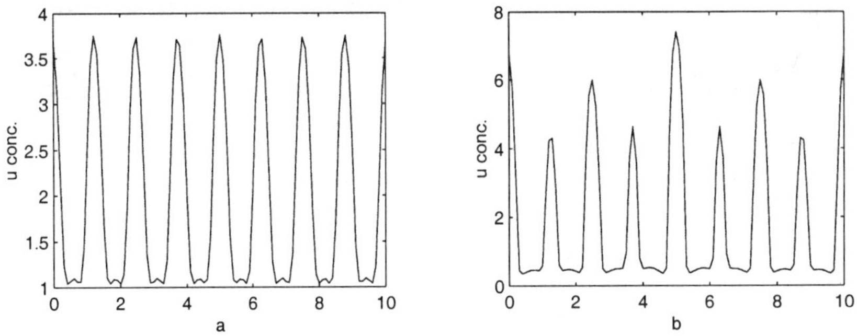

FIG. 6. *Comparison of the chemical concentration profiles at* $t = 4000$ *for the zero feedback model (a) and the feedback model (b), taken from the numerical simulations of Figure 3. The domain has grown from initial dimensions of 1.6* $\times$ *1.6 to 10.0* $\times$*10.0. See text for details.*

3.3. Geometry of the domain. To date, numerical simulations have been considered on either one-dimensional or two-dimensional domains of simple geometry (rectangular). As this is clearly a simplification of a real fish, we aim to understand the effects of realistic geometry on the patterning that develops.

3.3.1. Domain shape. We explore the effects of domain shape on patterning. The tail fin patterns of *Pomacanthus* provide a relatively simple geometry which can be solved numerically. In Figure 7(a) and (c), the results of two simulations under two sets of initial conditions are considered. Those for (a) create highly curved stripes which results in complexity of pattern at the edges of the tail fin. This is compared with the patterning on the tail fin of *P. imperator*, (b). The initial conditions in (c) give rise to "gently" curved stripes, resulting in a simple pattern of curved stripes, as seen on the tail fins of juvenile *P. semicirculatus*, (d). For both simulations, the geometry of the domain affects the pattern that develops by arranging stripes perpendicular to the boundaries. This forces the stripes into maintaining curved patterns.

The method of incorporating boundary shape has involved measuring dimensions of the fin from photographic images, calculating suitable functions to approximate the shape, and programming boundary conditions accordingly. This, of course, turns out to be a time-consuming and laborious process. A numerical scheme has been developed (see contribution by D. Bottino, this volume) which will solve a system of reaction-diffusion

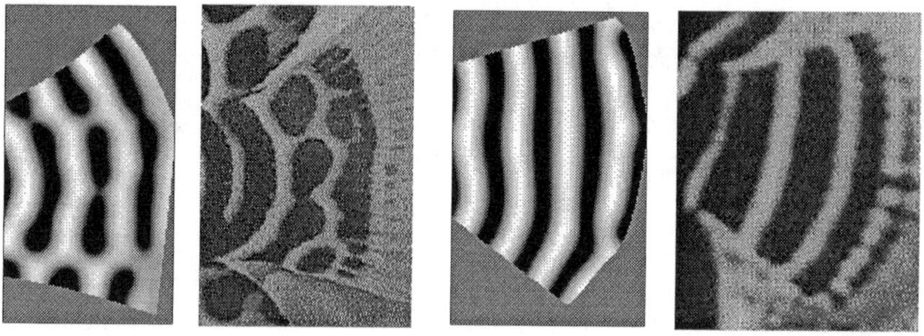

(a) (b) (c) (d)

FIG. 7. *Comparisons of patterns of the reaction-diffusion model with* Pomacanthus *tail fins on a domain geometry reflecting the actual fin. Simulation, (a), and* P. imperator *tail fin, (b). Simulation, (c), and* P. semicirculatus *tail fin (d). The simulations in (a) and (c) have been obtained using slightly different sets of initial conditions, with the same reaction-diffusion model as considered before. Initial conditions in (a) give rise to more highly curved stripes. See (Painter* et al., *1999) for details. Parameters:* $k_1 = 30.0$, $k_2 = 1.95$, $k_3 = 8.0$, $D_u = 1.0$, $D_v = 1.5$.

equations for a Voronoi tessellation. In Figure 8(a), the result of a numerical simulation is shown, which solves the reaction-diffusion system given by Equations (3.3) on an irregular domain using this method. The boundary itself was created by tracing an image of a juvenile *P. semicirculatus* at 6 months old (shown in Figure 8(b)) which is converted into a Voronoi grid by the scheme of Shewchuk [58].

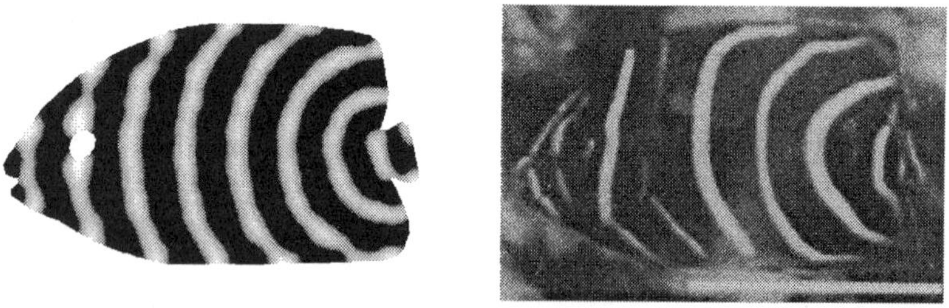

a b

FIG. 8. *(a) Pattern generated by solving the reaction-diffusion system on the irregular domain determined by tracing the boundary of the real fish shown in (b). Equations solved using the numerical scheme of Bottino (see this volume) on a Voronoi grid. Figure (b) taken from Kondo and Asai (1995) with kind permission of S. Kondo. (Reprinted by permission from Nature Vol. 376, No. 6543, pp. 765–768, August 31, 1995, Copyright Macmillan Magazines Ltd.) Parameters:* $k_1 = 30.0$, $k_2 = 1.9$, $k_3 = 8.0$, $D_u = 0.0025$, $D_v = 0.0125$. *Initial conditions consider a disturbance of the homogeneous steady state in the area of the tail fin. This forces a pattern of curved stripes.*

3.3.2. Three dimensional patterning. Fish chromatophore cells reside primarily in the dermis and, since body length and depth are much greater than the thickness of the skin (on the order of 0.1–1 mm, although there is wide variation between species), it is often adequate to model two dimensions only. In some species, however, patterning occurs on a fine scale: the first two stripes appearing in zebrafish are approximately 1 mm apart. In such cases it may be necessary to consider three dimensions.

Numerical simulations for Turing structures developing in a two species reaction-diffusion mechanism on a three-dimensional domain are plotted in Figure 9. The two standard patterns are shown on thin layers: stripes and spots. Here, depth is small and has no effect on the patterning.

Depth has negligible effect on pigmentation when the reaction occurs uniformly throughout, but may play a role when non-uniformity of reaction is introduced. For example, if reactants come from lower layers, a gradient in the supply may occur. The potential effect can be understood by imposing a gradient on a parameter through the depth. Jensen *et al.* [26] considered such effects in two dimensions for the kinetics given by Equations (3.3) under variation of the parameter k_2. It was noted that structure of the pattern moved through regions of spatial homogeneity to spotted patterns to stripes. Variation of this same parameter in three dimensions shows similar transitions, as we show in various slices of the three dimensional pattern plotted in Figure 10. At lower k_2 the pattern consists of stripes, however as this is increased we observe transition to spots and homogeneity.

In summary, when a parameter varies through depth it may disrupt the patterning. This may be significant if the parameter region for patterning is small, in which case small variations in the parameter may shift the reaction outside the parameter space for patterning.

4. Discussion. In this article, recent experimental data on the development of skin pigmentation patterns has been reviewed. The mathematical model presented here and in earlier work [51], has been shown to be capable of replicating many of the features observed during evolution of the pigment patterns in species such as *Pomacanthus semicirculatus*.

The understanding of patterns developing via a reaction-diffusion type mechanism on an irregular grid has received comparatively little attention to date. The development of numerical schemes in which a deforming irregular domain is considered will allow for simple incorporation of existing biological data on fish growth into the model and the understanding of pattern formation in such systems.

The modelling to date has considered a continuum approach to understand macroscopic patterning features. Experimental data in zebrafish and salamanders suggests that local interactions between pigment cell types may contribute to the patterning processes. One future direction is to employ a hybrid discrete/continuum approach whereby cells are consid-

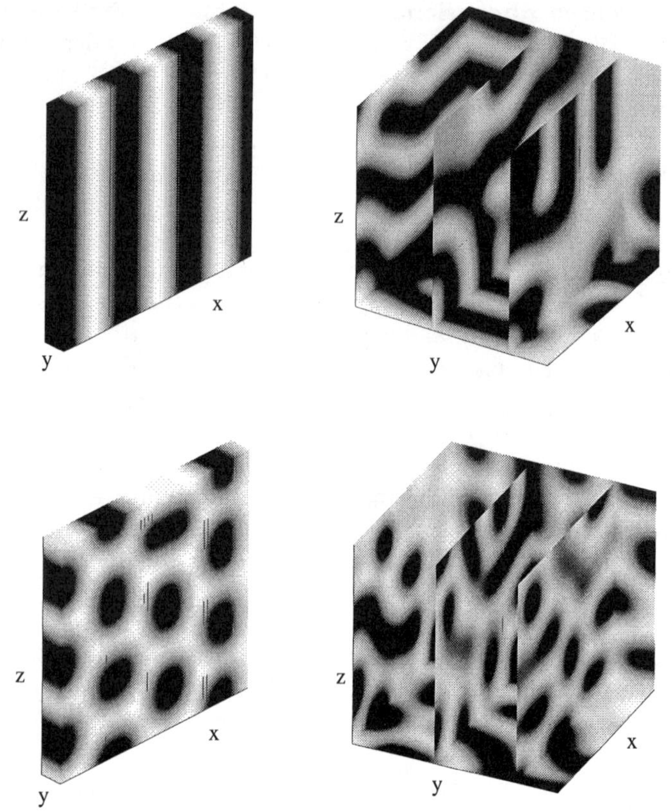

FIG. 9. *Three-dimensional patterns in a Turing system. Top left: On a thin domain, the pattern of stripes is maintained throughout the thickness of the domain. Bottom left: A similar effect occurs when the pattern is one of spots. Top and bottom right: Corresponding patterns for thicker layers. Slices through the pattern shows a large amount of structure and, in three dimensions, patterns show much less regularity than those in two dimensions. The regularity of the hexagonal patterns seen on the two dimensional layer is unlikely to develop when thickness becomes significant. Parameters use* $k_1 = 30.0$, $k_3 = 8.0$, $D_u = 1.0$, $D_v = 1.5$ *and* $k_2 = 1.4$ *for striped patterns,* $k_2 = 2.8$ *for spots. Equations solved using an Euler method with zero flux boundary conditions.*

ered as discrete particles. This approach would allow understanding of how different types of microscopic rules for movement give rise to different macroscopic patterning.

A number of areas are open to theoretical treatment. As mentioned above, the question of whether stripes or spots occur in a general reaction-diffusion model has yet to be answered satisfactorily. Patterns have also been observed to undergo break up from regular stripes into convoluted stripes under various forms and rates of growth [50, 51]. Whether these patterns represents a form of spatial chaos is an intriguing question.

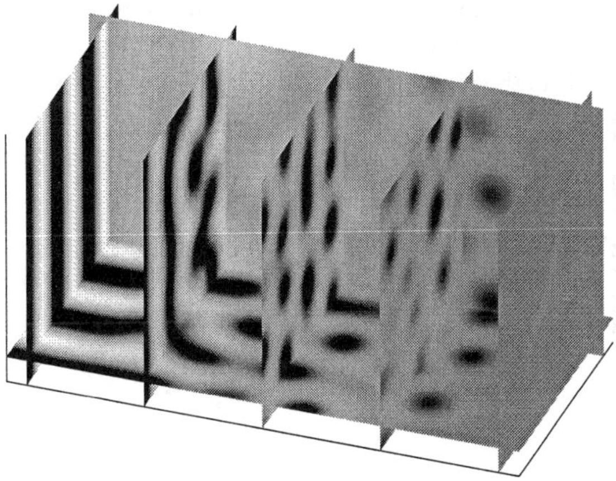

FIG. 10. *Simulations of three dimensional Turing patterns under linear variation of a kinetic parameter. Slices show how pattern varies along the direction of this variation. Patterning changes from stripes (left side) to spots (middle) and homogeneity (right). Simulations use the same kinetics as in previous simulations with* $k_1 = 30.0$, $k_3 = 8.0$, $D_u = 1.0$, $D_v = 1.5$ *and* k_2 *varies from 1.4 (left) to 5.6 (right). Domain dimensions are* $20 \times 40 \times 20$. *Solutions show pattern at* $t = 125$. *Simulations use an Euler method with zero flux boundary conditions.*

The model has been shown to reproduce many of the features associated with pigmentation in *Pomacanthus semicirculatus*. A principal prediction of this and previous models is that the appearances of the new "interstripes" occurs when the fish has approximately doubled its previous length. Growth of angelfish in an aquarium environment is limited by the size of the tank (e.g. see [9]) and the above prediction could therefore be tested by limiting the growth rate of the fish in this manner. However, due to the slow rate of growth (approximately one and a half years to reach adult), large size (15 inches) and the territorial nature of *Pomacanthus*, these fish are unsuitable as laboratory animals. A more widely studied species is the zebrafish, *Danio rerio*, which is small, easy to breed and has a transparent skin allowing for relatively straightforward observations of cell movement and pattern formation. We are currently applying the model to pattern formation in the zebrafish to develop a number of experimentally testable predictions. Excision and transplant experiments, whereby a fragment of skin tissue from a donor is removed and either replaced at a new orientation or transplanted onto a host fish (which has also had a fragment of skin removed), can be easily performed within the modelling framework. Comparison of the model results with existing transplant experiments (e.g. [29]) in zebrafish, together with the development of new predictions will be used to test the suitability of the model as a mechanism for pigmentation.

The mechanisms underlying pattern formation in areas such as pigmentation are still poorly understood. For this reason, modelling thus far has concentrated on developing simple models which capture basic features of the patterning. As more sophisticated techniques become available, and many of the current ambiguities are resolved, the model will be adapted accordingly.

Acknowledgments. I would like to thank D. Bottino for providing the numerical code and assisting me with simulations to solve the reaction-diffusion equations on an irregular grid. I would like to thank H.G. Othmer for useful comments on the manuscript. Supported by NIH grant GM 29123 to H.G. Othmer.

REFERENCES

[1] Arcuri, P. & Murray, J.D., 1986. Pattern sensitivity to boundary and initial conditions in reaction-diffusion models. *J. Math. Biol.*, **24**, 141–165.

[2] Bagnara, J.T. & Hadley, M.E., 1973. *Chromatophores and Color Change.* Eaglewood Cliffs, New Jersey: Prentice-Hall.

[3] Bagnara, J.T., Matsumoto, J., Ferris, W., Frost, S.K., Turner, W.A., Tchen, T.T., & Taylor, J.D., 1979. Common origin of pigment cells. *Science*, **182**, 1034–1035.

[4] Bard, J.B.L., 1981. A model for generating aspects of zebra and other mammalian coat patterns. *J. Theor. Biol.*, **93**, 363–385.

[5] Baynash, A. Greenstein, Hosoda, K., Giaid, A, Richardson, J.A., Emoto, N., Hammer, R.E., & Yanagisawa, M., 1994. Interaction of Endothelin-3 with Endothelin-B receptor is essential for development of epidermal melanocytes and enteric neurons. *Cell*, **79**, 1277–1285.

[6] Blume-Jensen, P., Claesson-Welsh, L., Siegbahn, A., Zsebo, K.M, Westermark, B., & Heldin., C.I., 1991. Activation of the human c-kit product by the ligand induced dimerization mediates circular actin reorganization and chemotaxis. *EMBO J.*, **10**, 4121–4128.

[7] Castets, V., Dulos, E., Boissonade, J., & Kepper, P. De., 1990. Experimental evidence of a sustained standing Turing-type nonequilibrium chemical pattern. *Phys. Rev. Lett.*, **64**, 2953–2956.

[8] Crampin, E., Gaffney, E., & Maini, P.K., submitted.

[9] Dakin, N., 1992. *The Macmillan book of the marine aquarium.* New York: Macmillan Publishing Company.

[10] Dillon, R., & Othmer, H.G., 1999. *A Mathematical Model for Outgrowth and Spatial Patterning of the Vertebrate Limb Bud.* To appear in J. Theor. Biol.

[11] Douarin, N.M. Le., 1982. *The Neural Crest.* Cambridge: CUP.

[12] Epperlein, H.-H. & Löfberg, J., 1990. The development of the larval pigment patterns in *Triturus alpestris* and *Ambystoma mexicanum. Adv. Anat. Embrol. Cell. Biol.*, **118**, 1–101.

[13] Erickson, C.A., 1993. From the crest to the periphery: Control of pigment cell migration and lineage segregation. *Pigment Cell Res.*, **6**, 336–347.

[14] Ermentrout, B., 1991. Stripes or spots? Nonlinear effects in bifurcation of reaction-diffusion equations on the square. *Proc. Roy. Soc. Lond. A.*, **434**, 413–417.

[15] Fraser-Brunner, A., 1933. A revision of the Chaetodont fishes of the subfamily *Pomacanthinae. Proc. Zool. Soc.*, **36**, 543–596.

[16] Fraser-Brunner, A., 1951. Pattern development in the chaetodont fish *Pomacanthus annularis* (Bloch), with a note on the status of *Euxiphipops. Copeia.*, **1**, 88–89.

[17] FRICKE, H.W., 1980. Juvenile-adult colour patterns and coexistence in the territorial coral reef fish *Pomacanthus imperator*. *Marine Ecology*, **1**, 133–141.

[18] FUJII, R., 1993a. Cytophysiology of fish chromatophores. *Int. Rev. Cytol.*, **143**, 191.

[19] FUJII, R., 1993b. *The Physiology of Fishes*. Marine Science Series. Boca Raton, Ann Arbor, London, Tokyo: CRC press. Chap. Coloration and Chromatophores, pp. 535–562.

[20] FUJII, R., KASUWAKA, H., , MIYAJI, K., & OSHIMA, N., 1989. Mechanism of skin coloration and its changes in the blue-green damselfish. *Zool. Sci.*, **6**, 477–486.

[21] GALLI, S.J., ZSEBO, K.M., & GEISSLER, E.N., 1993. The kit ligand, Stem cell factor. *Advances Immunol.*, **55**, 1–96.

[22] GEISSLER, E.N., RYAN, M.A., & HOUSMAN, D.E., 1988. The dominant-white spotting (W) locus of the mouse encodes the *c-kit* proto-oncogene. *Cell*, **55**, 185–192.

[23] GILBERT, S.F., 1997. *Developmental Biology*. Fifth edn. Sinauer Associates.

[24] HALABAN, R., GHOSH, S., & BAIRD, S., 1987. bFGF is the putative natural growth factor for human melanovytes. *In Vitro*, **23**, 47–52.

[25] HORIKAWA, T., NORRIS, D.A., YOHN, J.J., ZEKMAN, T., & MORELLI, J.B. TRAVERS J.G., 1995. Melanocyte mitogens induce both melanocyte chemokinesis and chemotaxis. *J. Invest. Derm.*, **104**, 256–259.

[26] JENSEN, O., MESEKILDED, E., BORCKMANS, P., & DEWEL, G., 1996. Computer-simulation of Turing structures in the Chlorite-Iodide-Malonic acid system. *Phys. Scripta.*, **53**, 243–251.

[27] KASUWAKA, H., OSHIMA, N., & FUJII, R., 1987. Mechanism of light reflection in blue damselfish motile iridophores. *Zool. Sci.*, **4**, 243–257.

[28] KELLEY, S., 1995. Pigmentation, squamation and the osteological development of larval and juvenile gray angelfish *Pomacanthus arcuatus* (Pomacanthidae: Pisces). *Bull. Mar. Sci.*, **56**(3), 826–848.

[29] KIRSCHBAUM, F., 1975. Untersuchungen uber dans Farbmuster der Zebrabarbe *Brachydanio rerio* (Cyprindae, Teleostei). *Roux's. Arch. Dev. Biol.*, **177**, 129–152.

[30] KONDO, S. & ASAI, R., 1995. A reaction-diffusion wave on the skin of the marine angelfish *Pomacanthus*. *Nature*, **376**, 675–768.

[31] KULESA, P.M., CRUYWAGEN, G.C., LUBKIN, S.R., MAINI, P.K., SNEYD, J., FERGUSON, M.W.J., & MURRAY, J.D., 1996. On a model mechanism for the spatial patterning of teeth primordia in the alligator. *J. Theor. Biol.*, **180**, 287–296.

[32] KUNISADA, T., YOSHIDA, H., YAMAZAKI, H., MIYAMOTO, A., HEMMI, H., NISHIMURA, E., SHULTZ, L.D., NISHIKAWA, S., & HAYASHI, S., 1998. Transgene expression of steel factor in the basal layer of the epidermis promotes survival, proliferation, differentiation and migration of melanocyte precursors. *Development*, **125**, 2915–2923.

[33] LENGYEL, I. & EPSTEIN, I.R., 1991. Modelling of Turing structures in the chlorite-iodide-malonic acid-starch reaction system. *Science*, **251**, 650–652.

[34] LYONS, M.J. & HARRISON, L.G., 1991. A class of reaction-diffusion mechanisms which preferentially select striped patterns. *Chem. Phys. Lett.*, **183**, 158–164.

[35] LYONS, M.J. & HARRISON, L.G., 1992. Stripe selection: An intrinsic property of some pattern-forming models with nonlinear dynamics. *Dev. Dyn.*, **195**, 201–215.

[36] MCCLURE, M., 1998. *Chapter3: Growth, Shape Change, and the development of pigment patterns in fishes of the genus* Danio *(Teleostei: cyprindae)*. Ph.D. thesis, Cornell University.

[37] MEINHARDT, H., 1989. Models for positional signalling with application to the dorsoventral patterning of insects and segregation into different cell types. *Development*, **supplement**, 169–180.

[38] MEINHARDT, H., 1995. Dynamics of stripe formation. *Nature*, **376**, 722–723.

[39] MEINHARDT, H., 1998. *The algorithmic beauty of sea shells.* 2nd edn. Berlin, New York: Springer.

[40] MORELLI, J.G., YOHN, J.J., LYONS, B., MURPHY, R.C., & NORRIS, D.A., 1989. Leukotrines C_4 and D_4 as potent mitogens for cultured human neonatal melanocytes. *J. Invest. Dermatol.*, **93**, 719–722.

[41] MURRAY, J.D., 1979. A pattern formation mechanism and its application to mammalian coat markings. Lecture Notes in Biomathematics, Vol. 39. Berlin, Heidelberg, New York.: Springer.

[42] MURRAY, J.D., 1981. A pre-pattern formation mechanism for animal coat markings. *J. Theor. Biol.*, **88**, 161–199.

[43] MURRAY, J.D., 1988. How the leopard got its spots. *Sci. Am.*, **258**, 80–87.

[44] MURRAY, J.D., 1993. *Mathematical Biology.* Second edition edn. Berlin, Heidelberg, New York: Springer-Verlag.

[45] NAGORCKA, B.N., 1992. From stripes to spots: Prepatterns which can be produced in the skin by a reaction-diffusion system. *IMA. J. Math. Appl. Med. & Biol.*, **9**, 249–267.

[46] NAITOH, T., MORIOKA, A., & OMURA, Y., 1985. Adaptation of a common freshwater goby, yoshinobori, *Rhinogobius brunneus* Temminck et Schlegel to various backgrounds including those containing different sizes of black and white checkerboard squares. *Zool. Sci.*, **2**, 59.

[47] NEWTH, D.R., 1956. On the neural crest of the lamprey embryo. *J. Embryol. Exp. Morphol.*, **4**, 358–375.

[48] OLSSON, L. & LÖFBERG, J., 1992. Pigment pattern formation in larval ambystomatid salamanders: *Ambystoma tigrinum tigrinum. J. Morphol.*, **211**, 73–85.

[49] OUYANG, Q. & SWINNEY, H.L., 1991. Transition from a uniform state to hexagonal and striped Turing patterns. *Nature*, **352**, 610–612.

[50] PAINTER, K.J., 1997. *Chemotaxis as a Mechanism for Morphogenesis.* Ph.D. thesis, University of Oxford.

[51] PAINTER, K.J., OTHMER, H.G., & MAINI, P.K., 1999. *Stripe formation in juvenile* Pomacanthus *explained by a generalized Turing mechanism with chemotaxis.* Proc. Natl. Acad. Sci. USA, Vol. **96**, 5549–5554.

[52] PARICHY, D.M., 1996a. Pigment patterns of larval salamanders (Ambystomatidae, Salamandridae): The role of the lateral line sensory system and the evolution of pattern-forming mechanisms. *Dev. Biol.*, **175**, 265–282.

[53] PARICHY, D.M., 1996b. When neural crest and placoses collide: Interactions between melanophores and the lateral lines that generate stripes in the salamander *Ambystoma tigrinum tigrinum* (Ambystomatidae). *Dev. Biol.*, **175**, 283–300.

[54] PARKER, G.H., 1948. *Animal Colour Changes and Their Neurohumours.* Cambridge: CUP.

[55] RUBIN, J.S., CHAN, A.M.L., BOTTARO, D.P., BURGESS, W.H., TAYLOR, W.G., CECH, A.C, HIRSCHFIELD, D.W., WONG, J., MIKI, T., FINCH, P.W., & AARONSON, S.T., 1991. A broad spectrum human lung fibroblast-derived mitogen is a variant of hepatocyte growth factor. *Proc. Natl Acad. Sci. USA*, **88**, 415–419.

[56] SCHLIWA, M., 1986. *Biology of the Integument 2: Vertebrates.* Berlin Heidelberg New York Tokyo: Springer-Verlag. Chap. Pigment Cells, pp. 65–77.

[57] SHANE, G.P. DU., 1934. The origin of pigment cells in Amphibia. *Science*, **80**, 620–621.

[58] SHEWCHUK, JONATHAN RICHARD, 1996. Triangle: Engineering a 2D Quality Mesh Generator and Delaunay Triangulator. *pp. 203–222 of:* Lin, Ming C., & Manocha, Dinesh (eds.), *Applied Computational Geometry: Towards Geometric Engineering.* Lecture Notes in Computer Science, Vol. 1148. Springer-Verlag. From the First ACM Workshop on Applied Computational Geometry.

[59] SUGIMOTO, M. 1993. Morphological colour changes in the medaka, *Oryzias latipes*, after prolonged background adaptation — I. Changes in the population and morphology of the melanophores. *Comp. Biochem. Physiol.*, **104**A, 513.

[60] TOSNEY, K.W., 1992. A long distance cue from emerging dermis stimulates neural crest migration. *Soc. Neurosci. Abs.*, **18**, 1284.

[61] TUCKER, R.P. & ERICKSON, C.A., 1986. Pigment patternformation in *Taricha torosa*: The role of the extracellular matrix in controlling pigment cell migration and differentition. *Dev. Biol.*, **118**, 268–285.

[62] TURING, A.M., 1952. The chemical basis for morphogenesis. *Phil. Trans. Roy. Soc. Lond. B.*, **237**, 37–72.

[63] VAREA, C., ARAGON, J.L., & BARRIO, R.A., 1997. Confined Turing patterns in growing systems. *Phys. Rev. E.*, **56**, 1250–1253.

[64] WEHRLE-HALLER, B. & WESTON, J.A., 1995. Soluble and cell-bound forms of steel factor activity play distinct roles in melanovyte precurso dispersal and survival on the lateral neuarl crest migration pathway. *Development*, **121**, 731–742.

[65] WILLIAMS, D.E., EISENMAN, J., BAIRD, A., RUACH, C., NESS, K. VAN, MARCH, C.J., PARK, L.S., MARTIN., U., MOCHIZUKI, D.Y., BOSWELL, H.S., BURGESS, G.S., COSMAN, D., & LYMAN, S.D., 1990. Identification of a ligand for the *c-kit* proto-oncogene. *Cell*, **63**, 167–174.

[66] YADA, Y., HIGUCHI, K., & IMOKAWA, G., 1991. Effects on endothelins on signal transduction and proliferation in human melanocytes. *J. Biol. Chem.*, **266**, 18352–18357.

[67] YOUNG, D.A., 1984. A local activator-inhibitor model of vertebrate skin patterns. *Math. Biosci.*, **72**, 51–58.

GENERIC MODELLING OF VEGETATION PATTERNS. A CASE STUDY OF *TIGER BUSH* IN SUB-SAHARIAN SAHEL

R. LEFEVER*, O. LEJEUNE*, AND P. COUTERON†

Abstract. The conditions underlying the mean field description of vegetation patterns are reviewed. A generic partial differential equation model describing vegetation propagation over space by reproduction under isotropic as well as anisotropic environmental conditions is discussed. An example of *Tiger Bush* in Burkina Faso is analyzed on the basis of this model.

1. Introduction. The modelling of vegetation patterns presented here finds its inspiration in the mean field theory of equilibrium phase transitions which has been used, notably, to study the spinodal decomposition of alloys and of polymer blends [1]. Pattern formation in these conservative systems is the result of nonlinear diffusion processes which allow for the occurrence of negative diffusion coefficients as a consequence of the existence of *nonlocal* dispersive forces. This situation contrasts with the by now classical "Turing inspired" reaction-diffusion theory and its applications to many pattern formation problems in biology where diffusion is approximated by linear fickian laws and where nonlinearities originate from nonconservative *local* chemical source terms [2].

The need for generalizations including nonlocal interactions has however been pointed out in biological contexts such as population dynamics, where more complex dispersal mechanisms than simple diffusion come into play (see, e.g., [3–5]). Botanical ecosystems clearly belong to this category of biological systems. Furthermore, they display an amazing variety of complex, large scale space-time organisations which so far have attracted relatively little attention from theoreticians. We would like to show that these behaviors can be appropriately addressed within the framework of the mean field approach. The latter indeed is well-suited to treat the space-time behavior of vegetal populations constituted of individuals which do not move, and which hence, can only propagate over space by reproduction. We will see that a generic formulation of this transport by reproduction phenomenon can be achieved in the form of a nonlinear fourth order partial differential equation model which encompasses isotropic as well as anisotropic environmental conditions. A case study based on this model will also be presented. In the next section we recall the properties of the vegetation patterns which will interest us in the following.

*Faculté des Sciences, CP 231, Université Libre de Bruxelles, B-1050 Bruxelles, BELGIUM.

†Ecole Nationale du Génie Rural des Eaux et Forêts, ENGREF – B.P. 5093, 34033 Montpellier Cedex 01, FRANCE.

1.1. Field observations. In many uninhabited arid regions of Africa, America and Australia, one observes vast territories the vegetation of which exhibits patterns having a spatial periodicity of the order of 10 to 500 meters (see Table 1). These large, botanical, ecological structures are called *vegetation patterns.* The most often reported structures are the most striking ones, i. e. vegetation stripes, but spotted patterns have also been described. Our analysis carried out in Burkina Faso [9] has established that bare spots within a wooded savanna display a periodicity and are spatially distributed according to a lattice of hexagonal symmetry.

TABLE 1

Characteristics of some vegetation patterns observed in Africa. Height = h, width = w, spacing = s, length = ℓ. Dimensions are given in meters. Rainfall is annual (in mm).

Region	Vegetation	Patterns		Slope, orientation	Rainfall
British Somaliland [7]	grass, shrubs, isolated trees	arcs, w:	30–40	1:400	70–430
		s:	120–130	perpendicular	
	h: 6–10	ℓ:	500–1000		
		lanes, w:	80–100	parallel	
		s:	80–400		
		ℓ:	3000–4000		
Sudan [8, 10]	grass	stripes, w:	8–12	1:160 to 1:300	100–400
		s:	20–30	perpendicular	
	trees	arcs, w:	40–50	1:200 to 1:50	
	h: 4–5	s:	50–60		
		ℓ:	150–300		
Burkina Faso [9]	shrubs and trees	stripes		1:100 or less	450–600
		w:	20–30	perpendicular	
	h: 3–5	s:	30–40		
		ℓ:	100–500		
		bare spots,			
		w: 15–25 (diameter)			
		s:	40–50 levelled		
		(no consistent slope)			
South Africa [11]	grass, shrubs and trees	stripes, w:	20–50	less than 1:100	about 600
		s:	20–50		
	h: 3–5				
Niger [12]	dense shrubs	stripes, w:	20–40	less than 1:100	400–750
	h: 4	s:	35–150	perpendicular	

More specific names have been given by different authors, like vegetation arcs, bands, stripes, lanes, groves, tiger bush, spotted bush, pearled bush, which describe different modes of patterning. The differences in aspect, symmetry, or orientation existing between these modes are phenotypic rather than fundamental. They represent various manifestations of a general botanical phenomenon which is characteristic of arid regions where rainfall is intense, rare and of short duration. Such climatic conditions prevail over one third of the earth surface.

As Table 1 also shows, vegetation patterns are not specific to particular plants or soils. They may consist of herbs and grasses as well as of shrubs and trees; they have been observed on soils which range from sandy and silty to clayey. Comparing different patterns of the same vegetation shows that on the average vegetation density diminishes as aridity increases. Simultaneously, the size of the spatial periodicities increases. Typically, the vegetated bands of tiger bush (also often called vegetation stripes), become less densely populated while, at the same time, the sum of the vegetated band width and of the sparse inter-band width, i.e., the pattern wavelength, increases [12–14]. Figure 1 represents a characteristic example of tiger bush in Niger; one sees that, in spite of the climatic fluctuations, over a time interval of forty years the pattern is remarkably invariant; Figure 2 represents an example of spotted bush in Burkina Faso.

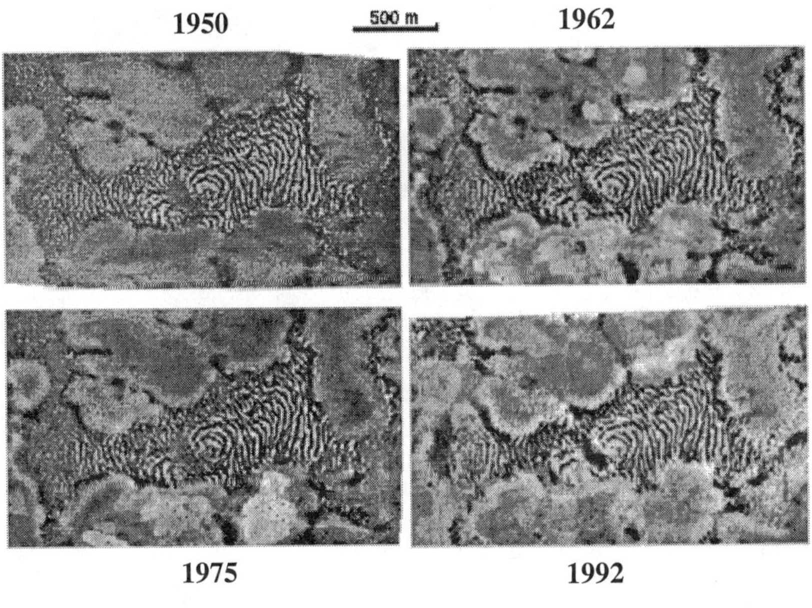

FIG. 1. *Example of Tiger Bush in south west Niger (13°30′ N, 2°40′ E, Banizoumbou site). The sequence of aerial photographs reveals no significant change in the structures over a period of 40 years. The bands of dense vegetation (dark) are approximately 50 m wide; the width of the separating lanes (bright) is of the same order of magnitude. The vegetation is dominantly constituted of Combretaceae. Annual rainfall: 300–600 mm (courtesy of J.-M. d'Herbès).*

Vegetation patterns are subjected to anisotropic environmental factors, like the direction of dominant winds, the sun's course in the sky, the slope of the ground surface, etc. None of these *extrinsic anisotropies*, displays a spatial periodicity which, *per se*, imposes a well-defined wavelength on

FIG. 2. *Example of spotted vegetation in north west Burkina Faso (*14° 14′ *N,* 2° 29′ *W). The vegetation displays a savanna physionomy characterized by an extended herbaceous layer (light grey on the photograph), with scattered trees and groves (darker dots). The most peculiar structure is due to spots of bare soil (bright) with an average diameter of 20–25 m and a wavelength of ca. 40–50m.*

vegetal communities. These anisotropies can however determine the symmetry and/or orientation of the patterns. There is a great deal of evidence indicating that the presence of a small slope, on which water flows without forming drainage channels, orients vegetation stripes [16, 17, 33]. The orientation adopted by the stripes can be either orthogonal to the ground slope, i. e., parallel to the contour lines, or parallel to the ground slope. In the former case, an upslope migration of the stripes has been observed for grass patterns [8]. This invalidates the hypothesis that an underlying soil pattern causes the formation of vegetation stripes: it is unrealistic to think that a soil's property moves upward and drives the stripes migration. Upslope motion has been advocated also with trees and bushes, but on a much slower time scale [18–20, 33], whilst several other authors remained skeptical [10, 12].

2. Mean-field propagator-inhibitor model. We have proposed the mean-field logistic equation

$$
\begin{aligned}
\partial_t \rho(\mathbf{r},t) &= F_1 \times F_2 - F_3 \\
&= \left[\int d\mathbf{r}' w_1(\mathbf{r}') f_1(\rho(\mathbf{r}+\mathbf{r}',t))\right] \\
&\quad \times \left[\int d\mathbf{r}' w_2(\mathbf{r}') f_2(\rho(\mathbf{r}+\mathbf{r}',t))\right] - f_3(\rho(\mathbf{r},t))
\end{aligned} \tag{1}
$$

as a starting point for modelling the space-time organisation of botanical ecosystems [21, 22]. In this section, we review the assumptions underlying this approach and we derive from them a partial differential equation formulation which is minimal, generic and can easily be used for interpreting the data concerning real systems. The case study performed in Section 4 will illustrate this practical aspect.

In Equation (1), the state variable $\rho(\cdot, t)$ represents the phytomass density (units: kg m^{-2}) of all vegetal species present at space point $\cdot$ and time t[1]. The position vectors $\{\mathbf{r}, \mathbf{r}'\}$ refer to 2D-territories supposed vast enough for their geometric shape and boundary conditions to be of no concern; in practice, we admit that $\{\mathbf{r}, \mathbf{r}'\} \in \mathbb{R}^2$ ($d\mathbf{r}' = dx'dy'$, x', y' being cartesian coordinates $\in \mathbb{R}$). The dynamics described by (1) is a birth and death competition. The *nonlocal birth term*, $F_1 \times F_2$ (units: kg m^{-2} s^{-1}), expresses that biomass production depends on different functional processes and plant structures. The nonlocal (integral) character of this term expresses that the variations in biomass taking place at a given point $\mathbf{r}$, are due not only to the plants living at that point, but also to the contributions coming from plants located at the neighboring points $\mathbf{r} + \mathbf{r}'$.

F_1 (units: kg m^{-2} s^{-1}) is the actual *propagation distribution function* of the vegetation: it stands for reproduction, seed dissemination, germination and the maturation of new plants; it mainly depends on plant aerial structures like the canopy. The integral's argument, $w_1(\mathbf{r}') f_1(\rho(\mathbf{r}+\mathbf{r}',t))$ (units: kg m^{-4} s^{-1}), represents the propagation density at point $\mathbf{r}$. The function F_2 (units: dimensionless), called therefore the *inhibition distribution function*, accounts for effects which inhibit the development of vegetation. These effects mainly involve the roots system and the uptake of such limited resources as water and nutrients. They depend on the availability of free soil, which new vegetation may colonize, and on soil resources which are in limited supply. The argument of the integral, $w_2(\mathbf{r}') f_2(\rho(\mathbf{r}+\mathbf{r}',t))$ (units: m^{-2}), represents the inhibition density at point $\mathbf{r}$. In choosing adequately the propagation and inhibition functions $f_1(\rho(\mathbf{r}+\mathbf{r}',t))$ (units: kg m^{-2} s^{-1}), $f_2(\rho(\mathbf{r}+\mathbf{r}',t))$ (units: dimensionless) and the weighting functions $w_1(\mathbf{r}')$, $w_2(\mathbf{r}')$ (units: m^{-2}), modulating the strength of nonlocal interactions existing between plants, the specificities of vegetal communities and of their environmental constraints can be modelled.

[1]Using a global state variable, ρ, is justified when a dominant species imposes its spatial distribution [23] and when genotypic differences as well as age classes can be neglected.

The second term in (1) is the *death term.* We admit that plants do not kill each other at a distance. Death is a local process the rate of which we assume is proportional to the vegetation density at the local point $\mathbf{r}$ considered. Thus, we set

$$f_3(\rho(\mathbf{r},t)) = d\ \rho(\mathbf{r},t).$$

Note that the validity of this assumption of locality in regard to the death term is not only biologically plausible, it also conditions the mathematical consistency of the mean field treatment adopted here [21]. The rate constant d (units: s^{-1}) represents the inverse of the vegetation lifetime. When the vegetation consists of several species, having different lifetimes, an effective $< d >$ can be defined by weighting the lifetime of each species according to its average phytomass density $< \rho_i >$:

$$(2) \qquad < d >= \sum_i d_i \frac{< \rho_i >}{S}, \qquad \text{where} \quad S = \sum_i < \rho_i > .$$

Let us consider more in detail the weighting functions w_i and the problem of modelling through them the nonlocal interactions by which plants are influenced.

2.1. Modellisation of weighting functions. Given the extreme diversity of vegetation ecosystems, the weighting functions, w_i, $i = 1, 2$, cannot be stated once and for all in a completely general form encompassing all possible situations. Whatever these situations are, however, the mathematical form of these functions must satisfy the following generic properties which guide for their choice:

1. The w_i are non-negative everywhere:

$$(3) \qquad w_i(\mathbf{r}') \geq 0, \qquad \forall\ \mathbf{r}' \in \mathbb{R}^2.$$

2. They are maximum at $\mathbf{r}' = 0$, monotonously decreasing with $|\mathbf{r}'|$, and tending to zero as this modulus tends to infinity:

$$(4) \qquad \lim_{|\mathbf{r}'| \to \infty} w_i(\mathbf{r}') = 0.$$

3. Since all interactions generated on the territory remain confined on it, they obey the normalisation condition

$$(5) \qquad \int d\mathbf{r}' w_i(\mathbf{r}') = 1.$$

4. In the limit $L_i \to 0$, where L_i is the effective isotropic operation range of propagative and inhibitive interactions, a strictly local dynamics should be recovered, implying that the w_i's must then reduce to δ-distributions:

(6) $$\lim_{L_i \to 0} w_i(\mathbf{r}') = \delta(\mathbf{r}').$$

The parameters L_1, L_2 control the steepness with which the weighting functions vary; they have the dimension of a length.

Given these conditions, let us deal specifically with the problem of choosing weighting functions modelling vegetation interactions under isotropic and anisotropic environmental conditions.

Case 1: The environment is isotropic. Such environments impose that the weighting functions w_i be chosen among the class of functions which possess the property of rotational invariance and are therefore unaffected by the symmetry transformation $\mathbf{r}' \longleftrightarrow -\mathbf{r}'$. These functions may only depend on the distance $|\mathbf{r}'| = \sqrt{x'^2 + y'^2}$ separating points; they may not depend on the spatial direction of $\mathbf{r}'$. Hence, the interaction ranges are independent of the spatial direction considered and can be fixed constant. Gaussian functions of half width L_i possess the required symmetry properties, fulfill conditions (3–6) and seem, at least qualitatively, appropriate for numerous systems. We adopt this choice for modelling isotropic systems and thus suppose that in that case the w_i functions are given by the expressions:

(7) $$w_i(\mathbf{r}') = \frac{1}{2\pi L_i^2}\, e^{-\frac{|\mathbf{r}'|^2}{2L_i^2}}.$$

Case 2: The environment is anisotropic. Since rotational invariance breaks down, the variations of the 2D-weighting functions $w_i^{(2D)}(\mathbf{r}') \equiv w_i(\mathbf{r}')$ depend upon the direction considered. To account of this feature, we set

(8) $$w_i^{(2D)}(\mathbf{r}') = w_i^{(1D)}(x')\, w_i^{(1D)}(y'),$$

where $w_i^{(1D)}(x')$ and $w_i^{(1D)}(y')$ are 1D-weighting functions associated with the x and y spatial directions. Conditions (3–6) must then be obeyed by these 1D-weighting function separately,

$$w_i^{(1D)}(x') \geq 0 \quad \text{and} \quad w_i^{(1D)}(y') \geq 0, \quad \forall\, x', y' \in \mathbb{R}$$

$$\lim_{|x'| \to \infty} w_i^{(1D)}(x') = \lim_{|y'| \to \infty} w_i^{(1D)}(y') = 0,$$

$$\int dx' w_i^{(1D)}(x') = \int dy' w_i^{(1D)}(y') = 1$$

$$\lim_{L_i \to 0} w_i^{(1D)}(x') = \delta(x'), \qquad \lim_{L_i \to 0} w_i^{(1D)}(y') = \delta(y'),$$

and the problem of modelling anisotropic constraints can be addressed by parameterizing differently the x and y 1D-weighting functions. Without loss of generality, we keep the $x' \longleftrightarrow -x'$ symmetry and consider that anisotropic effects operate in the y-direction. Accordingly, in the x-direction, as in the isotropic case, we adopt a Gaussian distribution function

$$w_i^{(1D)}(x') = \frac{1}{\sqrt{2\pi}L_i} e^{-\frac{x'^2}{2L_i^2}}, \tag{9}$$

while in the y-direction (see Figure 3), we define $w_i^{(1D)}(y')$ piecewisely: for $y' < 0$, as a Gaussian of half-width L_i^-; for $y' \geq 0$, as a Gaussian of half-width L_i^+. After normalisation, the asymmetric weighting functions $w_i^{(1D)}(y')$ constructed in this manner read:

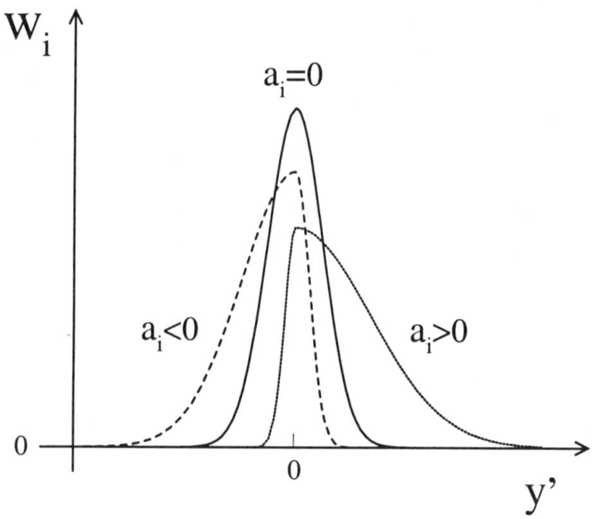

FIG. 3. *Modellisation of the $w_i^{1D}(y')$ weighting functions for L_i fixed, in the absence ($a_i = 0$) and presence ($a_i \neq 0$) of anisotropy oriented in the y-direction. Depending on whether a_i is positive or negative nonlocal interactions are more important in the positive or negative y-direction.*

$$w_i^{(1D)}(y') = \begin{cases} \dfrac{1}{\sqrt{2\pi}\,\dfrac{L_i^- + L_i^+}{2}} e^{-\frac{y'^2}{2L_i^{-2}}} & \text{if} \quad y' < 0 \\[2ex] \dfrac{1}{\sqrt{2\pi}\,\dfrac{L_i^- + L_i^+}{2}} e^{-\frac{y'^2}{2L_i^{+2}}} & \text{if} \quad y' \geq 0. \end{cases} \tag{10}$$

In order to explicitly define the anisotropy, we may simply express that the half-widths L_i differ from the interaction range L_i, corresponding to isotropic conditions, by a proportionality coefficient which is a function of some parameter a_i related to the degree of anisotropy of the environment. Accordingly, we set

$$L_i = g(a_i)\ L_i, \tag{11}$$

where the $g(a_i)$ are positive functions of the anisotropy parameter $a_i \in \mathbb{R}$. Furthermore, since we admit that reflexion symmetry w.r.t. the plane $y' = 0$ is broken and that one direction is favored over the other, the functions g^+ and g^- vary in opposite directions: when g^+ increases and tends asymptotically to infinity, g^-, on the contrary, decreases and tends to zero (and vice versa). When $a_i = 0$, the isotropic situation should be recovered and both functions g should be equal to one. The simplest function which fulfills all these conditions, and which we therefore adopt in the following, is the exponential:

$$\left.\begin{array}{ll} \dfrac{dg^-(a_i)}{da_i} < 0 & \dfrac{dg^+(a_i)}{da_i} > 0 \\ \lim\limits_{a_i \to -\infty} g^-(a_i) = +\infty & \lim\limits_{a_i \to -\infty} g^+(a_i) = 0 \\ \lim\limits_{a_i \to +\infty} g^-(a_i) = 0 & \lim\limits_{a_i \to +\infty} g^+(a_i) = +\infty \\ g^-(0) = 1 & g^+(0) = 1 \end{array}\right\} \longrightarrow g^{\pm}(a_i) = e^{\pm a_i}. \tag{12}$$

Replacing these expressions for the proportionality coefficients $g^{\pm}(a_i)$ in (10), yields:

$$w_i^{(1D)}(y') = \begin{cases} \dfrac{1}{\sqrt{2\pi}\cosh(a_i)L_i}\, e^{-\frac{y'^2}{2(e^{-a_i}L_i)^2}} & y' < 0 \\ \dfrac{1}{\sqrt{2\pi}\cosh(a_i)L_i}\, e^{-\frac{y'^2}{2(e^{a_i}L_i)^2}} & y' \geq 0. \end{cases} \tag{13}$$

Multiplying then the $w_i^{(1D)}(y')$ functions by the weighting functions (9) referring to the x' direction, we obtain that, under very general conditions, the 2D-anisotropic weighting functions $w_i^{(2D)}(\mathbf{r}')$ are given by:

$$w_i^{(2D)}(\mathbf{r}') = \begin{cases} \dfrac{1}{2\pi\cosh(a_i)L_i^2}\, e^{-\frac{x'^2+(e^{a_i}y')^2}{2L_i^2}} & y' < 0 \\ \dfrac{1}{2\pi\cosh(a_i)L_i^2}\, e^{-\frac{x'^2+(e^{-a_i}y')^2}{2L_i^2}} & y' \geq 0. \end{cases} \tag{14}$$

We shall see later that the results obtained in using the weighting functions (14) qualitatively agree with those reported earlier by the method of translating the Gaussians in the case of anisotropic systems [21, 22]. As this cruder treatment is not applicable to strongly anisotropic situations, we henceforth prefer to use (14).

2.2. Modellisation of propagation and inhibition densities. To specify the generic form of the propagation and inhibition densities, we consider their Taylor expansion in terms of the vegetation density ρ (it is superfluous to explicitly define the space and time dependency of ρ here). The dominant terms of $f_1(\rho)$ read (prime $'$ denotes derivative with respect to ρ):

$$
\begin{aligned}
f_1(\rho) &= f_1(0) + f_1'(0)\rho + \frac{1}{2} f_1''(0)\rho^2 \\
&= \rho \left[f_1'(0) + \frac{1}{2} f_1''(0)\rho \right].
\end{aligned} \tag{15}
$$

By definition, the propagation distribution function F_1 vanishes in the absence of vegetation which implies that $f_1(0) = 0$ in (15); $f_1'(0)$ determines the rate at which phytomass increases when its density is small. The quadratic term in the density, $\frac{1}{2} f_1''(0)\rho^2$, amounts to a feedback effect of the vegetation on its propagation rate. Clearly, this feedback is cooperative (anticooperative) if $f_1''(0)$ is positive (negative). The dominant terms in the Taylor expansion of $f_2(\rho)$ read

$$
f_2(\rho) = f_2(0) + f_2'(0)\rho. \tag{16}
$$

By definition, the inhibition distribution function F_2 is positive and decreases with ρ. The inequalities $f_2(0) > 0$ and $f_2'(0) < 0$ must hold, so that vegetation development be turned off as the phytomass density gets high and approaches the ratio

$$
K = -\frac{f_2(0)}{f_2'(0)}, \tag{17}
$$

which has the dimensions of a vegetation density, and represents the territory's carrying capacity. The notion of carrying capacity expresses that environmental resources put an upper bound on the phytomass density. The very fact that mature plants have a size and thus need a minimum of physical space puts an upper bound on their density. K is the *close packed density* of a fully occupied territory.

In practice, the quantities introduced in (15–17) can be estimated as follows. Knowing the average mass m and average area σ occupied by an isolated adult plant one has that

$$
K = \frac{m}{\sigma}.
$$

The product $f_1'(0)f_2(0)$, the units of which are the inverse of a time, determines the vegetation reproduction-maturation time scale. Setting

$$
f_1'(0)f_2(0) = \frac{\ln 2}{t_{1/2}}, \tag{18}
$$

where $t_{1/2}$ is the phytomass density doubling time, provides an estimate of it. Alternatively, $t_{1/2}$ could be identified with the time it takes for the vegetation density to recover its usual level after having been halved by some external cause such as, e.g., the action of man, grazing, fire, unusual drought, etc.

Taking (18) as unit of time, the inhibition range L_2 as unit of length and the carrying capacity K as unit of vegetation density, we redefine time, space, vegetation density and vegetation inverse lifetime by setting

$$\{t, \mathbf{r}, \rho, d\} = \left\{ \frac{t_{1/2}}{\ln 2} \tilde{t},\ L_2\, \tilde{\mathbf{r}},\ K\, \tilde{\rho},\ \frac{\ln 2}{t_{1/2}}\, \mu \right\}. \tag{19}$$

As a measure of (anti-) cooperative feedbacks and of the range of interactions, we introduce the dimensionless parameters

$$\Lambda = \frac{f_1''(0)}{2 f_1'(0)} K, \qquad \text{and} \qquad L = \frac{L_1}{L_2}. \tag{20}$$

Replacing (14, 15–20) in (1), we obtain for the evolution equation of the phytomass density the explicit expression:

$$\begin{aligned} \partial_t \rho(\mathbf{r}, t) = & \left[\frac{1}{2\pi \cosh(a_1) L^2} \int_{-\infty}^{+\infty} dx' e^{-\frac{x'^2}{2L^2}} \left(\int_{-\infty}^{0} dy' e^{-\frac{(e^{a_1} y')^2}{2L^2}} \right.\right. \\ & \left.\left. + \int_{0}^{+\infty} dy' e^{\frac{(e^{-a_1} y')^2}{2L^2}} \right) f_1(\rho(\mathbf{r} + \mathbf{r}', t)) \right] \\ & \times \left[1 - \frac{1}{2\pi \cosh(a_2)} \int_{-\infty}^{+\infty} dx' e^{-\frac{x'^2}{2}} \left(\int_{-\infty}^{0} dy' e^{-\frac{(e^{a_2} y')^2}{2}} \right.\right. \\ & \left.\left. + \int_{0}^{+\infty} dy' e^{-\frac{(e^{-a_2} y')^2}{2}} \right) \rho(\mathbf{r} + \mathbf{r}', t) \right] - \mu\, \rho(\mathbf{r}, t). \end{aligned} \tag{1'}$$

where for notational simplicity, the tildes over $\tilde{t}$, $\tilde{\mathbf{r}}^{(\prime)}$, and $\tilde{\rho}$ have been dropped and the propagation function is given by

$$f_1(\rho(\mathbf{r} + \mathbf{r}', t)) = \rho(\mathbf{r} + \mathbf{r}', t) \left[1 + \Lambda\, \rho(\mathbf{r} + \mathbf{r}', t) \right].$$

In summary, we note that within the framework of this mean field approach, the space-time organisation of the vegetation is described in terms of three sorts of parameters:

(i) The kinetic parameters μ and Λ which appear in the Verhulst type of logistic equation,

$$d_t \rho(t) = \rho(t) \left[1 + \Lambda\, \rho(t) \right] \left[1 - \rho(t) \right] - \mu\, \rho(t), \tag{21}$$

controlling the dynamics of homogeneous vegetation distributions. The parameter $\mu \in \mathbb{R}^+$ serves as a measure of the environment's aridity: in our case, increasing μ obviously diminishes the survival possibilities of the vegetation. The second parameter, $\Lambda \in \mathbb{R}$, expresses whether the propagation of plants is a cooperative ($\Lambda > 0$) or anti-cooperative ($\Lambda < 0$) process[2].

(ii) The ratio $L = L_1/L_2$ which fixes the range of propagative over inhibitive interactions.

(iii) The parameters a_1 and a_2 which express whether the nonlocal interactions involved in vegetation propagation and inhibition are affected by spatial anisotropies.

We shall see in Section 4 that these parameters can be estimated rather straightforwardly by quantifying pattern characteristics with the help of a Fourier analysis of their aerial photographic images. Before that let us present an approximate version of (1′) which applies when the characteristic length of spatial heterogeneities is large compared to the range of interactions (which is the case of interest in the study of vegetation patterns), and when the average density of the vegetation is small. A study of the bifurcation mechanisms and nonlinear solutions of Equation (1′) itself will be reported elsewhere [24, 25].

3. Weak gradient-low density approximation.

3.1. Isotropic case. To begin with, let us consider the case $a_1 = a_2 = 0$. Equation (1′) then reduces to its isotropic version

$$\begin{aligned} \partial_t \rho(\mathbf{r},t) &= \left[\frac{1}{2\pi L^2}\int d\mathbf{r}' e^{-\frac{|\mathbf{r}'|^2}{2L^2}}\,\rho(\mathbf{r}+\mathbf{r}',t)\big(1+\Lambda\,\rho(\mathbf{r}+\mathbf{r}',t)\big)\right] \\ &\quad \times \left[1-\frac{1}{2\pi}\int d\mathbf{r}' e^{-\frac{|\mathbf{r}'|^2}{2}}\rho(\mathbf{r}+\mathbf{r}',t)\right] - \mu\,\rho(\mathbf{r},t). \end{aligned} \tag{22}$$

Replacing $\rho(\mathbf{r}+\mathbf{r}',t)$ in (22) by its Taylor expansion,

$$\rho(\mathbf{r}+\mathbf{r}',t) = \sum_{n=0}^{\infty}\frac{1}{n!}\,(\mathbf{r}'\cdot\nabla)^n\rho(\mathbf{r},t), \tag{23}$$

yields the equivalent partial differential equation

$$\begin{aligned} \partial_t \rho(\mathbf{r},t) &= \left[\sum_{n=0}^{\infty}\frac{1}{(2n)!!}L^{2n}\Delta^n\left[\rho(\mathbf{r},t)(1+\Lambda\,\rho(\mathbf{r},t))\right]\right] \\ &\quad \times \left[1-\sum_{n=0}^{\infty}\frac{1}{(2n)!!}\,\Delta^n\rho(\mathbf{r},t)\right] - \mu\,\rho(\mathbf{r},t), \end{aligned} \tag{24}$$

[2] Only cooperative systems are of interest here. It has already been reported that vegetation patterns cannot form in anticooperative systems [22].

where $\Delta = \partial_x^2 + \partial_y^2$ and $(2n)!! = 2.4.6\ldots(2n)$, with $0!! = 1$. We have shown [21] that homogeneous vegetation distributions are stable when climatic (environmental) conditions are favorable, which corresponds in the model to values of μ close or equal to zero. Heterogeneous vegetation distributions, on the contrary, are stable only if some minimum level of aridity is reached; typically, they appear when the aridity parameter μ gets close to one and when, as a result, the average vegetation density $<\rho>$ gets small. On the other hand, pattern formation is the result of the ecosystem's *intrinsic instability* w.r.t. heterogeneous density fluctuations. In the vicinity of the bifurcation points where this instability appears, the heterogeneities which grow and dominate in the pattern finally formed generally correspond to spatial modes the wavelength of which is large compared to the size of the interactions between plants (space unit is L_2). These considerations suggest that the patterns correspond to smooth variations of $\rho(\mathbf{r}, t)$ and that, at least in the neighborhood of the bifurcation points where the density is small, it is possible to truncate the infinite series of partial derivatives appearing in (24). More precisely, using the scaling laws [24]

$$\begin{aligned} &1 - \mu \sim o(\epsilon^2) \\ &\Lambda - 1,\ \rho \sim o(\epsilon) \\ &L,\ \Delta \sim o(\epsilon^{\frac{1}{2}}) \\ &\partial_t \sim o(\epsilon^{\frac{3}{2}}) \end{aligned} \tag{25}$$

it can be shown that the first non trivial contributions in the ϵ-expansion of (24) amount to the fourth order partial differential equation

$$\begin{aligned} \partial_t \rho(\mathbf{r}, t) &= \rho(\mathbf{r}, t)\left[1 - \mu + (\Lambda - 1)\,\rho(\mathbf{r}, t) - \rho^2(\mathbf{r}, t)\right] \\ &\quad + \frac{1}{2}\left[L^2 - \rho(\mathbf{r}, t)\right]\Delta\rho(\mathbf{r}, t) - \frac{1}{8}\,\rho(\mathbf{r}, t)\,\Delta^2\rho(\mathbf{r}, t). \end{aligned} \tag{26}$$

This truncated evolution equation can be considered as minimal in the sense that it contains the necessary and sufficient terms for exhibiting the same homogeneous stationary state behaviors and symmetry breaking transitions as the integro-differential equation (22) studied in earlier works [21, 22]. Let us briefly review these properties in the case of (26).

The homogeneous stationary states ρ_s are the solutions of

$$\rho_s\left[1 - \mu + (\Lambda - 1)\ \rho_s - \rho_s^2\right] = 0.$$

Figure 4 sketches their behavior in terms of the kinetic parameters μ and Λ. Increasing μ amounts to a decrease of the plants life span, given by $1/<d>$ (cf. (2)), relatively to their normal generation-maturation time span, given by (18). This mimics the effect of an increasingly arid environment. Hence,

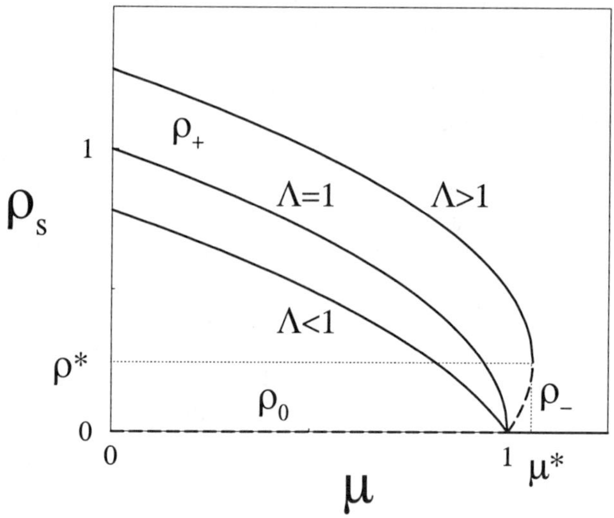

FIG. 4. *Uniform stationary distributions of vegetation ρ_s, and their stability w.r.t. homogeneous perturbations, as a function of environmental aridity measured by μ, for different (positive) values of the vegetation feedback constant Λ. The trivial uniform distribution, $\rho_0 = 0$, is always a solution of (26); it is unstable (dashed line) for $0 \leq \mu < 1$ and stable otherwise (full line). When the vegetation is weakly cooperative ($\Lambda \leq 1$), the finite, non-zero homogeneous stationary distribution ρ_+ exists for $0 \leq \mu \leq 1$. Strong inter-plant cooperative interactions ($\Lambda > 1$) give rise to a hysteresis loop allowing the survival of a stable vegetal population up to $\mu^* = 1 + (\Lambda - 1)^2/4$, $\rho^* = (\Lambda - 1)/2$, i. e. under environmental conditions harsher than those corresponding to $\mu = 1$. The bistable range $1 \leq \mu \leq \mu^*$ is characterized by the coexistence of two stable states, ρ_0 and ρ_+, separated by an intermediate unstable state, ρ_-.*

as already mentioned, the most favorable environmental condition is $\mu = 0$, for which one has a *pure birth* process the rate of which vanishes only if $\rho_s = \rho_0 = 0$, i. e., if the territory is a complete desert without any vegetation present at all, or on the contrary, if the vegetation has attained the highest state of density possible. With respect to homogeneous perturbations, the trivial state ρ_0 is then unstable while the most densely populated state is stable. For $\mu > 0$, a vegetal population can only survive if at least one of the non-trivial roots,

$$\rho_\pm = \frac{\Lambda - 1 \pm \sqrt{(\Lambda - 1)^2 + 4(1 - \mu)}}{2}, \tag{27}$$

is real and non-negative. For $0 < \Lambda \leq 1$, only the branch of uniform stationary solutions ρ_+ (curves drawn in full line in Figure 4) is meaningful insofar as $0 \leq \mu \leq 1$. When aridity (μ) increases, the vegetal population density ρ_+ monotonously decreases to zero, which it reaches for $\mu = 1$. At this point, ρ_+ intersects the trivial branch ρ_0 which becomes stable.

Subsequently, for $\mu > 1$, the trivial state ρ_0 is stable and remains the only stationary solution possible. When cooperative interactions are strong, $\Lambda > 1$, the branch of solutions ρ_+ extends beyond $\mu = 1$, up to the turning point (μ^*, ρ^*) given by

$$\mu^* = 1 + \frac{(\Lambda - 1)^2}{4}, \qquad \rho^* = \frac{\Lambda - 1}{2}. \tag{28}$$

Consequently, for $1 \leq \mu \leq \mu^*$, an hysteresis loop and a bistability phenomenon appear: both ρ_0 and ρ_+ are stable while ρ_-, which takes values in between these states (dashed curve in Figure 4) is unstable. When $\mu > \mu^*$, only the trivial state ρ_0 is possible.

The degree of aridity μ controls the switching of the trivial state $\rho_0 = 0$ from instability (for $0 \leq \mu < 1$) to stability (for $\mu > 1$). In agreement with the assumptions of Section 2 and the definition (19), it can be estimated by setting

$$\mu = \frac{< d > t_{1/2}}{\ln 2}. \tag{29}$$

The switching point $\mu = 1$ corresponds then simply to the situation where the vegetation average lifetime and generation-maturation time are equal. Regarding Λ, we shall discuss its determination later on, in Section 4.

Concerning the stability of the homogeneous stationary states $\rho_s \equiv \{\rho_+, \rho_-\}$, we observe that the "diffusion" coefficient, $1/2\,(L^2 - \rho)$, multiplying the Laplacian term in (26) is negative for $L < \sqrt{\rho_s}$. Given that ρ_s does not depend upon L, it is clear that this condition can always be satisfied by letting the values of L decrease. Heterogeneous perturbations $\delta\rho(\mathbf{r}, t)$ then destabilize the homogeneous stationary distribution ρ_s. In terms of two-dimensional Fourier modes $\delta\rho(\mathbf{r}, t)$ can be written as

$$\delta\rho(\mathbf{r}, t) = \frac{1}{2\pi} \int d\mathbf{k}\; \delta\tilde{\rho}(\mathbf{k}, t)\; e^{i\mathbf{k}.\mathbf{r}}, \tag{30}$$

where $\delta\tilde{\rho}(\mathbf{k}, t) = e^{\omega_k t}\delta\tilde{\rho}(\mathbf{k}, 0)$ is the amplitude of the Fourier mode of wave vector $\mathbf{k}$, or equivalently, of wavelength $\lambda = 2\pi/k$; the wavenumber $k = |\mathbf{k}|$ being the modulus of $\mathbf{k}$. Replacing (30) in (26) furnishes, after linearisation, the dispersion relation

$$\omega_k = 1 - \mu + 2\,(\Lambda - 1)\,\rho_s - 3\,\rho_s^2 - \frac{1}{2}\,(L^2 - \rho_s)\,k^2 - \frac{1}{8}\,\rho_s\,k^4, \tag{31}$$

obeyed by the eigenvalues ω_k which are necessarily real in the isotropic case. Depending on whether ω_k is negative or positive, the ring of modes $\mathbf{k}$ to which it refers is stable or unstable. As is well-known, a symmetry breaking transition occurs when there exists a critical wavenumber k_c for which

$$(\omega_k)_{k=k_c} = 0, \qquad \left(\frac{d\omega_k}{dk}\right)_{k=k_c} = 0, \qquad \text{and} \qquad \left(\frac{d^2\omega_k}{dk^2}\right)_{k=k_c} < 0. \tag{32}$$

Figure 5 represents the conditions for which relations (32) are verified in the case of the simplified model (26). One finds that k_c and the corresponding critical value L_c of L are given by:

$$k_c = \left[8\ (1 - \Lambda + 2\ \rho_+)\right]^{1/4} \tag{33}$$

$$L_c = \sqrt{(1 - \sqrt{2\ (1 - \Lambda + 2\ \rho_+)})\ \rho_+}, \tag{34}$$

while the fastest growing mode (see Figure 5(a)) simply is:

$$k_0 = \sqrt{2\ \left(1 - \frac{L^2}{\rho_+}\right)}. \tag{35}$$

The analysis of expressions (33–35) shows that:

(i) In agreement with experimental observations, the wavelength of the patterns (predicted to a first approximation by the critical wavelength $\lambda_c = 2\pi/k_c$) increases as the aridity of the environment increases, i.e., for ρ_+ decreasing (μ increasing) and Λ constant.

(ii) Decreasing L, for fixed values of μ and Λ, in general destabilizes the system and broadens the band of unstable modes.

(iii) Strictly speaking, Equation (26) only applies in the low density limit, $\rho_+ \ll 1$, which permits its derivation. This implies, in agreement with the conclusions of [21, 22], that the inequality $L_c < 1$ should hold in all instances.

(iv) When the homogeneous stationary distribution ρ_s exhibits no hysteresis in terms of μ (when $\Lambda < 1$), the domain of instability is bounded by an upper and lower critical point (see Figure 5(b)).

(v) Numerical simulations and nonlinear analysis show that different symmetries are possible for the patterns which form when ρ_+ becomes unstable. Typically, varying the conditions but keeping space isotropic, the patterns may switch from stripes to hexagonal symmetry [21, 22].

This latter result that isotropic systems, and thus flat territories, can support vegetation stripes, steps aside from the traditional view that a ground slope is necessary for tiger bush to form [8, 23, 33]. The second kind of pattern, consisting of densely or sparsely vegetated spots arranged on an hexagonal lattice, could be related to spotted [34] or pearly bush [35]. More recently, this kind of symmetry has been unambiguously identified in Burkina Faso [9].

3.2. Anisotropic case. Replacing in the general evolution Equation (1′), $\rho(\mathbf{r}+\mathbf{r}', t)$ by its Taylor expansion (23), and using the same scaling laws as in the isotropic case, a weak gradient-low density approximation describing the spatio-temporal evolution of vegetation in anisotropic environments can be derived [24]:

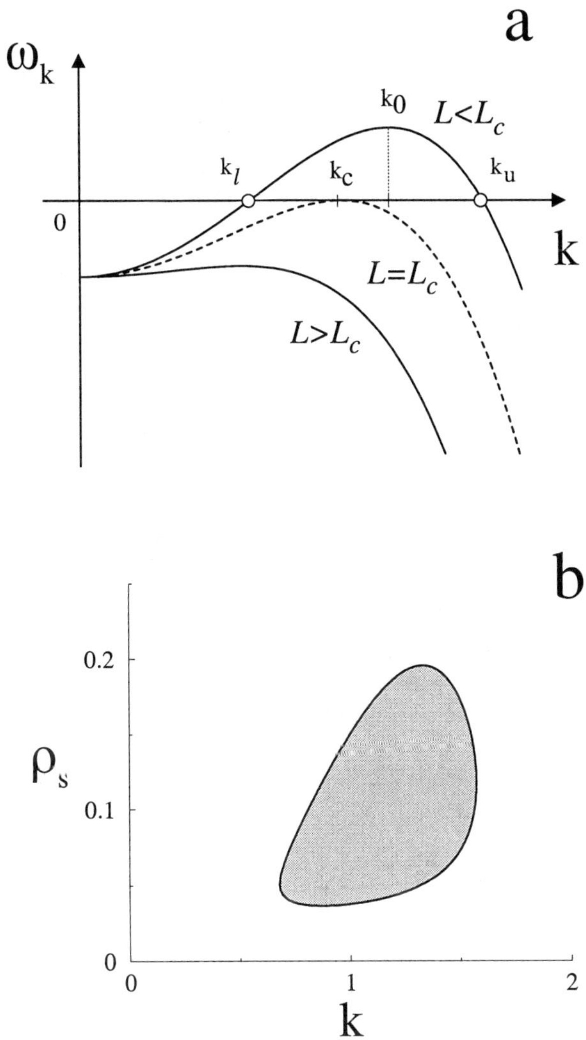

FIG. 5. *(a) Behavior of the eigenvalues* ω_k *as a function of the wave number* k, *for* Λ *and* μ *fixed and varying values of* L. *For* $L > L_c$, *all modes are stable;* $L = L_c$, *is the critical value at which* $\omega_k = 0$ *for* $k = k_c$ *(dashed curve); when* $L < L_c$, *the modes between* k_ℓ *and* k_u *are unstable. Vanishingly small modes and arbitrarily large modes are always stable. When the ratio of reproduction and inhibition ranges* L *tends to zero,* k_ℓ *decreases and* k_u *increases to a finite value different from zero. (b) Domain of instability (grey shade) in terms of* $\rho_s = \rho_+$ *and* k *for* $\Lambda = 1$ *and* $L = 0.15$. *As the value of* ρ_s *diminishes, i.e., as the aridity increases, the band of unstable wavenumbers* k *shifts towards smaller values, predicting, in agreement with field observations, that the wavelength of the periodic patterns produced by a given vegetation increases when it is submitted to harsher environmental stress.*

$$
\begin{aligned}
\partial_t \rho \; = \;& \rho\left[1-\mu+(\Lambda-1)\,\rho-\rho^2\right] \\
&+ \sqrt{\frac{8}{\pi}}\Big[\sinh(a_1)L(1+\rho)-\sinh(a_2)\rho\Big]\partial_y\rho+\frac{1}{2}\left(L^2-\rho\right)\partial_x^2\rho \\
&+ \frac{1}{2}\Big[\left(2\cosh(2a_1)-1\right)L^2-\left(2\cosh(2a_2)-1\right)\rho\Big]\partial_y^2\rho \\
&- \frac{8}{\pi}\sinh(a_1)\sinh(a_2)L(\partial_y\rho)^2 \qquad (36)\\
&- \sqrt{\frac{2}{\pi}}\sinh(a_2)\rho\left[\partial_x^2\partial_y+\frac{4}{3}\cosh(2a_2)\partial_y^3\right]\rho \\
&- \frac{1}{8}\rho\Big[\partial_x^4+2\left(2\cosh(2a_2)-1\right)\partial_x^2\partial_y^2 \\
&\qquad + \left(4\cosh^2(2a_2)-2\cosh(2a_2)-1\right)\partial_y^4\Big]\rho
\end{aligned}
$$

(the dependence of ρ upon the dimensional spatial coordinates x, y and time t is not explicitly stated). As before, this simplified version of (1′) is minimal in the form of a fourth order partial differential equation. As expected also, the breakdown of the $\mathbf{r}' \longleftrightarrow -\mathbf{r}'$ symmetry entails the appearance of odd spatial derivatives; we shall see later that their presence allows for the occurrence of non-stationary patterns of stripes which move parallel to a direction of anisotropy; the isotropic version (26) is recovered if the parameters a_1 and a_2 describing the influence of anisotropic factors on the processes of propagation and inhibition are set equal to zero.

The influence of anisotropy is best seen by investigating its effect on the linear stability of homogeneous stationary distributions. The eigenvalues $\omega_{\mathbf{k}}$, associated with the wave vectors $\mathbf{k} \equiv (k_x, k_y)$, are now in general complex. They obey the dispersion relation:

$$
\omega_{\mathbf{k}} = \Re\,\omega_{\mathbf{k}} + i\,\Im\,\omega_{\mathbf{k}}, \qquad (37)
$$

where

$$
\begin{aligned}
\Re\,\omega_{\mathbf{k}} \; = \;& 1-\mu+2\,(\Lambda-1)\,\rho_s-3\,\rho_s^2-\frac{1}{2}\left(L^2-\rho_s\right)k_x^2 \\
&- \frac{1}{2}\Big[\left(2\cosh(2a_1)-1\right)L^2-\left(2\cosh(2a_2)-1\right)\rho_s\Big]\,k_y^2 \\
&- \frac{1}{8}\rho_s\Big(k_x^4+2\left[2\cosh(2a_2)-1\right]k_x^2k_y^2 \qquad (38)\\
&\qquad + \left[4\cosh^2(2a_2)-2\cosh(2a_2)-1\right]k_y^4\Big)
\end{aligned}
$$

$$\Im\,\omega_{\mathbf{k}} = \sqrt{\frac{2}{\pi}}\,k_y\,\Big(2\big[\sinh(a_1)\,L + (\sinh(a_1)\,L - \sinh(a_2))\,\rho_s\big] + \sinh(a_2)\,\rho_s\left[k_x^2 + \tfrac{4}{3}\cosh(2a_2)\,k_y^2\right]\Big). \tag{39}$$

Figure 6(a) represents $\Re\,\omega_{\mathbf{k}}$ for a typical isotropic situation where there exists a finite band of unstable modes. Clearly, the value of $\Re\,\omega_{\mathbf{k}}$ only depends on the wave vector modulus $k = |\mathbf{k}| = \sqrt{k_x^2 + k_y^2}$. In Figure 6(b) we see, for identical values of the kinetic and interaction parameters, that the effect of an anisotropic factor acting upon the propagation density tends to stabilize the k_y component of unstable modes corresponding to small values of k_x. Accordingly, in the neighborhood of the bifurcation point, the pattern finally establishing itself will consist of stripes oriented in the y-direction, i. e., *parallel to the direction of the anisotropy.* On the contrary, as Figure 6(d) shows, anisotropy acting upon inhibition tends to destabilize further the k_y component of unstable modes corresponding to small values of k_x and to select patterns of stripes oriented *perpendicular to the direction of the anisotropy.* Remarkably, in the transition from the isotropic situation of Figure 6(a) to the one described by Figure 6(d), one passes through an intermediary state (see Figure 6(c)) where the most unstable modes are located at the summits of a rectangle, suggesting the possible appearance, at least transiently in time, of patterns displaying a rhomboidal symmetry. Such situations seems to have been identified recently [9].

4. A case study of tiger bush in sub-Saharian Sahel. Let us now study an example of tiger bush representative of the sub-Saharian Sahel region in Africa. Our objective is to explain the organisation of this vegetal ecosystem in terms of its dynamics as described by the generic model equations (26), (36). More precisely, we show that the information which can be extracted from aerial photographs by image treatment and Fourier analysis techniques[3], allows the estimate of the kinetic parameters Λ and μ, as well as the nonlocal interaction ranges L_1 and L_2 which constitute the basic phenomenological ingredients of our approach. Next, feeding these estimates in (26), (36) provides deeper biological understanding of the advantages associated with pattern formation for the vegetation. In this manner also, new capabilities are gained to make useful predictions which field observations may test; in particular, the variability of vegetation organisations w.r.t. environmental changes, notably w.r.t. anisotropies and changes in aridity, can be predicted and classified in a systematic way.

4.1. Experimental observations and data analysis. The tiger bush studied is represented in Figure 7. The climate in the region is tropical semi-arid; mean rainfall amounts to 490 mm year^{-1} and Potential

[3]Following procedures which could be automated.

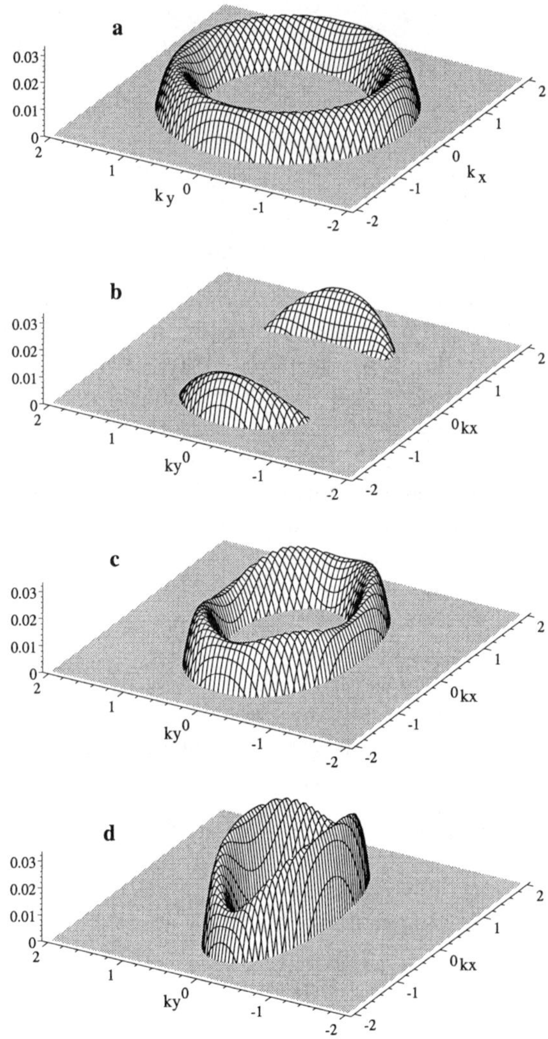

FIG. 6. *Influence of anisotropy on* $\Re\,\omega_{\mathbf{k}}$. *Only the positive real part of the eigenvalues is represented in terms of the wavevector components* k_x *and* k_y. *In all cases* $\rho_s = \rho_+$ *with* $\mu = 0.99$, $\Lambda = 1.2$ *and* $L = 0.2$. *Values of the anisotropy parameters: (a)* $a_1 = a_2 = 0$; *(b)* $a_1 = 0.5,\ a_2 = 0$; *(c)* $a_1 = 0,\ a_2 = 0.5$; *(d)* $a_1 = 0,\ a_2 = 1$. *All anisotropies operate in the direction of the* y *coordinate and preserve the* $x \longleftrightarrow -x$ *symmetry transformation.*

Evapotranspiration (Penman) is slightly under 2000 mm year^{-1} [26]. The vegetation grows on a gentle slope (around 1%) located on Paleozoic sandstones. Soils are shallow (between 10 cm and 20 cm), and poorly developed on ferruginised sandstone debris, overtopping the unweathered sandstone (encountered around 60–80 cm in depth). Vegetation belongs to the Sahel transition zone with most species related to the sudanian center of endemism [27]; it consists mainly of multistemmed shrubs/trees and of annual grasses. Dominant woody species are *Combretum micranthum* G. Don and *Pterocarpus lucens* Lepr. which account respectively for 50% and 30% of total basal area. Average height of woody individuals is around 3 m [9]. The site experiences a low grazing pressure (mainly goats) but neither wood-cutting nor cultivation. Using the image treatment methods described in references [9, 22], the Fourier analysis of the pattern contained in the square area limited by a white stroke in Figure 7, yields the periodogram, and the radial and angular spectrum reported in Figure 8.

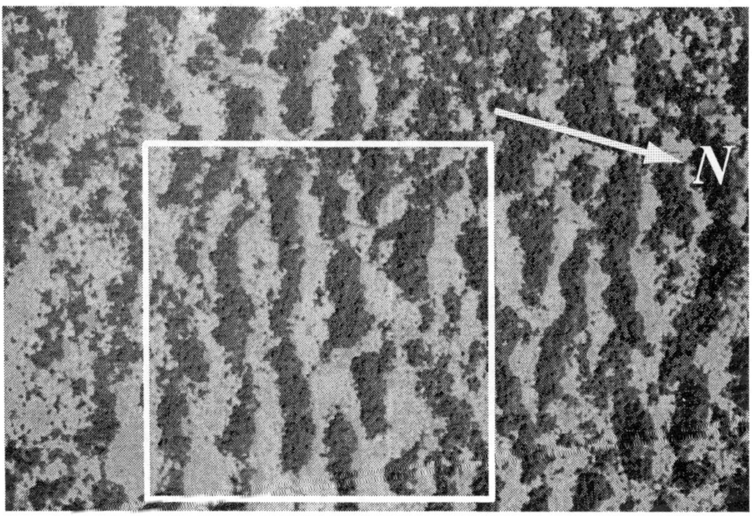

FIG. 7. *Aerial photograph of a tiger bush pattern located at and around 14° 10′ N and 2° 28′ W in the North-West part of Burkina Faso. It was obtained on October 10, 1995 around 10h30 (U.T.), i. e., with a zenithal solar angle of about 38° from an elevation of 750 m. The camera was a Pentax ILX (50 mm focus and 35 mm lens). The film (Kodak Gold 100 ASA) was machine-processed into colored* 7.5×5 *cm printed outlooks, which have been numerised (grey-scale values in the range 0–255) at a resolution of 300 dots per inch (DPI) through a HP Scanjet scanner (pixel side of 0.8 m in the field). The square window of* 400×400 *pixels (i.e.* 320×320 *m in the field) was extracted for analysis. Bright pixels correspond to bare soils, whereas dark ones are dominated by woody vegetation; intermediate grey-scale values can mainly be interpreted as being dense grass cover. Since continuous grass and woody vegetation have respective phytomass averaging 1,500 kg ha*$^{-1}$ *and* 2×10^4 *kg ha*$^{-1}$ *[28] (Couteron unpl. data), grey-scale values can be seen as a monotone function of the phytomass. The picture is oriented according to the main slope, with vegetation bands roughly following the contour.*

As expected, the periodogram is clearly dominated by two groups of entries (Figure 8(a)). A well-defined spike in the radial spectrum (Figure 8(b)) for wave numbers 4 and 5 reveals a dominant wavelength ranging between 64 m and 70 m. A main orientation between 90° and 100° was

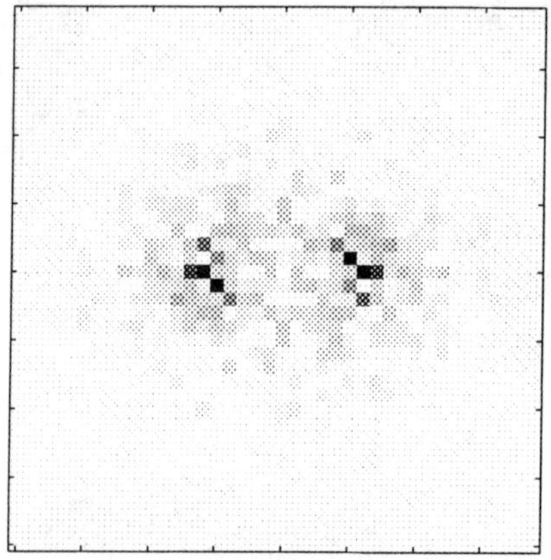

(a) Periodogram

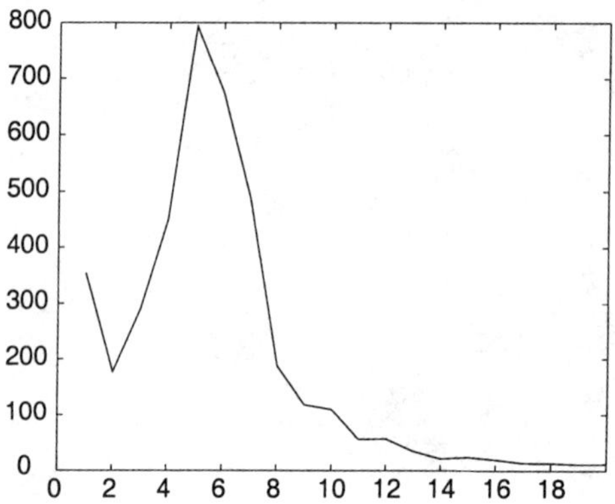

(b) Radial spectrum

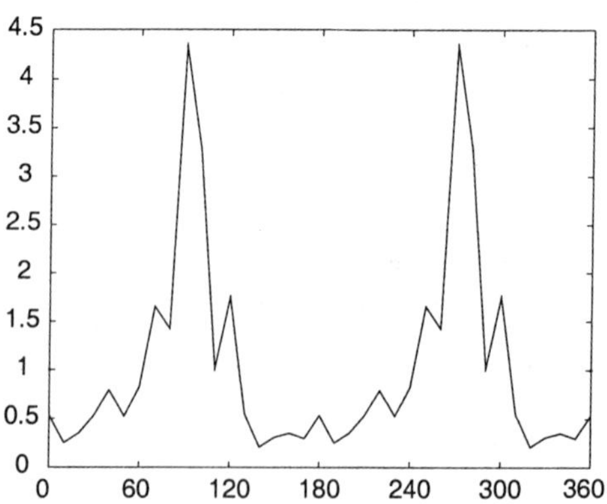

(c) Angular spectrum

FIG. 8. *Spectral analysis of the tiger bush represented in figure 7.*

deduced from the angular spectra (see Figure 8(c)). On the other hand, to estimate the homogeneous distribution ρ_+ introduced above, we note that to a first approximation patterning can be viewed as a phenomenon of redistribution of the vegetation over space. In other words the value of ρ_+ is likely to be approximately equal to the average density obtained by redistributing the vegetation patterns over the entire territory, i. e., by setting:

$$\rho_+ \approx <\rho> = \frac{1}{A}\int d\mathbf{r}\ \rho(\mathbf{r}), \tag{40}$$

where A is the territory's surface area. Hence, assuming a linear dependence of vegetation density upon grey tone scale, we obtain $\rho_+ \approx 0.3$. In summary, the spectral analysis of Figure 7 allows us to conclude that the vegetation pattern corresponds to

$$\lambda = 70 \text{ m} \qquad \text{and} \qquad \rho_+ \approx <\rho> \approx 0.3. \tag{41}$$

From these values, the propagation and inhibition ranges L_1 and L_2 of the vegetation, as well as its cooperativity Λ can be deduced. Transforming back Equations (33) and (34) to real space, we rewrite them as

$$\frac{L_2}{\lambda_c} = \frac{1}{\pi}\Big[\frac{1}{2}\,(1-\Lambda+2\,\rho_+)\Big]^{1/4} \tag{42}$$

$$\frac{L_1}{L_2} = \sqrt{(1-\sqrt{2\,(1-\Lambda+2\,\rho_+)})\,\rho_+}, \tag{43}$$

where λ_c is now expressed in physical space units. Since the value of ρ_+ is fixed, cf. (41), the right hand sides of these expressions are functions of Λ only. Furthermore, in (42), we may equate λ_c to the measured wavelength λ given by (41). This procedure is exact at the critical point. Below this point, when there exists a finite band of unstable modes (for $L < L_c$), the value of the fastest growing mode k_0 is generally close to that of k_c and the relationship $\lambda = \lambda_c$ remains even then a good fit. On the other hand, the existence on the terrain of a small ground slope means that in reality the environment is anisotropic. The influence on the pattern wavelength of this anisotropy can be neglected in a first approach[4]. Hence, setting in (42), (43) $\rho_+ = 0.3$ and $\lambda_c = 70$ m, we solve these equations for L_1 and L_2 in terms of Λ. Figure 9 reports the values of L_1 and L_2 obtained in this manner. One finds that the values of Λ and L_1 for which the generic version (26) of the model predicts the existence of patterns are given by the inequalities

$$\frac{11}{10} < \Lambda < 1.6 \qquad \text{and} \qquad 0 \text{ m} < L_1 < 4 \text{ m}. \tag{44}$$

[4]Passing from the isotropic situation to the cases where the propagation and inhibition distributions are lowly anisotropic produces no significant variation of the modulus k of unstable wave vectors.

In agreement with the general discussion presented above, in this domain L_2 ranges between 0 and 16 meters and the inequality $L_2 > L_1$ always holds. One sees also that for a fixed value of L_1, Figure 9 predicts the possible existence of two distinct values for Λ, and hence for L_2.

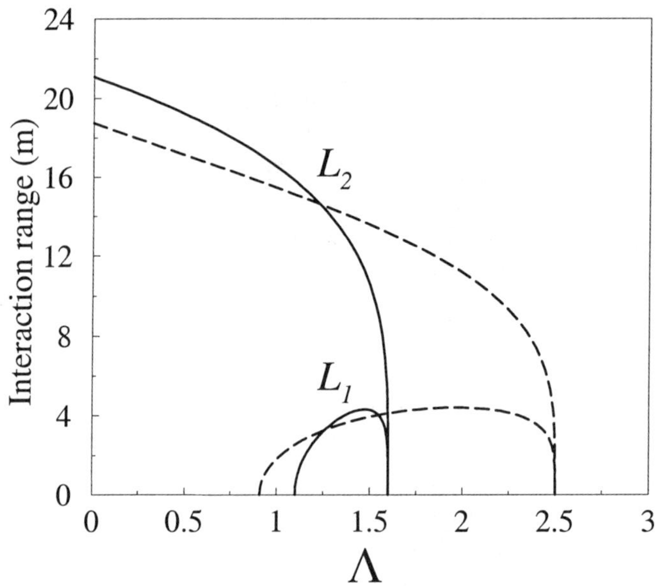

FIG. 9. *Range of interactions L_1 and L_2 as a function of Λ for $< \rho >= 0.3$ and $\lambda = 70$ m. The full lines represent the results obtained within the framework of the low density limit which conditions the derivation of the generic isotropic model (26) and equations (42), (43). The dashed curve permits a comparison, for the same values of λ and ρ_+, with the results reported on the basis of Equation (22) by using as here the weak gradient approximation but not the low density approximation [22]. Qualitatively the results of both cases are in agreement. The orders of magnitude for L_1 and L_2 are comparable and biologically reasonable. One sees however that the low density approximation considerably decreases the upper bound of the acceptable Λ values, from approximately 2.5 to 1.6. The magnitude of this difference explains itself by the fact that 0.3 is not a very small $< \rho >$.*

4.2. Discussion. The choice of values possible for L_1 and L_2 can be further discussed in the light of the field data. In semi-arid and arid situations, short-range cooperativity is likely to result from the favorable influence exerted by the crown of a mature woody individual on the surrounding vegetation. Such influences have been extensively studied with respect to herbaceous vegetation (see [29], for a review), but the results are likely to hold also for woody seedlings. On the other hand, long-range inhibition may result from adverse influences exerted by the root system of a pre-existing mature individual. Data regarding root systems are scarce, but it is well established that lateral roots do extend far beyond the limit of the crown [30, 31]. Furthermore, having lateral roots extending outside the crown (i.e. $L_2 > L_1$) may be an outcome of aridity [32], a recording that may explain why periodic vegetation patterns are so frequent in arid and semi-arid zones.

On our field site, we found an average value of 1.2 m for the radius of the crown of mature *C. micranthum* (the dominant species), and 1.3 m for all mature individuals, irrespective to species [9]. Since favorable influences are supposed to extend significantly up to $2L_1$, due to the properties of the Gaussian distribution, it is reasonable to consider $L_1 < 1$ m. Such a range of values yields two contrasting sets of solutions for Λ and L_2, which correspond either to low ($\Lambda \approx 1.1$) or to high ($\Lambda \approx 1.6$) cooperativity. Low cooperativity determines high values for L_2, around 15 m, whilst high cooperativity implies $L_2 < 4$ m. In the first case, the range of significant inhibition, $2L_2$, and hence of lateral root extension, should extend up to 30 m. So a large extension, though not impossible, is not reported in the literature, even for more arid conditions [30], whilst a lateral root extension between 2 m and 8 m is fully consistent with most published results. As a consequence, the high cooperativity situation appears more realistic in the light of the available data. This suggests looking for a value of Λ in this range: tentatively, we set it equal to 1.5. Reading from Figure 9 the corresponding values of L_1 and L_2, we see that the theoretical propagation/inhibition ratio $L = L_1/L_2 = 0.4$ is then in agreement with the data in the literature.

A robust prediction, supported by all versions of the model studied so far (see the comparison reported in Figure 9), is thus that small values for Λ, i.e., $\Lambda \lesssim 1$ can be excluded. This implies that the ecosystem's stationary state behavior in terms of the aridity parameter μ involves a phenomenon of hysteresis. Let us consider in more detail this behavior in terms of μ.

For $\Lambda = 1.5$, Equation (28) predicts an hysteresis domain when $1 \leq \mu \leq \mu^* = 1.0625$. The corresponding homogeneous stationary state curve, calculated from (27), is represented in Figure 10(a). Setting $\rho_+ = 0.3$, one finds that $\mu = 1.06$. Remarkably, this value lies in the domain of hysteresis, close to the turning point ($\mu^* = 1.0625$, $\rho^* = 0.25$) beyond which the only homogeneous stationary states possible are the ones of the trivial branch, $\rho_0 = 0$. Strikingly however, the domain of existence of the patterns extends well beyond μ^*. By numerical integration of Equation (26), we have verified that it extends, at least, up to $\mu = 1.066$ (cf. the black circles corresponding to this value of μ in Figure 10(a) and referring to the pattern of Figure 10(c)). This survival is only possible because short range cooperative effects competing with long range inhibitive effects allows *a switch from a homogeneous to a heterogeneous density distribution.* Clearly, the pattern is a selected "collective behavior" permitting vegetation to survive in spite of environmental harshness. Note also that the patterns appear subcritically and their amplitude increases with μ, as the spatial extrema values reported in Figure 10(a) indicate.

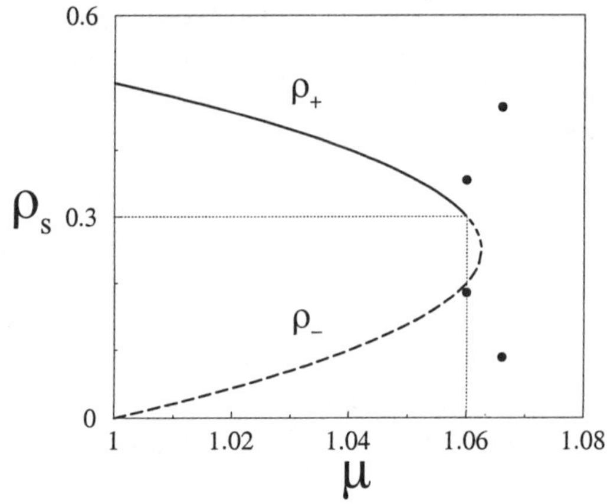

(a) Bifurcation diagram

(b) π-hexagons

(c) 0-hexagons

FIG. 10. *(a) Homogeneous stationary state curve in terms of* μ *for* $\Lambda = 1.5$. *Only the domain of hysteresis* $\mu > 1$ *relevant for Figure 7 is shown. The part of the upper branch* ρ_+ *drawn as a dashed line is unstable w.r.t. inhomogeneous perturbations. The black circles indicate the maximum and minimum density of the patterns represented in (b) and (c), obtained by integrating (26) numerically for* $\Lambda = 1.5$, $L = 0.4$, $\mu = 1.06$ *and* $\mu = 1.066$. *The integration domain is a square-shaped territory subjected to periodic boundary conditions. For pattern (b), the initial condition is a slight heterogeneous perturbation of the uniform stationary distribution* $\rho_+ = 0.3$. *For pattern (c), the initial condition is the pattern (a) itself, since the non trivial branch of solutions* $\rho_\pm$ *does not exist anymore. Black corresponds to the highest phyto-density.*

The comparison of Figures 10(b) and 10(c), shows that when μ increases from 1.06 to 1.066, an exchange of stability takes place: the π-hexagonal symmetry is replaced by the 0-hexagonal symmetry. The existence of patterns possessing the π-hexagonal symmetry in the sub-Saharian Sahel region, is established [9]. It is related to patterns often described in the ecological literature as *spotted bush*. We shall discuss their occurrence and properties more in depth elsewhere. Furthermore, we would like to mention also the existence of stable stripe-symmetric patterns, even under the isotropic conditions considered here [36]. The bifurcation diagram of the amplitude equations describing the nonlinear solutions of (26) confirms analytically this behavior [25]. As a consequence, the question of whether the banded patterns of Figure 7 would form in the absence of a ground slope (i.e., in an isotropic environment) is an open question which cannot be settled completely and definitively for the moment. The fact however that the orientation of the bands in the field studied is far from random, i.e., clearly perpendicular to the contour lines of the existing slight ground slope, indicates at minimum that this slope is an important element of the orientation selection mechanism. Using the model, the *modus operandi* of this mechanism has been identified. We know indeed (see Figure 6 and [21, 22]) that anisotropic inhibition favors the emergence of stripe-symmetric patterns orientated perpendicularly to the direction of anisotropy. This suggests that in the tiger bush of Figure 7, the slope predominantly affects inhibitive rather than cooperative interactions. This conclusion is illustrated in Figure 11(b) which shows that the hexagonal symmetry obtained under the isotropic conditions of Figure 10(a) is replaced by a pattern of stripes when inhibition is sufficiently anisotropic. For $a_2 = 0.3$, the switching from hexagons to stripes is complete, indicating that rather weak deformations of the Gaussian weighting function w_2 (see Figure 11(a)) can do the job.

We would like to draw attention to another interesting prediction concerning the effect of anisotropy on inhibition, namely, that since it implies the appearance in evolution equation (36) of first order derivatives w.r.t. the direction of anisotropy y, the final patterns are not stationary but move upward in the y-direction (see the space-time map of Figure 11(c)). Though such motion, given the vegetation slow turn-over time, is necessarily slow and thus difficult to observe, its occurrence is a robust feature unaffected by changes in boundary conditions, e.g., replacing periodic boundaries by Neumann or Dirichlet boundaries does not suppress this motion.

To conclude, in regard to the mean field approach used here in the context of botanical ecosystems, let us still express the opinion that it is susceptible to find broader applications, notably in the theoretical description of cellular tissues growth and/or bacterial colonies growth and spatial differentiation. We hope to come back to these problems in the future.

Acknowledgments. The support of the Instituts Internationaux de Physique et de Chimie Solvay and of the Centre for Nonlinear Phenomena and Complex Systems (U.L.B.) is acknowledged.

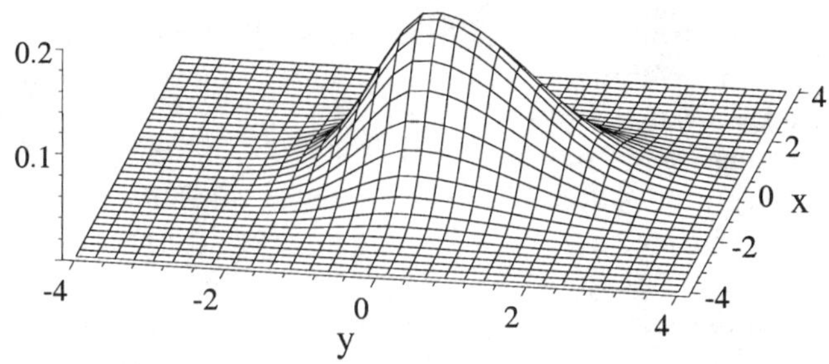

(a) Anisotropic inhibition

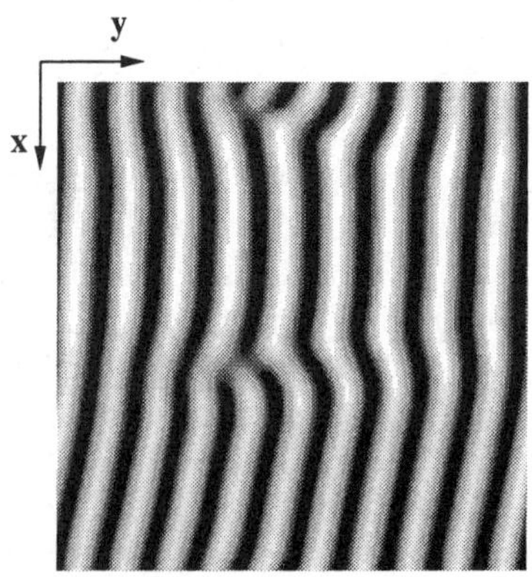

(b) Orthogonal stripes

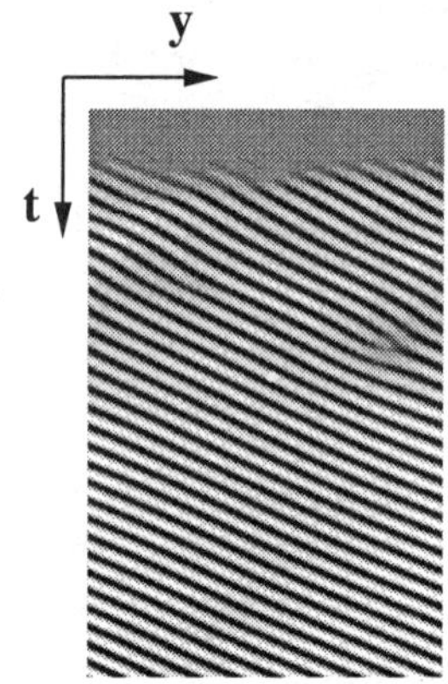

(c) Upward migration

FIG. 11. *(a) Plot of the anisotropic weighting function* w_2 *(*$a_2 = 0.3$*) used in simulations (b) and (c). (b) Pattern obtained by integrating the anisotropic version (26) of the model for* $\mu = 1.06$, $\Lambda = 1.5$, $L = 0.4$ *(same values as in Figure 10(a)) and* $a_1 = 0$, $a_2 = 0.3$. *The y-coordinate corresponds to the horizontal direction so that the bands be orientated as in Figure 7. (c) Space-time map of a linear transect parallel to the anisotropy direction, showing the stripes movement in the positive y-direction.*

REFERENCES

[1] For reviews, see e.g., J.D. GUNTON, M. SAN MIGUEL AND P.S. SAHNI, in: *Phase Transitions and Critical Phenomena*, Vol. **8**, edited by C. Domb and J.L. Lebowitz (Academic Press 1983). K. BINDER, in: Phase Transformations of Materials (Material Science and Technology 5), edited by P. Haasen (Weinheim: VCH 1991).

[2] cf. the numerous examples described in, e.g., J.D. MURRAY (1989). *Mathematical Biology.* Biomathematics Texts **19**. Berlin/New York: Springer-Verlag. H. MEINHARDT (1982). *Models of Biological Pattern Formation.* New York: Academic Press.

[3] H. OTHMER (1969). *Interactions of reaction and diffusion in open systems.* Doctoral Thesis, University of Minnesota (USA).

[4] D.S. COHEN AND J.D. MURRAY (1981). *A generalized diffusion model for growth and dispersal in a population.* J. Math. Biol. **12**, 237–249.

[5] L.A. Segel (1985). *Pattern generation in space and aspect.* SIAM Rev. **27**, 45–67.

[6] W.A. Macfadyen (1950a). *Soil and vegetation in British Somaliland.* Nature **165**, p. 121.

[7] W.A. Macfadyen (1950b). *Vegetation patterns in the semi-desert plains of British Somaliland.* Geogr. J. **116**, 199–211.

[8] G.A. Worrall (1959). *The Butana grass patterns.* J. Soil Sci. **10** $n^{\circ}1$, 34–53.

[9] P. Couteron (1998). *Relations spatiales entre individus et structure d'ensemble dans des peuplements ligneux soudano-sahéliens au nord-ouest du Burkina Faso.* Doctoral Thesis, University of Toulouse (France).

[10] G.E. Wickens and F.W. Collier (1971). *Some vegetation patterns in the Republic of the Sudan.* Geoderma **6**, 43–59.

[11] F. Van Der Meulen and J.W. Morris (1979). *Striped vegetation patterns in a Transvaal savanna.* Geo-Eco-Trop **3**(4), 253–266.

[12] L.P. White (1970). *Brousses Tigrées Patterns in Southern Niger.* J. Ecol. **58**, 549–553.

[13] M. Gavaud (1966). *Etude pédologique du Niger Occidental.* Editions de l'ORSTOM de Dakkar-Hann.

[14] J.-M. d'Herbès, C. Valentin, and J.M. Thiéry (1997). *La brousse tigrée au Niger: Synthèse des connaissances acquises. Hypothèse sur la génèse et les facteurs déterminant les différentes structures contractées.* In *Fonctionnement et gestion des écosystèmes contractés sahéliens* (J.-M. d'Herbès, J.M.K. Ambouta and R. Peltier, eds.), 131–152. John Libbey Eurotext, Paris.

[15] S.B. Boaler and C.A.H. Hodge (1964). *Observations on vegetation arcs in the northern region, Somali Republic.* J. Ecol. **52**, 511–544.

[16] J.A. Mabbutt and P.C. Fanning (1987). *Vegetation banding in arid Western Australia.* J. Arid Env. **12**, 41–59.

[17] C. Montaña, J. Lopez-Portillo, and A. Mauchamp (1990). *The response of two woody species to the conditions created by a shifting ecotone in an arid ecosystem.* J. Ecol. **78**, 789–798.

[18] C. Montaña (1992). *The colonization of bare areas in two-phase mosaics of an arid ecosystem.* J. Ecol. **80**, 315–327.

[19] A.F. Cornet, J.P. Delhoume, and C. Montaña (1988). *Dynamics of striped vegetation patterns and water balance in the Chihuahuan desert.* In *Diversity and Pattern in Plant Communities* (H.J. During, M.J.A. Werger, and J.H. Willems, eds.), 221–231. SPB Academic Publishing: The Hague.

[20] D.J. Tongway and J.A. Ludwig (1990). *Vegetation and soil patterning in semi-arid mulga lands of Eastern Australia.* Aust. J. Ecol. **15**, 23–34.

[21] R. Lefever and O. Lejeune (1997). *On the origin of tiger bush.* Bull. Math. Biol. **59**, 263–294.

[22] O. Lejeune, P. Couteron, and R. Lefever (1999). *Short range cooperativity competing with long range inhibition explains vegetation patterns.* Acta Oecologica **20**, 171–183.

[23] P. Greig-Smith (1979). *Pattern in vegetation.* J. Ecology **67**, 755–779.

[24] O. Lejeune, June 1999. *Une théorie champ moyen de l'organisation spatio-temporelle des écosystmes végétaux.* Doctoral Thesis, University of Brussels (Belgium).

[25] O. Lejeune and R. Lefever. *Propagator-inhibitor model of vegetal ecosystems: amplitude equations in the isotropic case.* In preparation.

[26] P. Couteron and K. Kokou (1997). *Woody vegetation spatial patterns in a semi-arid savanna of Burkina Faso, West Africa.* Plant Ecol. **132**, 211–227.

[27] F. White (1983). *The vegetation of Africa.* A descriptive memoir to accompany the UNESCO/AEFTAT/UNSO vegetation map. (UNESCO/AEFTAT/UNSO. Paris), p. 356.

[28] H.N. Le Houerou (1989). *The grazing land ecosystems of the African Sahel.* (Springer-Verlag. Berlin), p. 282.

[29] O.R. VETAAS (1992). *Micro-site effects of trees and shrubs in dry savannas.* J. Veg. Sci. **3**, 337–344.
[30] P.E. GLOVER (1951). *The root systems of some British-Somaliland plants-IV.* East Afr. Agr. J. **17**, 38–50.
[31] H. POUPON (1980). *Structure et dynamique de la strate ligneuse d'une steppe sahélienne au nord du Sénégal.* (ORSTOM. Paris), p. 351.
[32] A.J. BELSKY (1994). *Influence of trees on savanna productivity: tests of shade, nutrients, and tree-grass competition.* Ecol. **75**, 922–932.
[33] J.E.G.W. GREENWOOD (1957). *The development of vegetation in Somaliland Protectorate.* Geogr. J. **123**, 465–473.
[34] A. CLOS-ARCEDUC (1964). *La géométrie des associations végétales en zone aride.* Act. Conf. UNESCO: Explorations aériennes et études intégrées, Toulouse, 419–421.
[35] S.B. BOALER AND C.A.H. HODGE (1962). *Vegetation stripes in Somaliland.* J. Ecol. **50**, 465–474.
[36] O. LEJEUNE AND M. TLIDI (1999). *A model for the explanation of vegetation stripes (tiger bush).* J. Veg. Sci. **10**, 201–208.

CHEMICAL TURING PATTERNS: A MODEL SYSTEM OF A PARADIGM FOR MORPHOGENESIS*

DAVID J. WOLLKIND† AND LAURA E. STEPHENSON‡

Abstract. The development of one- and two-dimensional Turing patterns characteristic of the chlorite-iodide-malonic acid/indicator reaction occurring in an open gel continuously fed unstirred reactor is investigated by means of various weakly nonlinear stability analyses applied to the appropriately scaled governing chlorine dioxide-iodine-malonic acid/indicator reaction-diffusion model system. Then the theoretical predictions deduced from these pattern formation studies are compared with experimental evidence relevant to the diffusive instabilities under examination. The latter consist of stripes, rhombic arrays of rectangles, and hexagonal arrays of spots, nets, or black-eyes. Here, starch, for the case of a polyacrylamide gel, or the gel itself, for a polyvinyl alcohol gel, serves as the Turing pattern indicator. The main purpose of these analyses is to explain more fully the transition to such stationary symmetry-breaking structures when the malonic acid or iodine reservoir concentrations are varied.

1. Introduction and historical review. Almost half a century ago Turing (1952) proposed *the chemical basis of morphogenesis* in a landmark paper with that title. In particular he postulated the existence of chemical morphogens which formed the basis of embryomorphogenesis through the development of prepatterns. That concept along with the one of positional information regarding subsequent differentiation have made Turing theory a fundamental paradigm for explaining developmental processes ranging from embryology to limb formation and coat patterning (reviewed by Murray, 1990). Specifically he investigated the possibility of an instability occurring in purely dissipative systems involving chemical reactions far from equilibrium and the transport process of diffusion but no hydrodynamic motion. Diffusive instabilities of this sort differ from hydrodynamic ones which involve both convective and dissipative processes. When restricted to two chemical species, an activator and an inhibitor, the existence of such instabilities requires an autocatalytic reaction for the activator and a diffusive advantage for the inhibitor as necessary conditions. Then an initially homogeneous state which would be stable in the absence of diffusion can be destabilized resulting in a re-equilibrated nonhomogeneous symmetry breaking pattern. The need for the activator species to diffuse significantly less rapidly than the inhibitor posed a major obstacle for designing an experiment which exhibited chemical Turing instability patterns since in aqueous media nearly all simple molecules and ions have diffusion coefficients within a factor of two of 1.5×10^{-5} cm^2/sec.

---

*This research was supported by National Science Foundation grant DMS-9531797 and the Institute for Mathematics and its Applications at the University of Minnesota.

†Department of Pure and Applied Mathematics, Washington State University, Pullman, Washington 99164-3113.

‡United Defense LP, Armament Systems Division, 4800 E. River Road, Minneapolis, MN 55421.

Indeed twenty years after Turing's original theoretical prediction of such chemical patterns it was felt that if diffusive instabilities were ever to be encountered in natural systems one would have to seek them elsewhere: Namely in an ecological context. Toward that end Segel and Jackson (1972) considered a general predator-prey interaction-diffusion model system and demonstrated by means of linear stability theory that spatially uniform steady states which would have been stable for homogeneously distributed populations could be destabilized through the introduction of dispersal effects. As usual the occurrence of these ecological diffusive instabilities depended on the activator prey species exhibiting an autocatalytic or Allee effect at equilibrium and having much lower motility than the inhibitory predaceous one. Then Segel and Levin (1976) performed a weakly nonlinear stability analysis on a certain interaction-diffusion system of this type and showed that a new stable nonuniform stationary pattern would emerge following the destabilization of its spatially-uniform steady state. The method they employed, while incorporating the nonlinearities of the relevant model system, basically pivoted a perturbation procedure about the critical point of linear stability theory (reviewed by Wollkind *et al.*, 1994). The advantage of such an approach over strictly numerical procedures is that it allows one to deduce quantitative relationships between system parameters and stable patterns which are valuable for experimental design and difficult to accomplish using simulation alone. Levin and Segel (1976) suggested that diffusive instabilities might explain instances of spatial irregularity for natural communities in which the prey population survives in a clumped pattern forced upon it by the predator's more rapid dispersion that caused the initial breakdown of the uniform state. Although the latter authors had plankton communities in mind, differential dispersal ability of this sort has been documented in arthropod predator-prey systems characterized by patchy distribution patterning both in laboratory (Huffaker *et al.*, 1963) and field (Kareiva and Odell, 1987) experiments.

Finally, almost forty years after its theoretical prediction, there has been experimental confirmation in a chemical activator-inhibitor/immobilizer laboratory system of Turing pattern formation. Castets *et al.* (1990) and Ouyang and Swinney (1991a,b) managed to overcome the aqueous solution difficulty by conducting their experiments involving the chlorite-iodide-malonic acid (CIMA) system in a gel reactor with a Turing pattern indicator which besides preventing convection also resulted in a marked reduction of the effective diffusion coefficient of the activator iodide species. A multitude of Turing structures consisting of parallel stripes, rhombic arrays of rectangles, and hexagonal arrays of spots, net-like honeycombs, or black-eye patterns appeared or disappeared as the system control parameters of temperature or pool species reservoir concentrations were tuned (Ouyang and Swinney, 1991a,b, 1995; Ouyang *et al.*, 1993; Gunaratne *et al.*, 1994). Turing (1952) predicted the formation of stationary two-dimensional structures which developed from a uniform homogeneous base state. In these experiments however there was an inherent concentration gradient for each of the pool species because of the imposed differences at the boundary reservoirs. The observed structures formed perpendicular to these concentration gradients and were shown by Ouyang *et al.* (1992) to be quasi-two-dimensional in that they occupied a single thin layer.

For the very first time, theoreticians were presented with the exciting prospect of an experimental Turing system which invited the challenge of quantification. A number of activator-inhibitor/immobilizer model reaction-diffusion systems were developed to quantify these CIMA gel experiments and various bifurcation analyses relevant to the linear problem and numerical simulations relevant to the nonlinear one have been conducted on those model systems (Lengyel and Epstein, 1991, 1992; Lengyel *et al.*, 1992, 1993; Guslander and Field, 1991; Pearson, 1992; Jensen *et al.*, 1993, 1996). In particular the CIMA system is one of the reactions commonly employed in a well-stirred (homogeneous) batch (closed) experimental environment to illustrate the occurrence of chemical oscillations. After an initial rapid consumption of most of the chlorite and iodide ions to generate chlorine dioxide and iodine, such oscillations result from the interaction of the latter molecules with the reactant malonic acid (Lengyel and Epstein, 1995). An ordinary differential equation model for the chemistry of this CDIMA reaction was proposed by Lengyel *et al.* (1990a,b). Numerical integration of that five-component system yielded oscillatory behavior strikingly similar in nature to the experimental evidence. Further this CDIMA model lent itself readily to an additional simplification. The calculated concentrations of the reactants and intermediate species of this system suggested that, under a wide range of oscillatory conditions, the chlorite and iodide concentrations changed rapidly by several orders of magnitude while the chlorine dioxide, iodine, and malonic acid concentrations varied much more slowly. This allowed Lengyel *et al.* (1990b) to reduce their five-component system to a two-component model by treating these three slowly varying concentrations as constants and to identify the resulting Hopf bifurcation to a limit cycle occurring in the latter system with the transition to chemical oscillations observed experimentally. Lengyel and Epstein (1991) coupled this CDIMA reaction mechanism with diffusion in one spatial dimension, performed a stationary bifurcation analysis on the simplified two-component version of that model, and produced an iodide concentration profile for a particular set of parameter values in the Turing-structure regime by integrating the original five-component form of this system numerically. Guslander and Field (1991) extended that numerical simulation to two spatial dimensions. In each of these cases the mechanism suggested for the reduction of the effective iodide diffusion coefficient was the interaction of iodide with the gel and/or the Turing pattern indicator to form binding sites which acted as traps. Lengyel and Epstein (1992) modified their previous approach by introducing a Turing pattern indicator which reversibly forms an unreactive immobile complex with the activator iodide species rapidly enough to allow them in essence to circumvent the differential diffusivity requirement. They then demonstrated by means of the same linear stability analysis employed in Lengyel and Epstein (1991) that Turing instabilities could be generated over a parameter range where their two-component CDIMA model system would ordinarily exhibit oscillatory behavior in the absence of the indicator. Jensen *et al.* (1996) performed a numerical simulation of that model system for two spatial dimensions. In order to ascertain which of these activator-inhibitor/immobilizer two-component CDIMA reaction-diffusion models proposed by Lengyel and Epstein (1991,1992) was more appropriate for representing CIMA/indicator gel reactor experiments, Noszticzius *et al.* (1992) determined the effect of various Turing pattern indicators on oscillations occurring in a homogeneous batch CIMA system. They found that starch and polyvinyl alcohol suppressed all but the last few large-amplitude oscillations and increased the period of the latter whereas glucose, ethanol, and proponal had no effect on this well-mixed closed CIMA system. Hence Noszticzius *et al.* (1992) concluded that the modified model (Lengyel and Epstein, 1992) was more appropriate for the first group of complex forming Turing pattern indicators while the original one (Lengyel and Epstein, 1991) could still be used to represent the second group. Lengyel *et al.* (1992) formulated a mathematical model which was a quasi-two-dimensional extension of Lengyel and Epstein (1992) and developed a linear stability method to determine the position of Turing structures along the gradient direction and the layer thickness wherein such structures could form. They then compared these theoretical predictions and a numerically simulated two-dimensional pattern with their experimental results. Armed with this knowledge Lengyel *et al.* (1993) devised a closed gradient-free aqueous analog to that gel experiment involving a starch indicator which produced transient Turing patterns and compared them to numerical simulations obtained by using the Lengyel and Epstein (1992) model. All of these analyses dealt with supercritical Turing bifurcations. Jensen *et al.* (1993) examined the possibility of a subcritical transition to Turing structures by numerically integrating the Lengyel and Epstein (1992) model for the relevant parameter range.

In addition there also have been several weakly nonlinear stability analyses performed on these two-component CDIMA reaction-diffusion Turing pattern indicator model systems. Specifically, Rovinsky and Menzinger (1992) considered the interaction of Turing and Hopf bifurcations in the Lengyel and Epstein (1991) model for both one and two spatial dimensions by performing a weakly nonlinear stability analysis about the degenerate point where those bifurcations occur simultaneously. Stephenson and Wollkind (1995) investigated the development of one-dimensional Turing patterns characteristic of CIMA/indicator gel reactor experiments by performing a weakly nonlinear stability analysis on the appropriately scaled Lengyel and Epstein (1992) model system into which had been incorporated the temperature dependence of the reaction rates and the gel axial-coordinate dependence of the pool species concentrations. Their CDIMA/indicator model system also contained a uniform rate parameter which was taken to be either a fixed constant or a function of the malonic acid reservoir concentration depending on whether Turing patterns emerged upon increase or decrease of the latter quantity, respectively, in isothermal experiments. Wollkind and Stephenson (2000) extended that analysis to two spatial dimensions by considering both rhombic and hexagonal basic planforms. They investigated the possibility of occurrence of two-dimensional Turing patterns consisting of rhombic arrays of rectangles and hexagonal arrays of spots or nets, respectively, versus stripes by performing the appropriate weakly nonlinear stability analyses of the homogeneous solution to that model. Each of these analyses employed amplitude functions and the hexagonal one, phase functions as well, the sizes of which are governed by an associated system of differential equations containing constant coefficients. In both cases Wollkind and Stephenson (2000) were primarily concerned with evaluating those Landau constants and cataloguing the stability of the critical points of the amplitude or amplitude-phase equations for a parameter range that permitted quantitative comparison of theory with experiment when Turing patterns emerged upon decrease of the malonic acid reservoir concentration. Our main emphasis will be on interpreting these results so that they may be compared with other classes of experiments such as ones in which Turing patterns emerge upon increase of that concentration. Here starch serves as the Turing pattern indicator for a polyacrylamide gel, or the gel itself, for polyvinyl alcohol.

Although the experiments we wish to quantify involve chemical rather than biochemical reactions, they may still have biological implications given the role that the Turing pattern mechanism has played in developmental biology. Since this experimental evidence represents the first laboratory verification of chemical Turing patterns, its quantification should perhaps be of some interest to mathematical biologists. As such it is also a continuation of Harrison's (1993) recent pioneering effort to bridge the cultural gap between the physical and biological sciences with regard to kinetic theory and living patterns.

We begin in Section 2 with the formulation of our CDIMA/indicator reaction-diffusion problem and a review of those one-dimensional results of Stephenson and Wollkind (1995) involving the development of stripes for this system which are relevant to the two-dimensional analyses of it that follow. In Section 3 we investigate the possibility of occurrence of two-dimensional Turing patterns consisting of rhombic arrays of rectangles and hexagonal arrays of spots or nets, respectively, versus stripes by summarizing the appropriate weakly nonlinear stability results of Wollkind and Stephenson (2000). We conclude in Section 4 with comparisons of the sort mentioned above, a commentary placing our contributions in the context of some recent pattern formation studies in alloy solidication which includes an explanation for black-eye hexagons, and a final discussion of those requisite extensions still needed to examine the interaction of rhombic with hexagonal patterns.

2. The CDIMA/indicator reaction-diffusion model system and its one-dimensional results. Let us first consider the reaction scheme proposed by Lengyel *et al.* (1990b) and Lengyel and Epstein (1992) as a two-component simplification for the CDIMA/indicator system

$$(2.1\text{a})\quad \begin{aligned} &R \xrightarrow{k_1} X, \quad V_1 = k_1' = k_1[MA];\\ &X \xrightarrow{k_2} Y, \quad V_2 = k_2'[X], \quad k_2' = k_2[ClO_2]; \end{aligned}$$

$$(2.1\text{b})\quad 4X + Y \xrightarrow{k_3} P, \quad V_3 = k_3' \ [X][Y]/(u^2 + [X]^2), \quad k_3' = k_3[I_2];$$

$$(2.1\text{c})\quad X + S + I_2 \underset{k_r}{\overset{k_f}{\rightleftarrows}} SI_3^-;$$

which is obtained from the latter by taking the chlorine dioxide (ClO_2), iodine (I_2), malonic acid (MA), and pattern indicator (S) concentrations constant where in the above a bracketed character represents the concentration of that species. Here $X = I^- \equiv$ iodide and $Y = ClO_2^- \equiv$ chlorite, the concentrations of which are our dynamical variables and may be regarded as functions of space and time denoted by $\boldsymbol{s}$ and τ, respectively. Further the latter species have self-diffusion coefficients D_1 and D_2 taken to be constant as is the case for the reaction rates k_1, k_2, k_3, k_f, and k_r while u is a uniform shaping concentration selected historically to provide agreement with experiment, all of which will be assigned later. Then introducing the following dimensionless variables and parameters

$$(2.2\text{a})\quad t = k\tau, \quad \boldsymbol{r} = \frac{\boldsymbol{s}}{(D_2/k)^{1/2}}, \quad x = \frac{[X]}{u}, \quad y = \frac{k_3'[Y]}{k_2'u^2},$$

$$(2.2\text{b})\quad \alpha = \frac{k_1'}{5k_2'u}, \quad \beta = \frac{k_3'}{k_2'u}, \quad \mu = \frac{D_1}{D_2}, \quad K = \frac{k_f}{k_r}[S][I_2], \quad k = \frac{k_2'}{1+K},$$

and employing the law of mass action and Fick's second law in conjunction with this scaling, we deduce the nondimensional governing activator-inhibitor/immobilizer reaction-diffusion system defined on an unbounded flat domain (the r_1-r_2 plane)

$$(2.3\text{a})\quad \frac{\partial x}{\partial t} = F(x, y; \alpha) + \frac{\mu}{1+K}\nabla_2^2 x, \quad \frac{\partial y}{\partial t} = \beta(1+K)G(x, y) + \nabla_2^2 y,$$

where

$$(2.3\text{b})\quad \begin{aligned} &F(x, y; \alpha) = 5\alpha - x - 4xy/(1+x^2),\\ &G(x, y) = x - xy/(1+x^2), \qquad \nabla_2^2 = \sum_{i=1}^{2} \partial^2/\partial r_i^2, \end{aligned}$$

with

$$F(x_0, y_0; \alpha) = G(x_0, y_0) = 0 \implies \qquad (2.3c)$$
$$x_0 = x_0(\alpha) = \alpha, \quad y_0 = y_0(\alpha) = 1 + \alpha^2.$$

In the above we have implicitly made use of the fact that the tri-iodide complex, (SI_3^-), does not diffuse and satisfies the chemical quasi-equilibrium condition (Stephenson and Wollkind, 1995)

$$x^* = x \quad \text{where} \quad x^* = \frac{[SI_3^-]}{Ku}. \qquad (2.3d)$$

We have also employed a quasi-two-dimensional approximation (Stephenson and Wollkind, 1995) which allows us to consider the axial coordinate z, scaled with the height of the gel disk, as a parameter and to introduce the pool species concentration gradient relations for $0 < z < 1$ given by

$$[ClO_2] = [ClO_2]_0(1 - z), \; [MA] = [MA]_0 z, \; [I_2] = [I_2]_0 \; ; \qquad (2.3e)$$

analogous to the laboratory reservoir configurations of Ouyang and Swinney (1991a,b). In this context we observe that the latter authors reported $[ClO_2^-]_0$ and $[I^-]_0$ concentrations rather than $[ClO_2]_0$ and $[I_2]_0$. Indeed Pearson (1992) applied an immobilizer appended version of a CDIMA system of this sort to such CIMA/starch experiments in gels and presented his predicted bifurcation behavior in $[MA]$-$[I_2]$ phase space for temperature $T = 288°$ K and $[ClO_2] = 10^{-4}$M.

We note that by necessity the existing numerical simulations of Dufiet and Boissonade (1992) for the Schnackenberg (1979) reaction-diffusion model (Murray, 1989; Ouyang *et al.*, 1992) and Lengyel *et al.* (1992,1993) for the CDIMA/starch system were performed on a square array with periodic and zero-flux boundary conditions, respectively. Given that the experimental patterns investigated by Ouyang and Swinney (1991a,b) typically had a gel disk diameter to characteristic wavelength ratio on the order of 100, it seems reasonable as a first approximation for us to consider our activator-inhibitor/immobilizer equations on an unbounded spatial domain. Indeed this effect was even more pronounced in the experiments of Gunaratne *et al.* (1994) which, although having zero flux at its boundaries, involved a system about 160 times longer in extent than the characteristic wavelength and consequently those boundaries did not significantly influence the patterns (Graham *et al.*, 1994).

The equilibrium point (2.3c) to our model system (2.3a,b) represents a uniform steady-state spatially homogeneous exact solution to these governing equations. It was the stability of this solution to one-dimensional perturbations with which Stephenson and Wollkind (1995) were concerned and hence they considered solutions of (2.3) of the form

$$x(\boldsymbol{r}, t) \sim x_0(\alpha) + A_1(t)\cos(q_c r_1) + A_1^2(t)\big[x_{20} + x_{22}\cos(2q_c r_1)\big]$$

$$+ A_1^3(t)\left[x_{31}\cos(q_c r_1) + x_{33}\cos(3q_c r_1)\right] \tag{2.4a}$$

with an analogous expansion for $y(\boldsymbol{r},t)$ where the amplitude function $A_1(t)$ satisfied the equation

$$\frac{dA_1}{dt} \sim \sigma(\beta;\alpha,\mu,K)A_1 - a_1(\alpha;\mu,K)A_1^3 \tag{2.4b}$$

and $q_c = q_c(\alpha;\mu,K)$ was the critical wavenumber of linear stability theory while σ denoted the growth rate associated with that most dangerous mode and a_1, the corresponding Landau constant. They found that

$$q_c^2(\alpha;\mu,K) = \frac{5(1+K)(3\alpha^2-5)}{\mu(1+\alpha^2)[5+2\alpha(10)^{1/2}(1+\alpha^2)^{-1/2}]} \tag{2.5}$$

and diffusive instabilities ($\sigma > 0$) occurred whenever

$$0 < \beta_1(\alpha;K) < \beta < \beta_2(\alpha;\mu) \tag{2.6a}$$

where

$$\beta_1(\alpha;K) = (3\alpha - 5/\alpha)/(1+K) \tag{2.6b}$$

and

$$\beta_2(\alpha;\mu) = \frac{(3\alpha^2-5)^2}{\alpha\mu[13\alpha^2+5+4\alpha(10)^{1/2}(1+\alpha^2)^{1/2}]}, \tag{2.6c}$$

which re-equilibrated ($a_1 > 0$) to form a striped pattern of characteristic dimensional wavelength

$$\lambda_c^*(\alpha) = 2\pi/q_c^*(\alpha) \tag{2.7a}$$

such that

$$q_c^*(\alpha) = (k/D_2)^{1/2}q_c(\alpha;\mu,K) = (k_2'/D_1)^{1/2}q_c(\alpha;1,0) \tag{2.7b}$$

provided

$$\alpha_1 < \alpha < \alpha_2 \tag{2.8a}$$

where

$$\alpha_1 = 1.40, \quad \alpha_2 = 2.77. \tag{2.8b}$$

Further under this condition on α the uniform state was stable for $\beta > \beta_2(\alpha;\mu)$.

Finally Stephenson and Wollkind (1995) took

$$D_1 = D_2 = \chi^2 D_x \tag{2.9a}$$

with

$$\chi = 0.40 \quad \text{and} \quad D_x = 7 \times 10^{-6}\ \text{cm}^2/\ \text{sec} \ . \tag{2.9b}$$

The constitutive relation of (2.9) reflects the fact that a fully hydrolyzed saturated gel will result in an ionic diffusion coefficient which has been uniformly reduced from its common aqueous solution value D_x (in this case associated with $T = 280°$ K), the amount of that reduction being dependent on the characteristic pore diameter of the gel itself (Ouyang *et al.*, 1995). In this context we note that when Pearson (1992) assigned u the value of 0M in his basic dimensional system and then took $\chi = 1$ in (2.9a), he predicted a $\lambda_c^* \cong 0.40$mm along the relevant line in his bifurcation diagram instead of the observed value of $\lambda_c^* \cong 0.17$mm (DeKepper *et al.*, 1991), an overprediction which would also be adjusted correctly upon adoption of the χ of (2.9b).

3. Two-dimensional patterns: Rhombic and hexagonal planform analyses. In order to investigate the possibility of occurrence for our CDIMA/indicator model system of either those rhombic-type patterns observed by Ouyang *et al.* (1993) and Gunaratne *et al.* (1994) or those hexagonal patterns observed by Ouyang and Swinney (1991b, 1995) and Ouyang *et al.* (1992), Wollkind and Stephenson (2000) sought weakly nonlinear solutions of these equations which to lowest order satisfied either

$$\begin{aligned} &x(\boldsymbol{r}, t) \sim x_0(\alpha) + A_1(t)\cos(q_c r_1) + B_1(t)\cos(q_c r_3)\ , \\ &\qquad r_3 = r_1\cos(\psi) + r_2\sin(\psi) \end{aligned} \tag{3.1a}$$

such that

$$\begin{aligned} \frac{dA_1}{dt} &\sim \sigma A_1 - A_1(a_1 A_1^2 + b_1 B_1^2)\ , \\ \frac{dB_1}{dt} &\sim \sigma B_1 - B_1(a_1 B_1^2 + b_1 A_1^2) \end{aligned} \tag{3.1b}$$

or

$$\begin{aligned} x(\boldsymbol{r}, t) - x_0(\alpha) \sim\ & A_1(t)\cos[q_c r_1 + \phi_1(t)] \\ &+ A_2(t)\cos[q_c(r_1 - \sqrt{3}r_2)/2 - \phi_2(t)] \\ &+ A_3(t)\cos[q_c(r_1 + \sqrt{3}r_2)/2 - \phi_3(t)] \end{aligned} \tag{3.2a}$$

where

$$\frac{dA_i}{dt} \sim \sigma A_i - 4a_0 A_j A_k \cos(\phi_i + \phi_j + \phi_k) - A_i[a_1 A_i^2 + 2a_2(A_j^2 + A_k^2)] \ , \tag{3.2b}$$

$$A_i \frac{d\phi_i}{dt} \sim 4a_0 A_j A_k \sin(\phi_i + \phi_j + \phi_k) \ , \quad (i,j,k) = \text{ even permutation of } (1,2,3) \ , \tag{3.2c}$$

respectively, with analogous expansions for $y(\boldsymbol{r}, t)$.

Amplitude equations (3.1b) possess the following equivalence classes of critical points: I: $A_1 = B_1 = 0$; II: $A_1^2 = \sigma/a_1, B_1 = 0$; V: $A_1 = B_1$ with $A_1^2 = \sigma/(a_1 + b_1)$. Assuming that $a_1, a_1 + b_1 > 0$ and investigating the stability of these critical points, one finds that I is stable for $\sigma < 0$; II, for $\sigma > 0, b_1 > a_1$; and V, for $\sigma > 0, a_1 > b_1$. Noting that I and II, as in the one-dimensional analysis, represent the homogeneous and striped states, respectively, while V can be identified with a rhombic array of rectangles of characteristic angle ψ, Wollkind and Stephenson (2000) next used these criteria to refine those predictions of Stephenson and Wollkind (1995) relevant to the former states due to the presence of the latter. Toward this end they obtained the explicit formulae

$$a_1 = a_1(\alpha; \mu, K) \ , \qquad b_1 = b_1(\alpha, \psi; \mu, K) \tag{3.3}$$

by employing Fredholm-type solvability conditions, where the expression for a_1 is identical to that appearing in Stephenson and Wollkind (1995), and examined the signs of $a_1 + b_1$ and $b_1 - a_1$ for $\alpha_1 < \alpha < \alpha_2$ and $0 < \psi \leq \pi/2$, with $\psi = \pi/2$ (or equivalently 90°) representing a square planform, after assigning μ and K the typical values

$$\mu = 1 \ , \qquad K = 100 \ . \tag{3.4}$$

These results are summarized in the chart which comprises Fig. 1. This figure has been drawn for the extended interval $\pi/2 < \psi \leq \pi$ in order to demonstrate graphically the symmetry about $\psi = \pi/2$ characteristic of rhombic patterns. Restricting ourselves to the interval of interest $0 < \psi \leq \pi/2$, we see that for $\alpha = 1.9$ there are two narrow bands of stable rhombic patterns flanking $\psi = \pi/3$ with no pattern between these bands and stable stripes outside of them. Observe from Fig. 1 that there exist no stable rhombic patterns of characteristic angle $\psi = \pi/3$. Wollkind and Stephenson (1999) conjectured that this angle was reserved for hexagonal arrays and stable patterns of the latter sort occurred in those parameter ranges where their rhombic planform analysis predicted no pattern.

In cataloguing the critical points of the amplitude-phase equations (3.2b,c) and summarizing their orbital stability behavior it is necessary to employ the quantities

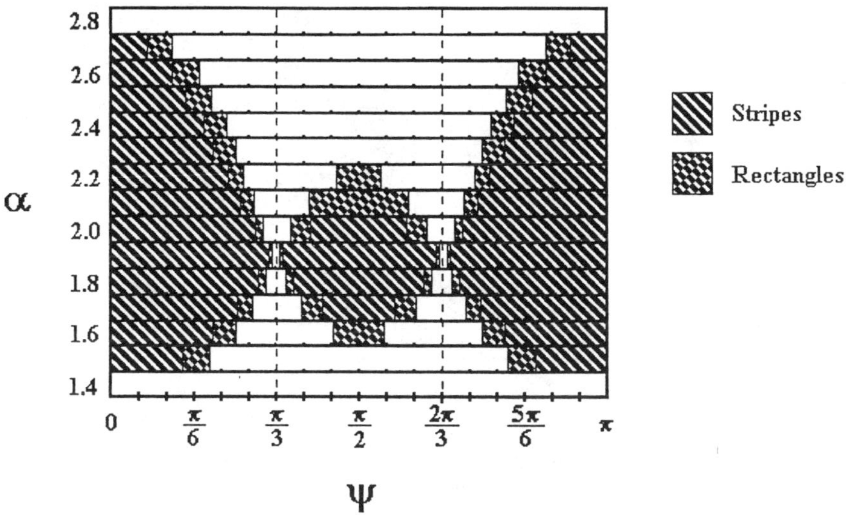

FIG. 1. *Chart in ψ-α parameter space summarizing rhombic versus striped pattern predictions with $\mu = 1$ and $K = 100$ for $\alpha = 1.4, 1.5, \cdots, 2.7, 2.8$.*

$$\tag{3.5} \begin{aligned} \sigma_{-1} &= -4a_0^2/(a_1 + 4a_2), \qquad \sigma_1 = 16a_1a_0^2/(2a_2 - a_1)^2, \\ \sigma_2 &= 32(a_1 + a_2)a_0^2/(2a_2 - a_1)^2 . \end{aligned}$$

There exist equivalence classes of critical points of (3.2b,c) given by $\phi_1 = \phi_2 = \phi_3 = 0$ and I: $A_1 = A_2 = A_3 = 0$; II: $A_1^2 = \sigma/a_1$, $A_2 = A_3 = 0$; III$^{\pm}$: $A_1 = A_2 = A_3 = A_0^{\pm} = \{-2a_0 \pm [4a_0^2 + (a_1 + 4a_2)\sigma]^{1/2}\}/(a_1 + 4a_2)$; IV: $A_1 = -4a_0/(2a_2 - a_1)$, $A_2^2 = A_3^2 = (\sigma - \sigma_1)/(a_1 + 2a_2)$; where it is assumed that $a_1, a_1 + 4a_2 > 0$. The orbital stability conditions for these critical points can be posed in terms of σ. Thus critical point I is stable in this sense for $\sigma < 0$ while the stability behavior of II and III$^{\pm}$ which depends upon the signs of a_0 and $2a_2 - a_1$ as well has been summarized in Table 1. Here, when stable, II and III$^{\pm}$ represent one- and two-dimensional periodic structures, respectively, the latter pattern exhibiting hexagonal symmetry in the plane such that $A_0^+ > 0$ and $A_0^- < 0$. Finally critical point IV, which reduces to II for $\sigma = \sigma_1$ and to III$^{\pm}$ for $\sigma = \sigma_2$ and hence called a generalized cell, is not stable for any value of σ. The stable critical points described above were identified with the following Turing patterns generated during CIMA/indicator experiments in the comparison of such observations with theoretical prediction included in Wollkind and Stephenson (2000): I, uniform or homogeneous; II, stripes or bands; III$^+$, nets or honeycombs; III$^-$, spots or dots. Further given the hexagonal close-packed nature of the arrays associated with III$^{\pm}$, we shall also refer to them collectively as hexagons.

TABLE 1
Orbital stability behavior of critical points II and III$^{\pm}$.

a_0	$2a_2 - a_1$	Stable Structures
$+$	$-, 0$	III$^-$ for $\sigma > \sigma_{-1}$
$+$	$+$	III$^-$ for $\sigma_{-1} < \sigma < \sigma_2$, II for $\sigma > \sigma_1$
0	$-$	III$^{\pm}$ for $\sigma > 0$
0	$+$	II for $\sigma > 0$
$-$	$+$	III$^+$ for $\sigma_{-1} < \sigma < \sigma_2$, II for $\sigma > \sigma_1$
$-$	$-, 0$	III$^+$ for $\sigma > \sigma_{-1}$

Wollkind and Stephenson (2000) used these criteria to refine those one-dimensional predictions of Stephenson and Wollkind (1995) relevant to states I and II due to the presence of the two-dimensional hexagonal states III$^{\pm}$. Toward this end they obtained the explicit formulae

$$a_0 = a_0(\alpha; \mu, K) \ , \ a_1 = a_1(\alpha; \mu, K) \ , \ a_2 = a_2(\alpha; \mu, K) \tag{3.6}$$

by employing Fredholm-type solvability conditions, where the expression for a_1 is again identical to that appearing in Stephenson and Wollkind (1995), and examined the signs of $a_0, 2a_2 - a_1$, and $a_1 + 4a_2$ for $\alpha_1 < \alpha < \alpha_2$ after assigning μ and K the typical values of (3.4). From these results they observed that besides α_1 and α_2 there exist the following other significant values of α

$$\alpha_3 = 1.44 \ , \ \alpha_5 = 1.48 \ , \ \alpha_c = 1.88 \ , \ \alpha_6 = 2.36 \ , \ \alpha_4 = 2.53 \tag{3.7a}$$

such that

$$a_1 + 4a_2 = 0 \text{ for } \alpha = \alpha_3 \text{ or } \alpha_4 \ , \ a_1 + 4a_2 > 0 \text{ for } \alpha_3 < \alpha < \alpha_4 \ ; \tag{3.7b}$$

$$\begin{aligned} &2a_2 - a_1 = 0 \text{ for } \alpha = \alpha_5 \text{ or } \alpha_6 \ , \ 2a_2 - a_1 > 0 \text{ for } \alpha_5 < \alpha < \alpha_6 \ , \\ &2a_2 - a_1 < 0 \text{ for } \alpha < \alpha_5 \text{ or } \alpha > \alpha_6 \ ; \end{aligned} \tag{3.7c}$$

$$a_0 = 0 \text{ for } \alpha = \alpha_c \ , \ a_0 > 0 \text{ for } \alpha > \alpha_c \ , a_0 < 0 \text{ for } \alpha < \alpha_c \ . \tag{3.7d}$$

We note that this behavior is independent of μ and K as was demonstrated explicitly by Stephenson and Wollkind (1995) for a_1. The same thing was also true for b_1 of (3.3).

Finally upon determining the functions

$$\sigma_1 = \sigma_1(\alpha; 1, 100) \ , \ \sigma_2 = \sigma_2(\alpha; 1, 100) \tag{3.8a}$$

obtained from the definitions of (3.5) in conjunction with (3.6) for these fixed values of μ and K, Wollkind and Stephenson (2000) produced the loci

$$\beta = \beta_{\sigma_i}(\alpha) = \beta_c[\alpha; 1, 100, \sigma_i(\alpha; 1, 100)] \ , \ i = 1 \text{ and } 2 \ , \tag{3.8b}$$

where

$$\beta = \beta_c(\alpha; \mu, K, s) \tag{3.8c}$$

represents the generalized marginal curve corresponding to $\sigma = s$ in the α-β plane and thus as is to be expected

$$\beta_c(\alpha; \mu, K, 0) = \beta_2(\alpha; \mu) \ . \tag{3.8d}$$

Since all the quantities required for the identification of the Turing patterns of Table 1 had been evaluated, Wollkind and Stephenson (2000) could represent graphically the regions corresponding to these patterns in the α-β plane of Fig. 2, where the loci of (3.8b) are denoted by $\sigma = \sigma_i, i = 1$ and 2, in that figure. Then from Fig. 2 we observe that for $a_1 + 4a_2 > 0$ all (when $2a_2 - a_1 < 0$) or part (when $2a_2 - a_1 > 0$) of the region $(\sigma, a_1 > 0)$ where the one-dimensional analysis of Stephenson and Wollkind (1995) predicted striped Turing patterns is further divided into two subregions characterized by hexagonal patterns consisting of either dots (when $a_0 > 0$) or honeycombs (when $a_0 < 0$), respectively. In the overlap regions satisfying

$$\beta_1(\alpha; 100) < \beta_{\sigma_2}(\alpha) < \beta < \beta_{\sigma_1}(\alpha) < \beta_2(\alpha; 1) \tag{3.9}$$

where stripes and nets $(\alpha_c^- < \alpha < \alpha_c)$ or stripes and spots $(\alpha_c < \alpha < \alpha_c^+)$ are predicted, the initial conditions determine which stable equilibrium structure of each pair will be selected. Here $\alpha_c^{\pm}$ are defined implicitly by

$$\beta_{\sigma_1}(\alpha_c^{\pm}) = \beta_1(\alpha_c^{\pm}; 100) \tag{3.10a}$$

which from Fig. 2 implies

$$\alpha_c^- = 1.58 \ , \ \alpha_c^+ = 2.19 \ . \tag{3.10b}$$

There also exists a region of bistability corresponding to $\sigma_{-1} < \sigma < 0$, the uniform state being stable for $\sigma < 0$ and hexagons for $\sigma_{-1} < \sigma < \sigma_2$. Given that $\sigma_{-1} = -4a_0^2/(a_1 + 4a_2) < 0$ for $a_1 + 4a_2 > 0$ and $a_0 \neq 0$, the hexagons persisting in this overlap region would be subcritical in nature. However as can be seen from Fig. 3

$$|a_0| << a_1 + 4a_2 \tag{3.11a}$$

in the parameter range of interest and thus the loci $\sigma = \sigma_{-1}$ and $\sigma = 0$ are virtually indistinguishable over that range. Hence unlike the type between

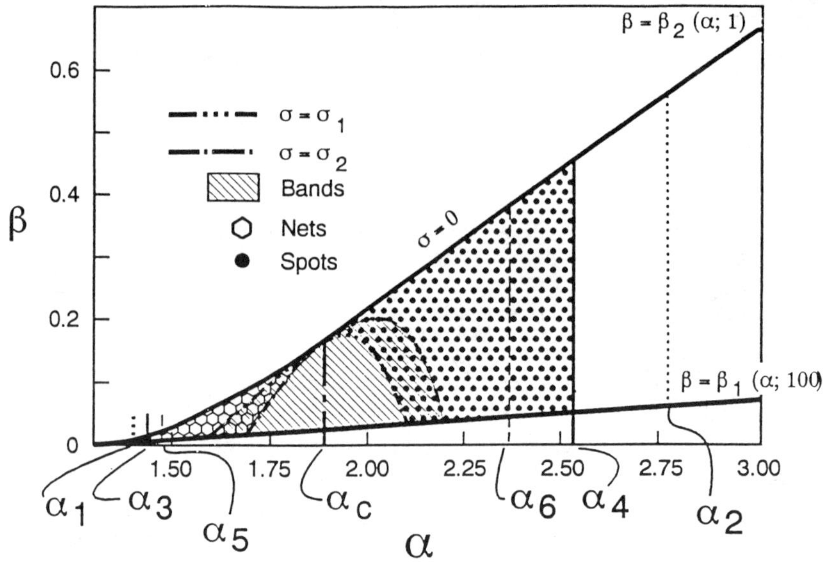

FIG. 2. *Stability diagram in the α-β plane for the CDIMA/indicator model system with $\mu = 1$ and $K = 100$ denoting the predicted Turing patterns summarized in Table 1.*

hexagons and stripes, this bistability is beyond experimental resolution. Further to justify the truncation procedure inherent to the asymptotic representation of (3.2) it is necessary that its coefficients satisfy the additional size constraint (Wollkind *et al.*, 1994)

$$|a_0|/(a_1 + 4a_2)^2 << 1 \ . \tag{3.11b}$$

Noting that the inequality condition (3.11a) also guarantees the satisfaction of this constraint, we can conclude that such a truncation procedure is valid for our hexagonal planform weakly nonlinear stability analysis of the CDIMA/indicator model system.

4. Comparisons, extensions, and conclusions. We are now ready to compare these theoretical predictions summarized in Section 3 with relevant experimental observations. We shall proceed by first considering those experiments which only involved stripes and hexagonal dot or net patterns that emerged upon increase of the $[MA]_0$ reservoir concentration. Since Fig. 2 represents a two-dimensional refinement of Stephenson and Wollkind's (1995) one-dimensional results, we shall examine the possible succession of Turing patterns predicted when a member of the one-parameter family of curves $\beta \equiv \beta_0$ is traversed in the direction of increasing α.

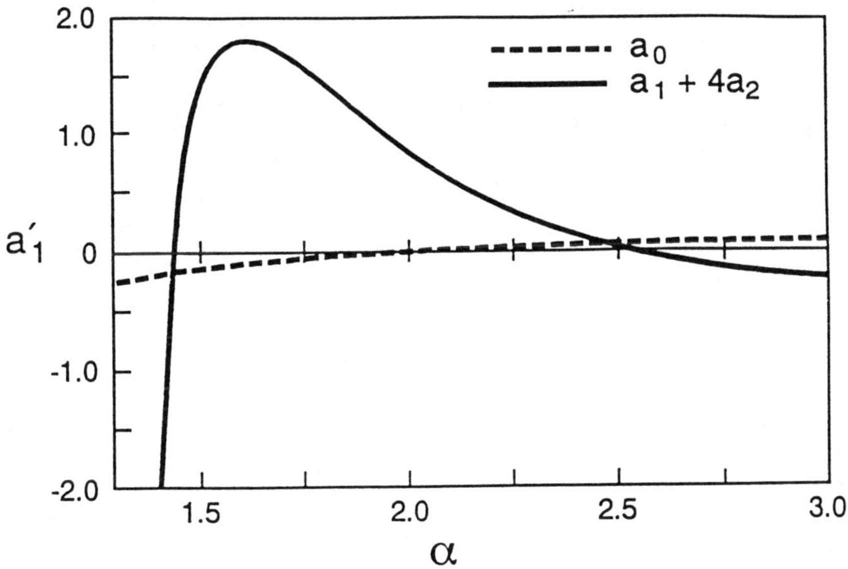

FIG. 3. *Plots of* a_0 *and* $a_1 + 4a_2$ *of (3.6) versus* α *for* $\mu = 1$ *and* $K = 100$.

In particular to produce a theoretical prediction consistent with these experimental observations, it is necessary for us to assume the constant u condition

$$u = u_0 \ , \tag{4.1a}$$

which yields the transit curve

$$\alpha = m[MA]_0 \ , \ \beta = \beta_0 \tag{4.1b}$$

where

$$m = \frac{k_1 z}{5k_2[ClO_2]_0 u_0(1-z)} \ , \qquad \beta_0 = \frac{k_3[I_2]_0}{k_2[ClO_2]_0 u_0(1-z)} \ . \tag{4.1c}$$

For fixed values of the other parameters including z this horizontal line is traversed in the direction of increasing α as $[MA]_0$ increases. Adjusting these parameters appropriately we conclude from an examination of Fig. 2 that such a transit line is capable of generating all those Turing pattern sequences catalogued in Table 2 as $[MA]_0$ increases.

Here after Borckmans *et al.* (1995) we are using the notation A/B to indicate the bistability of structures A and B. Note that Table 2 includes as its first entry the complete succession of Turing patterns depicted by the latter authors in their numerical study of the Brusselator reaction-diffusion

TABLE 2
Predicted Turing pattern sequence versus β_0.

β_0	Predicted Sequence
$(0.05, 0.16)$	I, III^+, III^+/II, II, II/III^-, III^-
0.16	I, II, II/III^-, III^-
$(0.16, 0.18)$	I, III^-, III^-/II, II, II/III^-, III^-
$(0.18, 0.20)$	I, III^-, III^-/II, III^-
$(0.20, 0.44)$	I, III^-

model system. Borckmans *et al.* (1995) numerically computed bifurcation diagrams for this system and found a particular set of parameter values which produced such a complete succession of Turing patterns consisting of the uniform state, nets, nets/stripes, stripes, stripes/spots, and spots as the bifurcation parameter was increased. From the comparison between the Brusselator and CDIMA reaction-diffusion systems included in Stephenson and Wollkind (1995), we can conclude that their scenario was analogous to the one which would be generated in our problem by allowing α to increase in Fig. 2 along the horizontal transit curve $\beta \equiv \beta_0$, where $\beta_0 \in (0.05, 0.16)$. Indeed Borckmans *et al.* (1995) described the occurrence of both types of hexagons in the same problem as surprising and hence stated that interestingly the complete scenario represented in the bifurcation diagram for the Brusselator under these conditions had been observed experimentally in the CIMA reaction by Ouyang and Swinney (1995). Finally they listed the identical set of sequences as that contained in Table 2 when compiling possible scenarios obtained from numerical simulations of the Schnackenberg simplification of the Brusselator. Since Stephenson and Wollkind (1995) showed that both the Brusselator and Schnackenberg model systems when appropriately scaled can be placed in the generic form of (2.3a), Borckmans *et al.*'s (1995) study demonstrates the robustness of our hexagonal stability results to other choices for the nonlinear chemical kinetic mechanism.

In addition to pure stripes and hexagons, Ouyang and Swinney (1991b) also observed what they originally termed a distinct stationary mixed state which appeared as if it were an overlap of hexagons and stripes. Subsequently Ouyang *et al.* (1993) reclassified this structure as a rhombic array of characteristic angle 58°. Upon examining Figs. 1 and 2, we see that $\psi = 58°$ (or equivalently 1.012 radians) is a typical allowable rhombic angle for α in the neighborhood of α_c relevant to the parametric curve $\beta \equiv 0.16$. Further the characteristic angles of the stable rhombic arrays observed by Gunaratne *et al.* (1994) occurred in two bands flanking either side of the 60° (or $\pi/3$ radians) of regular hexagons with $\psi = 66°$ (or 1.152 radians) being the representative angle portrayed in that figure depicting their results. These observations are in agreement with our predictions of Fig. 1 while from the latter we note that $\psi = 66°$ is an allowable rhombic angle for $\alpha = 2$. In this context we can now conclude that hexagons actually occur in those regions of Fig. 1 previously identified as having no pattern. Making this interpretation and, after Ouyang *et al.* (1993) and Gunaratne *et al.* (1994), classifying both stable rhombic and hexagonal arrays together where $\psi = \pi/3$ corresponds to the latter, we can deduce that the band width of the angular distribution for such structures tends to be narrow near the onset of patterns and broad farther away from the transition as one proceeds along the parametric curve under examination consistent with the experimentally generated histograms appearing in those references.

We complete this comparison with some comments regarding a topic usually only handled implicitly in current pattern formation literature (Walgraef, 1997): Namely the observational implications of various methodologies for selecting the dark and light regions in contour plots of Turing patterns. Ouyang and Swinney (1991b) and Ouyang *et al.* (1992) performed series of experiments involving a CIMA/starch reaction occurring in a polyacrylamide disc gel reactor. A variety of stationary one- and two-dimensional spatial Turing structures were observed to form from an initially uniform state as a control parameter was varied. These Turing patterns as depicted in their photographic reproductions appear as a yellow (light) oxidized (low $[I_3^-]$) region on a blue (dark) reduced (high $[I_3^-]$) background generated by the color change of the starch indicator during the redox reaction. Ouyang *et al.* (1993) and Gunaratne *et al.* (1994) performed similar experiments involving a disc reactor but instead employing a polyvinyl alcohol gel which served as its own Turing pattern indicator by turning dark red in the reduced state. In Section 3 the threshold triiodide concentration for that color change was implicitly chosen to coincide with the homogeneous state value of α. Hence all spatial regions characterized by $x \geq \alpha$ would appear dark and by $x < \alpha$, light. Thus given this morphological interpretation the homogeneous state $x = \alpha$ would appear uniformly dark. The adoption of such a protocol has also been standard operating procedure for representing two-dimensional patterns obtained by numerical analyses of reaction-diffusion systems (Lengyel *et al.*, 1992) although in some instances the light and dark regions were interchanged (Dufiet and Boissonade, 1992). That an interchange of this sort will result in Turing patterns which preserve stripes and rectangles while switching the two hexagon types is a distinguishing feature of all protocols based on a zero-deviation threshold. In fact any critical transition threshold x_c which satisfies $x_c \leq \alpha$ will guarantee that the homogeneous state appears uniformly dark. This being the case it is instructive to examine precisely what effect the adoption of a different protocol based on another member of that class would have on the appearance of our Turing patterns. Should we choose a lower threshold protocol satisfying $x_c < \alpha$ then the dark regions would predominate. Having investigated these lower threshold protocols there is also merit in examining the morphological consequences of the adoption of a higher threshold one satisfying $x_c > \alpha$. For this situation the homogeneous solution $x = \alpha$ would appear uniformly light. Hence it seems reasonable under the circumstances to view Turing patterns as being generated by the formation of dark regions against a light background. Given such a morphological interpretation III^+ now represents "spots" and III^-, "honeycombs", which reverses the roles played by these critical points heretofore. In particular a higher threshold protocol of this sort would give rise to Turing patterns for which the light regions now predominate. It is of interest to note that examples of all the Turing patterns described in this paragraph actually appear in the photographic reproductions of Ouyang and Swinney (1991b,1995) and Gunaratne *et al.* (1994).

That the adoption of such a protocol to replicate patterns of this sort may not be as well-known a theoretical pattern generation mechanism as one might have suspected is attested to by the following: Recently Kondo and Asai (1995) investigated the formation of stripes in marine angelfish by employing an activator-inhibitor reaction-diffusion model which incorporated the kinetics of Turing (1952). Analogous to our contour plots, high and low concentrations of the activator species were represented by dark and light colors, respectively, in their computer simulated patterns. When commenting on this work in conjunction with the actual appearance of the angelfish, Meinhardt (1995) stated that the light stripes on the fish still required further explanation since those shown by Kondo and Asai (1995) were very narrow with respect to the dark spaces in between while all the models of which he was aware could only produce stripes and interstripes of the same size.

We next place our CDIMA/indicator model system results in the context of some recent pattern formation studies in alloy solidification and discuss those requisite extensions of these analyses still needed to examine the interaction of rhombic with hexagonal patterns. We begin this commentary with a comparison of our predicted patterns to equilibrium structures characteristic of systems used to model interfacial morphologies observed during dilute binary alloy solidification (reviewed by Coriell and McFadden, 1993). In the process of doing so we shall also offer an explanation for the occurrence of black-eye hexagonal patterns within the framework of our quasi-two-dimensional results.

Wollkind *et al.* (1984) extended the two-disturbance results of Sriranganathan *et al.* (1983) by performing a six-disturbance hexagonal planform weakly nonlinear stability analysis of their model diffusion system of equations for the controlled solidification of a dilute binary alloy under the influence of a temperature gradient. The stable critical points of the amplitude-phase equations (3.2b,c), which can potentially have a variety of phenomenological interpretations, in this instance corresponded to the following interfacial morphologies: I, the planar interface; II, elongated cells or bands; III$^+$, dome-shaped regular hexagonal cells or nets; and III$^-$, a hexagonal close-packed array of circular depressions of liquid into the solid or nodes (see Fig. 4). Here the roles of the hexagonal cell and node structures are correlated with those of the nets and spots of the chemical Turing instability pattern since for the alloys under examination the interfacial planar deviation and solute impurity concentration were inversely related while the experimental Turing patterns have been characterized by low $[I_3^-]$. The interfacial morphological stability criteria relevant to this metallurgical problem can be represented in a plot analogous to Fig. 2 when α and β are associated with the appropriate nondimensionalized rate of solidification U and liquid temperature gradient H, respectively (see Fig. 4). In particular $\beta_1 \equiv 0$ for this situation while the inequalities of (3.7d) are replaced by

(4.2) $$a_0 > 0 \text{ for } \alpha < \alpha_c \ , \ a_0 < 0 \text{ for } \alpha > \alpha_c$$

which results in an interchange of the regions identified as III$^\pm$ upon comparison with Fig. 2. Since the inequality condition (3.12b) was also satisfied by the Landau constants for the solidification problem (Wollkind *et al.*, 1984), the sequence of interfacial morphologies consisting of planar interface, nodes, nodes/bands, bands, bands/cells, and cells predicted theoretically when α is increased along the horizontal line $\beta \equiv \beta_0$, which agrees with experimental observation (Morris and Winegard, 1969), can be considered a conclusive result. Specifically this complete sequence of morphologies is dependent upon β_0 being in the range

(4.3) $$\beta_2(\alpha_3) < \beta_0 < \beta_2(\alpha_c) \ .$$

Further if β is decreased along the vertical lines $\alpha \equiv \alpha_0^\pm$ where $\alpha_{cr}^- < \alpha_0^- < \alpha_c, \alpha_c < \alpha_0^+ < \alpha_{cr}^+$, and $\alpha_{cr}^\pm$ are defined implicitly by

(4.4) $$\beta_{\sigma 2}(\alpha_{cr}^\pm) = \beta_1(\alpha_{cr}^\pm) \ ,$$

the morphological behavior is reminiscent of the sequence of convection patterns predicted in the Rayleigh–Bénard problem when the Rayleigh number Ra, a nondimensional temperature difference, is increased (Wollkind, 1986). In particular, for $\alpha \equiv \alpha_0^-$ where $a_0 > 0$, the solidification sequence is planar interface, nodes, nodes/bands, and bands while for $\alpha \equiv \alpha_0^+$ where $a_0 < 0$, it consists of planar interface, cells, cells/bands, and bands. Upon examination of Fig. 2, we see that

(4.5) $$\alpha_{cr}^- = 1.68 \ , \ \alpha_{cr}^+ = 2.09 \ .$$

Hence when $\alpha_0^+ = 2$, we predict the Turing pattern sequence of spatial homogeniety, spots, spots/stripes, and stripes as β decreases, a prediction

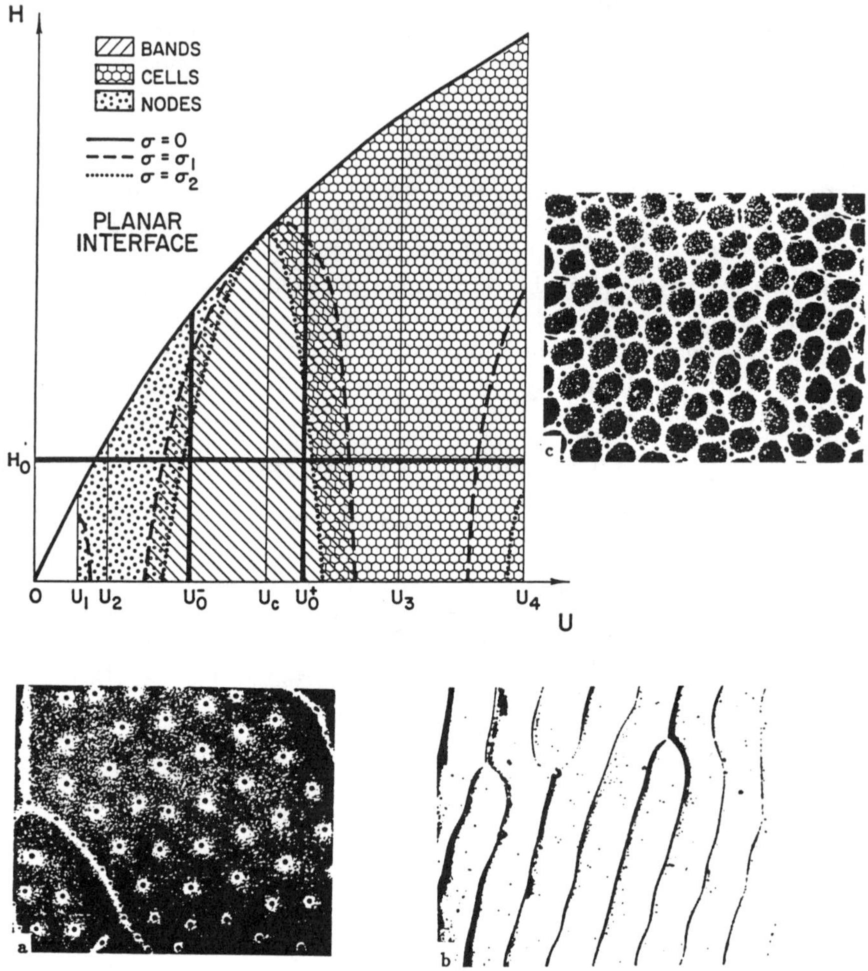

FIG. 4. *A plot in the U-H plane denoting the predicted two-dimensional interfacial morphologies of Wollkind (1986). Here the photographs labeled (a), (b), and (c) correspond to nodes, bands, and cells, respectively.*

which is confirmed directly for this CDIMA reaction-diffusion model system by the numerical simulations of Jensen *et al.* (1996). For fixed values of the other parameters such a transition should occur as $[I_2]_0$ varies which within the framework of our model system is consistent with experimental observation (Ouyang and Swinney, 1991b). More generally the transit line $\alpha \equiv \alpha_0$ is capable of generating all those Turing pattern sequences catalogued in Table 3 as $[I_2]_0$ decreases.

TABLE 3
Predicted Turing pattern sequence versus α_0.

α_0	Predicted Sequence
$(1.44, 1.58)$	I, III^+
$(1.58, 1.68)$	I, III^+, III^+/II
$(1.68, 1.88)$	I, III^+, III^+/II, II
1.88	I, II
$(1.88, 2.09)$	I, III^-, III^-/II, II
$(2.09, 2.19)$	I, III^-, III^-/II
$(2.19, 2.53)$	I, III^-

The morphological stability investigation of Wollkind *et al.* (1984) was an extension of Segel's (1965) six-disturbance analysis of the Rayleigh–Bénard problem of buoyancy-driven convection. In particular Wollkind *et al.* (1984) extended Segel's (1965) analysis to include the possibility of $2a_2 - a_1 < 0$ given that this quantity is identically positive for the Rayleigh–Bénard problem. Since the equivalence class of critical points of (3.2b,c) designated as II in Section 3 actually contains the three solutions

$$A_i^2 = \sigma/a_1 \ , \ A_j = A_k = 0 \ ,$$
$$(i,j,k) = \text{even permutation of } (1,2,3) \ , \tag{4.6}$$

the region of Fig. 2 identified with bands or stripes is itself a locus of multiple stable states. These represent a family of bands aligned parallel to the r_2-axis, plus two similar families of bands making angles of $\pm 60°$ to them for which stable co-existence with a member of either the original family or one another is impossible. Then, as initial conditions varied from point-to-point on the interfacial surface, Wollkind *et al.* (1984) concluded that such families of bands could give rise to polygonal arcs the boundaries of which would appear quite random in orientation. Indeed a number of the Turing patterns classified as stripes by Ouyang and Swinney (1995) and Boissonade *et al.* (1995) have the appearance of such curved elongated cells in the relevant photographic reproductions contained therein. Upon examination of Figs. 2 and 4 as well as Table 3 it can be seen that the vertical line $\alpha \equiv \alpha_c$ on which $a_0 = 0$ lies totally within the region where $2a_2 - a_1 > 0$. Hence we may conclude that for such combinations of parameter values corresponding to a vanishing coefficient of the quadratic terms in the amplitude equations only stripes or bands but never hexagonal solutions can be stable. Therefore in spite of the potentiality of bistability existing between the two types of hexagonal states when $a_0 = 0$ and $2a_2 - a_1 < 0$ (see the appropriate entry of Table 1) this particular possibility is precluded for our specific model. Further, we note that the identical value α_c playing a central and consistent role with respect to stripe formation in both Figs. 1 and 2 serves as a partial but independent check on those analyses which generated them. Finally, observe that in Fig. 2 the degenerate point (α_K, β_K) where the Turing $\beta = \beta_2$ and Hopf $\beta = \beta_1$ boundaries intersect, satisfies $\alpha_K = 1.36$ for $K = 100$ and thus is in the subcritical bifurcation region relevant to Turing instabilities. Hence, this codimension-two bifurcation point lies outside our parameter range of interest. Therefore, Rovinsky and Menzinger's (1992) predicted spatio-temporal patterns, occurring in the neighborhood of such a point when that bifurcation is supercritical, have no bearing on the scenario considered here.

We have deferred until now a discussion of the black-eye hexagonal patterns observed by Gunaratne *et al.* (1994) and Ouyang and Swinney (1995). For those experiments these authors discovered a complex stationary periodic black-eye array which they felt was unexpected from general pattern formation theory. Such structures were only obtained in experiments involving polyvinyl alcohol gel disks of high concentration. This pattern formed from a normal hexagonal one, which was the initial instability to the uniform state, when $[MA]_0$ increased slowly. It consisted of two hexagonal lattices: One of white spots and the other of black spots at the center of each white spot and at the center of the dark region in each equilateral triangle with three neighboring white spots at its vertices. Then upon further increase of $[MA]_0$ this pattern disappeared and was replaced by stripes. Specifically, the transitions from the uniform state to the normal hexagonal dot pattern and from the latter to the black-eye array, which occurred at $[MA]_0 = 7$ and 8mM, respectively, were nonhysteretic while that from the black-eye pattern to stripes was hysteretic with this hysteresis occurring over the 11.6–13.8mM range.

To reproduce the black-eye hexagonal array sequence described above, we next need to relax the patterned layer infinitesimal thickness constraint which was implicitly assumed when taking $z = z_0$ during all the interpretations of our quasi-two-dimensional model results presented previously. We do so by introducing a layer of thickness $\triangle z$ located in the interval $z \in (z_1, z_2)$ where $z_{2,1} = z_0 \pm \triangle z/2$. This replaces our transit curve $\beta \equiv \beta_0$ of (4.1) particularized to $z = z_0$ by a band of width

(4.7a)
$$\Delta\beta \cong \frac{\beta_0}{1 - z_0}\Delta z$$

centered about β_0 such that for a fixed value of $[MA]_0$ the locus of interest in α-β space becomes the line segment through (α_0, β_0) joining the end points $(\alpha_0 \pm \Delta\alpha/2, \beta_0 \pm \Delta\beta/2)$ for $z \in (z_1, z_2)$ where

(4.7b)
$$\Delta\alpha \cong \frac{\alpha_0}{z_0(1 - z_0)}\Delta z\ , \qquad \alpha_0 = m_0[MA]_0$$

with m_0 denoting m of (4.1c) evaluated at z_0. Finally considering a band of this sort, flanking $\beta_c = 0.16$ in Fig. 2, and superposing those patterns predicted in the top and bottom portions of the layer as $[MA]_0$ is increased, we can obtain the observed sequence under investigation. Let us describe that procedure in some detail. The optical transmission technique inherent to these experimental observations permits us for comparison purposes to superpose the predicted Turing patterns for the top $z \cong z_2$ and bottom $z \cong z_1$ surfaces of the thin layer. Upon examination of Fig. 2 and Table 2, we can conclude that the patterns in question are represented by the predicted sequences listed as the third and first entries of the latter, respectively, with Turing patterns for $z \cong z_2$ emerging from the uniform state before those for $z \cong z_1$ as $[MA]_0$ is increased. Moreover we shall truncate these sequences by limiting the $[MA]_0$ range so that each terminates in the region where stripes are the only stable pattern and further assume that the critical $[MA]_0$ value at which the polyvinyl alcohol color change occurs corresponds to a point in this region as well. Hence the $\text{III}^{\pm}$ hexagonal states in these sequences are of the higher threshold variety while the uniform state I appears clear. The superposition of those patterns in these sequences that can coexist for a particular value of $[MA]_0$ in the allowable range results in the following superposed combinations {I,I}, $\{\text{III}^-, \text{I}\}$, $\{\text{III}^-, \text{III}^+\}$ and {II,II} as $[MA]_0$ increases, which are identifiable with a homogeneous, "honeycomb", black-eye, and striped pattern, respectively. To justify our identifications we must examine the resolution of our hexagonal structures in more detail. Only regions of relatively high tri-iodide concentration appear dark in those patterns. From the definition of III^-, we can deduce that the highest such concentrations for the "honeycomb" are located in the circular regions about the vertices of the hexagons of Fig. 4c or equivalently inscribed within the equilateral triangles described above. Should the lighting conditions be sensitive enough for their resolution, these regions would appear as black dots standing out against the dark background of the "honeycomb" pattern. When superposed with the black "spots" of Fig. 4a this configuration yields the black-eye pattern described by Gunaratne *et al.* (1994) and Ouyang and Swinney (1995). When superposed with the uniform state it yields the normal "honeycomb" hexagonal pattern since the latter situation unlike the former fails to provide sufficient illumination at the key vertices to make them visible. This circumstance is analogous to the leopard coat patterning phenomenon of the black on black spots of its melanistic panther color phase only being visible in direct sunlight but not in the shade (Baker and Bridges, 1948). A resolution of this sort for the "honeycomb" and "spot" patterns is consistent with the appearance in Fig. 4 of the cells and nodes of the metallurgical problem (Wollkind *et al.*, 1984), respectively, to which they are now correlated. As one last detail we observe that the interplay of the hysteresis regions of the pattern sequences characteristic of the top and bottom of the layer gives rise to the hysteretic behavior between black-eyes and stripes reported by Gunaratne *et al.* (1994).

In order to make these qualitative comparisons between theoretical prediction and experimental observation more quantitative in nature, we assign the parameters in the model the typical values, appropriate for $T = 280°$ K, of

(4.8a)
$$k_1 = 9.0 \times 10^{-4}/\text{sec}\ , \quad k_2 = 1.2 \times 10^3/(\text{M sec})\ ,$$
$$k_3 = 1.5 \times 10^{-4}/\text{sec}\ ,$$

(4.8b)
$$[I_2]_0 = 7.2 \times 10^{-4}\text{M}\ , \quad [ClO_2]_0 = 1.6 \times 10^{-3}\text{M}\ ,$$
$$u_0 = 3.5 \times 10^{-6}\text{M}\ , \quad z_0 = 0.9\ ,$$

the latter condition being consistent with the location of the chemical front near the malonic acid reservoir boundary $z = 1$ during these experiments. For the polyvinyl alcohol gel sequence just described we then find that the intersection point (α_0, β_0) between the transit curve $\beta \equiv \beta_0$ and the Turing boundary $\beta = \beta_2$ satisfies

(4.9a)
$$\alpha_0 = \alpha_c\ , \quad \beta_0 = \beta_c$$

when $[MA]_0 = 7\text{mM}$ in agreement with the scenario proposed above and corresponds to

(4.9b)
$$\lambda_c^* = .20\text{mm}$$

from (2.7) and (2.9) in accordance with experimental measurement.

We close this discussion by pointing out that Gunaratne *et al.* (1994) in offering their explanation for the periodic black-eye array associated the black dots at the hexagonal vertices with the black "spot" pattern and postulated such a black hexagonal lattice was a spatial harmonic of the primary white spotted one, the former being generated as a secondary mode by the resonant interaction of the basic modes of the latter. Employing this hypothesis however they were unable to explain why that secondary mode did not grow continuously beyond the onset of the primary instability. Gunaratne *et al.* (1994) then stated that they did not understand this difference between theory and experiment while suggesting that either there might not be sufficient sensitivity to detect this harmonic closer to the onset of normal hexagons or perhaps the secondary modes were not slaved to the primary ones in the sense of Boissonade *et al.* (1995).

So far after Kuske and Matkowsky (1994) and Hoyle *et al.* (1995), who studied the behavior of a premixed flame anchored on a flat burner and the effect of surface free energy anisotropy on interfacial morphology during the controlled solidification of a dilute binary alloy, respectively, by both square and hexagonal planform weakly nonlinear perturbation analyses, we have investigated separately the stability of either rhombic or hexagonal arrays versus stripes but not considered the stability of these two-dimensional Turing patterns versus each other. To determine directly the outcomes of interactions of this sort it is necessary to introduce extensions of our two-dimensional analyses which would allow us simultaneously to consider the stability of both rhombic and hexagonal patterns. Although these extensions are beyond the scope of our present work, we conclude with a brief description of this related topic not only for the sake of completeness but also because it complements much of the material discussed already.

Two different methods of pattern selection have been developed for examining the competition between rhombic and hexagonal arrays. The first in essence is a synthesis of both our rhombic and hexagonal planform approaches which enlarges the class of perturbations allowed for either analysis by including members from the other one as well. The second is a Ginzburg–Landau formulation involving spatio-temporal amplitude equations in which rhombic patterns are obtained by stretching an array of regular hexagons along an axis of symmetry with that distortion occurring as a consequence of the action of the spatial derivatives contained in those equations.

The method of synthesis was originally devised by Kuznetsov and Spektor (1976) to study interfacial patterns on the surface of a dielectric fluid. Golovin *et al.* (1994) used a method of this type particularized to $\psi = \pi/2$ and, having normalized our $-4a_0$ to unity by appropriately scaling their amplitude equations, deduced stability criteria for hexagons versus squares and squares versus hexagons involving $\sigma, a_1, a_2, b_1(\pi/2)$, and $b_1(\pi/6)$ relevant to the Bénard–Marangoni surface-tension driven convection problem with poorly conducting boundaries. They also deduced stability criteria for hexagons or squares versus rolls equivalent to those of Kuske and Matkowsky (1994). From these criteria Golovin *et al.* (1994) constructed a pattern selection diagram in a gravity number-capillary number parameter space by identifying regions where various types of bistability could occur. They found squares to be stable in those regions for which a strictly hexagonal planform analysis would have predicted stable hexagons alone, the latter retaining their stability to the enlarged class of perturbations.

The Ginzburg–Landau method as proposed by Ouyang *et al.* (1993) and Gunaratne *et al.* (1994) adds second-order spatial operators which are equivariant with respect to arbitrary rotations to the hexagonal planform complex amplitude equations of Kuske and Matkowsky (1994), where the latter are related to (3.2b,c) by the transformation

$$\mathcal{R}_1 = A_1 e^{-i\phi_1}\ , \quad \mathcal{R}_2 = A_2 e^{i\phi_2}\ , \quad \mathcal{R}_3 = A_3 e^{i\phi_3}\ . \tag{4.10}$$

Here the terms in these equations involving those particular operators contain the proportionality constant (Cross and Hohenberg, 1993)

$$\lambda_0 = -\frac{1}{2}\frac{\partial^2 \sigma_c}{\partial q^2}(q_c, \beta_2; \alpha, \mu, K) \tag{4.11a}$$

where

$$\sigma_c(q_c, \beta; \alpha, \mu, K) = \sigma(\beta; \alpha, \mu, K)\ . \tag{4.11b}$$

Introducing

$$\mathcal{R}_1 = W_1\ , \quad \mathcal{R}_2 = W_2 e^{i\delta q_c r_2}\ , \quad \mathcal{R}_3 = W_3 e^{-i\delta q_c r_2}\ , \tag{4.12}$$

we note that the transformed system of amplitude equations involving the W_n's has a critical point corresponding to III$^{\pm}$ when $\delta = 0$. Ouyang *et al.* (1993) and Gunaratne *et al.* (1994) made the morphological interpretation that to lowest order this critical point could be associated with a rhombic array produced by stretching our hexagonal array along its r_2-axis of symmetry. In this context, δ represents a measure of that distortion with $\triangle\psi$, the deviation of the characteristic angle of such a rhombic array from $\pi/3$, related to it by geometrical considerations. Then they deduced criteria governing the stability of this critical point to long-wavelength perturbations and from these criteria in conjunction with the existence condition for that critical point found the allowable range of $\triangle\psi$ over which such structures could occur. In particular, Ouyang *et al.* (1993) and Gunaratne *et al.* (1994) analyzed an idealized set of their Ginzburg-Landau equations with the arbitrary assignment of

$$-4a_0 = a_1 = a_2 = \lambda_0 = q_c = 1 \tag{4.13}$$

and deduced a single criterion governing the stability of rhombic distortions to a hexagonal array, that criterion involving σ and $\triangle\psi$. Next they plotted the marginal stability locus associated with this criterion in the σ-$\triangle\psi$ plane. That curve formed the right-hand boundary of a stability region possessing the marginal locus relevant to the existence condition for the critical point as its left-hand boundary. When $\triangle\psi = 0$, or equivalently $\delta = 0$, there was a finite range of σ given in this instance by $\sigma_{-1} = -0.05 < \sigma < 4 = \sigma_2$ for which hexagons were stable. Their model predicted that for values of σ in this range different rhombic arrays could coexist with the hexagons, the characteristic angles of those arrays lying within an interval about

$\triangle\psi = 0$ determined from the stability domain described above. Ouyang *et al.* (1993) and Gunaratne *et al.* (1994) expected these results, which were in qualitative accord with their experimental observations, to be valid for sufficiently small values of δ or equivalently $\triangle\psi$. They stated that a quantitative comparison of experiment with theory would require the evaluation of the coefficients in the Ginzburg–Landau equations from the chemical kinetics and diffusion coefficients of the reaction, which has been our philosophy from the outset of this endeavor. The latter is compatible with our long range goal of developing the simplest reasonable natural science models which preserve the essential features of pattern generation and are still consistent with observation. Perhaps the best rationale for performing our two-dimensional analyses of this chemical Turing instability model system was offered by Murray (1989), who stated that, although there were a number of numerical studies, a general nonlinear analysis of the evolution to the finite amplitude steady-state spatial pattern for such diffusive instabilities was still lacking. We believe a procedure that employs a synergism between analytical stability techniques and experimental data to establish the parameter range of interest for pattern formation, such as our weakly nonlinear methods, and only then explores this regime more fully numerically, to have both a scientific and an economic advantage over one that attempts to use numerical techniques alone to accomplish the same end.

As pointed out at the end of Section 1, pattern formation processes of this sort may play a significant role in developmental biology particularly as related to embryology. Lengyel *et al.* (1993) felt that if patterns in developing embryos do arise by Turing bifurcation, it was likely to be through their gradient-free scenario rather than one in which different reactants entered the embryo from opposite ends. They imagined that prepatterns which formed transiently in the embryo from nearly uniform concentrations were then frozen into place as a result of developmentally determined changes in its chemical environment. Recent experiments involving meso dermal patterning in amphibian embryos induced by means of a chemical concentration gradient (Niehrs *et al.*, 1994) would seem to make as good a case for the gradient-imposed scenarios of Ouyang and Swinney (1991a,b; 1995), Ouyang *et al.* (1992), Ouyang *et al.* (1993), and Gunaratne *et al.* (1994) in this context.

REFERENCES

M. Baker and W. Bridges (1948), *Wild Animals of the World*, Garden City Publishing, Garden City, N.Y.

J. Boissonade, E. Dulos, and P. De Kepper (1995), *Turing patterns: Myth to reality*, in Chemical Waves and Patterns, R. Kapral and K. Showalter, eds., Kluwer, Dordrecht, pp. 221–268.

P. Borckmans, G. Dewel, A. Dewit, and D. Walgraef (1995), *Turing bifurcations and pattern selection*, in Chemical Waves and Patterns, R. Kapral and K. Showalter, eds., Kluwer, Dordrecht, pp. 323–363.

V. CASTETS, E. DULOS, J. BOISSONADE, AND P. DE KEPPER (1990), *Experimental evidence of a sustained standing Turing-type nonequilibrium chemical pattern*, Phys. Rev. Lett., **64**, pp. 2953–2956.

S.R. CORIELL AND G.B. MCFADDEN (1993), *Morphological stability* in Handbook of Crystal Growth, Vol. 1, (Ed. D.T.J. Hurle), Elsevier, Amsterdam, pp. 785–858.

M.C. CROSS AND P.C. HOHENBERG (1993), *Pattern formation outside of equilibrium*, Rev. Mod. Phys., **65**, pp. 851–1112.

P. DE KEPPER, V. CASTETS, E. DULOS, AND J. BOISSONADE (1991), *Turing-type chemical patterns in the chlorite-iodide-malonic acid reaction*, Physica, 49D, pp. 161–169.

V. DUFIET AND J. BOISSONADE (1992), *Conventional and unconventional Turing patterns*, J. Chem. Phys., **96**, pp. 664–672.

A.A. GOLOVIN, A.A. NEPOMNYASHCHY, AND L.M. PISMEN (1994), *Pattern formation in large-scale Marangoni convection with deformable interface*, preprint.

M.D. GRAHAM, I.G. KEVREKIDIS, K. ASAKURA, J. LAUTERBACH, K. KRISCHER, H.-H. ROTERMUND, AND G. ERTL (1994), *Effects of boundaries on pattern formation: Catalytic oxidation of CO on platinum*, Science, **264**, pp. 80–82.

G.H. GUNARATNE, Q. OUYANG, AND H.L. SWINNEY (1994), *Pattern formation in the presence of symmetries*, Phys. Rev. E, **50**, pp. 2802–2820.

J. GUSLANDER AND R.J. FIELD (1991), *Modeling of an observed Turing structure in the* ClO_2^-*-*I^-*-Malonic Acid system*, Int. J. Bifurcation and Chaos, **1**, pp. 929–931.

L.G. HARRISON (1993), *Kinetic Theory of Living Patterns*, Cambridge University Press, Cambridge.

R.B. HOYLE, G.B. MCFADDEN, AND S.H. DAVIS (1995), *Pattern selection with anisotropy during directional solidification*, Appl. Math. Tech., Report No. **9404**, Northwestern University.

C.B. HUFFAKER, K.P. SHEA, AND S.G. HERMAN (1963), *Experimental studies on predation (III). Complex dispersion and levels of food in an acarine predator-prey interaction*, Hilgardia, **34**, pp. 305–330.

O. JENSEN, E. MOSEKILDE, P. BORCKMANS, AND G. DEWEL (1996), *Computer Simulation of Turing structures in the chlorite-iodide-malonic acid system*, Phys. Scripta, **53**, pp. 243–251.

O. JENSEN, V.O. PANNBACKER, G. DEWEL, AND P. BORCKMANS (1993), *Subcritical transition to Turing structures*, Phys. Lett A, **179**, pp. 91–96.

P. KAREIVA AND G.M. ODELL (1987), *Swarms of predators exhibit "prey taxis" if individual predators use area search*, Am. Nat., **130**, pp. 233–270.

S. KONDO AND R. ASAI (1995), *A reaction-diffusion wave on the skin of the marine angelfish* Pomacanthus, Nature, **376**, pp. 765–768.

R. KUSKE AND B.J. MATKOWSKY (1994), *On roll, square, and hexagonal cellular flames*, Euro. Jnl. Appl. Math., **5**, pp. 65–93.

E.A. KUZNETSOV AND M.D. SPEKTOR (1976), *Existence of a hexagonal relief on the surface of a dielectric fluid in an external electric field*, Sov. Phys. JETP, **44**, pp. 136–141.

I. LENGYEL AND I.R. EPSTEIN (1991), *Modeling of Turing structures in the chlorite-iodide-malonic acid-starch reaction system*, Science, **251**, pp. 650–652.

——— (1992), *A chemical approach to designing Turing patterns in reaction-diffusion systems*, Proc. Natl. Acad. Sci. USA, **89**, pp. 3977–3979.

——— (1995), *The chemistry behind the first experimental chemical examples of Turing patterns*, in Chemical Waves and Patterns, R. Kapral and K. Showalter, eds., Kluwer, Dordrecht, pp. 297–322.

I. LENGYEL, S. KÁDÁR, AND I.R. EPSTEIN (1992), *Quasi-two-dimensional Turing patterns in an imposed gradient*, Phys. Rev. Lett., **69**, pp. 2729–2732.

——— (1993), *Transient Turing structures in a gradient-free closed system*, Science, **259**, pp. 493–495.

I. LENGYEL, G. RÁBAI, AND I.R. EPSTEIN (1990a), *Batch oscillation in the reaction of chlorine dioxide with iodine and malonic acid*, J. Amer. Chem. Soc., **112**, pp. 4606–4607.

——— (1990b), *Experimental and modeling study of oscillations in the chlorine dioxide-iodine-malonic acid reaction*, J. Amer. Chem. Soc., **112**, pp. 9104–9110.

S.A. Levin and L.A. Segel (1976), *Hypothesis for origin of planktonic patchiness*, Nature, **259**, p. 659.

H. Meinhardt (1995), *Dynamics of stripe formation*, Nature, **376**, pp. 722–723.

L.R. Morris and W.C. Winegard (1969), *The development of cells during the solidification of a dilute Pb-Sb alloy*, J. Crystal Growth, **5**, pp. 361–375.

J.D. Murray (1989), *Mathematical Biology*, Springer-Verlag, Berlin.

——— (1990), *Discussion: Developmental biology: Turing theory of morphogenesis - its influence on modelling biological pattern and form*, Bull. Math. Biol., **52**, pp. 119–152.

C. Niehrs, H. Steinbeisser, and E.M. De Robertis (1994), *Mesodermal patterning by a gradient of the vertebrate homeobox gene goosecoid*, Science, **263**, pp. 817–820.

Z. Noszticzius, Q. Ouyang, W.D. McCormick, and H.L. Swinney (1992), *Effect of Turing pattern indicators on CIMA oscillators*, J. Phys. Chem., **96**, pp. 6302–6307.

Q. Ouyang, G.H. Gunaratne, and H.L. Swinney (1993), *Rhombic patterns: Broken hexagonal symmetries*, Chaos, **3**, pp. 707–711.

Q. Ouyang, R. Li, G. Li, and H.L. Swinney (1995), *Dependence of Turing pattern wavelength on diffusion rate*, J. Chem. Phys., **102**, pp. 2551–2555.

Q. Ouyang, Z. Noszticzius, and H.L. Swinney (1992), *Spatial bistability of two-dimensional Turing patterns in a reaction-diffusion system*, J. Chem. Phys., **96**, pp. 6773–6776.

Q. Ouyang and H.L. Swinney (1991a), *Transition from a uniform state to hexagonal and striped Turing patterns*, Nature, **352**, pp. 610–612.

——— (1991b), *Transition to chemical turbulence*, Chaos, **1**, pp. 411–420.

——— (1995), *Onset and beyond Turing pattern formation,* in Chemical Waves and Patterns, R. Kapral and K. Showalter, eds., Kluwer, Dordrecht, pp. 269–295.

J.E. Pearson (1992), *Pattern formation in a (2 + 1)-species activator-inhibitor-immobilizer system*, Physica A, **188**, pp. 178–189.

A. Rovinsky and M. Menzinger (1992), *Interaction of Hopf and Turing bifurcations in chemical systems*, Phys. Rev. A, **46**, pp. 6315–6322.

J. Schnackenberg (1979), *Simple chemical reaction systems with limit cycle behavior*, J. Theor. Biol., **81**, pp. 389–400.

L.A. Segel (1965), *The nonlinear interaction of a finite number of disturbances in a layer of fluid heated from below*, J. Fluid Mech., **21**, pp. 359–384.

L.A. Segel and J.L. Jackson (1972), *Dissipative structure: An explanation and an ecological example*, J. Theor. Biol., **37**, pp. 545–592.

L.A. Segel and S.A. Levin (1976), *Applications of nonlinear stability theory to the study of the effects of diffusion on predator-prey interactions* in Topics in Statistical Mechanics and Biophysics: A Memorial to Julius L. Jackson, AIP Conf. Proc. No. 27, R.A. Piccirelli, ed., Amer. Inst. Phys., New York, pp. 123–152.

R. Sriranganathan, D.J. Wollkind, and D.B. Oulton (1983), *A theoretical investigation of the development of interfacial cells during the solidification of a dilute binary alloy: Comparison with the experiments of Morris and Winegard*, J. Crystal Growth, **62**, pp. 265–283.

L.E. Stephenson and D.J. Wollkind (1995), *Weakly nonlinear stability analyses of one-dimensional Turing pattern formation in activator-inhibitor/immobilizer model systems*, J. Math. Biol., **33**, pp. 771–815.

A.M. Turing (1952), *The chemical basis of morphogenesis*, Phil. Trans. R. Soc. Lond., B**237**, pp. 37–72.

D. Walgraef (1997), *Spatio-Temporal Pattern Formation*, Springer-Verlag, New York.

D.J. Wollkind (1986), *A new prototype problem for nonlinear stability theory: Plane-front alloy solidification versus free-surface Bénard convection* in Mathematics Applied to Fluid Mechanics and Stability, J.E. Flaherty and D.A. Drew, eds., SIAM, Philadelphia, pp. 205–217.

D.J. Wollkind, V.S. Manoranjan, and L. Zhang (1994), *Weakly nonlinear stability analyses of prototype reaction-diffusion model equations*, SIAM Review, **36**, pp. 176–214.

D.J. Wollkind, R. Sriranganathan, and D.B. Oulton (1984), *Interfacial patterns during plane front alloy solidification*, Physica, **12**D, pp. 215–240.

D.J. Wollkind and L.E. Stephenson (2000), *Chemical Turing pattern formation analyses: Comparison of theory with experiment*, SIAM J. Appl. Math., in press.

BEYOND SPOTS AND STRIPES: GENERATION OF MORE COMPLEX PATTERNS BY MODIFICATIONS AND ADDITIONS OF THE BASIC REACTION

HANS MEINHARDT*

1. Introduction. The question of how a complex organism can develop from a single fertilized egg has fascinated biologist for more than two centuries. In earlier times, basic insights have been obtained from experiments in which normal development has been perturbed. Like all other biological processes, development must be accomplished by interactions of molecules. From the regulations observed after experimental perturbations one cannot deduce directly the molecular interactions involved, however, these observations allow one to work out hypothetical interactions that have the same dynamic properties. A theory of development has to describe concentration changes of the relevant substances as functions of space and time. We have worked out several mathematically formulated models for different developmental situations that where able to describe many of the observations rather precisely. Since the molecular basis of development was completely unknown at that time, these models made firm predictions on the general types of interactions on which development is based.

Around 1984, the new molecular-genetic techniques opened a second approach. It became possible to clone relevant genes and to isolate the corresponding gene products. In the meantime, an overwhelming amount of data has been accumulated. It has turned out that the earlier proposed models have predicted essential elements of the actual interactions quite well. This shows that only a very limited set of molecular interactions is compatible with the observed dynamic behavior and that mathematical modeling is an appropriate tool to find the general character of these interactions. The following list summarizes the most essential elements of these models and gives some key reference to the corresponding observations:

1. **Primary pattern formation**: Unique organizing regions and periodic structures require the formation of local concentration maxima. They emerge if a self-enhancing reaction is coupled with an antagonistic reaction of long range (Gierer and Meinhardt, 1972, Meinhardt, 1982). The selection of sensory mother cells in *Drosophila* (see Culi and Modolell, 1998, Sun et al., 1998, Lee et al., 1996a) is an example.
2. **Gene activation**: Cells have to remember the signals they (or their progenitors) have received by forming stable states of differentiation. Such cell states result by a feedback of a gene product

*Max-Planck-Institut für Entwicklungsbiologie, Spemannstraße 35 IV, D-72076 Tübingen, GERMANY.

onto the activation of its own gene. Once activated, the activity of a gene is maintained by this positive feedback loop (Meinhardt, 1976, 1978). Many such autoregulatory genes are meanwhile known (Regulski et al., 1991; Leptin 1991). If genes responsible for alternative cell states compete with each other for becoming active, only one of these genes remains active within one cell: the cells have to make a choice.

3. **Activation of several genes under control of a gradient**: A position-dependent gene activation can result from an appropriate coupling of gene activation to a gradually distributed signaling substance. It was proposed that cells measure a particular concentration by becoming stepwise and irreversibly promoted to higher cell states until the actually achieved state corresponds to the local morphogen concentration. After this determination is completed, the signal is no longer required to maintain a particular differentiation. A later increase of the signal can lead to a further promotion ('distal transformation'), while a decrease is without effect (Meinhardt, 1978). An example is the activation of the *brachyury* and the *goosecoid* genes in *Xenopus* by different concentrations of *Activin* (Gurdon et al., 1995).
4. **Segmentation**: Segmentation was proposed to depend on the formation of a sequence of cell states with a predictable neighborhood. This requires a mutual long-range activation of feedback loops (genes) that locally exclude each other. Neighboring cell states need to interact in a symbiotic manner. If more than two cell states are involved, the resulting structure has an intrinsic polarity (a periodic pattern ... ABCABC ... has a polarity, a pattern ... ABABAB ... doesn't). Missing elements can be intercalated (Meinhardt and Gierer, 1980). The predicted complex molecular network has been fully confirmed by the elucidation of the *engrailed-wingless-hedgehog* interaction in *Drosophila*: the cell state characteristic for the posterior compartment requires the *engrailed (en)* activation. As expected, *en* is autocatalytic. The genes required for the neighboring anterior cell state are locally suppressed but activated on long range by the secreted molecule *hedgehog*. In turn, the secreted *wingless* protein, produced in the anterior compartment, stabilizes the *engrailed* activation in the posterior compartment (see Ingham 1991, Pfeifer and Bejsovec, 1992).
5. **Somites**: Somites are the most obviously-segmented structure in vertebrates. It was predicted that they are generated by a stepwise conversion of a periodic pattern in time into a periodic pattern in space (Meinhardt, 1982, 1986). Although somites are separated from the presomitic mesoderm in an anterior-to-posterior sequence, the counter-intuitive prediction was made that the specification of anterior and posterior half-somites occur by wave-like processes

that are initiated at the posterior end of the presomitic mesoderm and move toward the anterior until they come to rest at the correct distance from the last formed half-somite (Meinhardt, 1982; Meinhardt, 1986). The prediction has been confirmed by the observation of the *c-hairy1* gene in the chicken (Palmeirim et al., 1997) that behaves as expected for a signal generating the posterior half-somite. Still open is the prediction that this oscillation is used to advance the particular specification of the individual somites in a very reliable way. Each more posterior somite undergoes precisely one additional cycle of the anterior-posterior-anterior oscillation. This would allow to count the number of periodic elements on the gene level. Such a stop and go mechanism provides a rational as to why the homeotic genes crucial for the anterior-posterior axis and the real structures are co-linearly arranged.

6. **Legs and wings**: The initiation of substructures such as legs or wings was proposed to occur at borders between regions in which different genes are active (Meinhardt, 1983a,b). If the production of a new morphogen requires a cooperation of two adjacent differently determined cell types, the new signaling center will be formed at the common border. The local morphogen concentration provides a measure for the distance from this border. The intersection of two such borders, one resulting from a patterning along the anteroposterior (head-to-tail) axis, the other from a patterning in the dorsoventral (or mediolateral, back-to-belly) dimension, defines unique points and complete new coordinate systems for the initiation of the new structures. For a tube-like embryo, the initiation of legs, wings, etc. in pairs is a straightforward consequence of such a mechanism. For instance, digits have been proposed to arise at a dorsoventral (back-to-belly) border that give rise first to the apical ectodermal ridge. The digits are formed along this ridge and appear, therefore, in a plane. The inner side and outer side of a hand are necessarily different from the beginning. The type of the digits is determined by the distance from a border located at a posterior margin of the limb bud. The intersection of the two borders gives rise to the zone of polarizing activity (ZPA). This boundary model has found direct support for vertebrates (Martin, 1995) and for insects (Vincent and Lawrence, 1994).
7. **Net-like structures**: Filamentous branched structures result from moving signals that cause local elongation of the filaments. Long differentiated structures emerge as a trace behind shifting signals (Meinhardt, 1976). All these elements have been found, for instance, in the formation of the branching network of tracheae in insects (see below).

In the present article, after a brief description of the basic mechanism that allows pattern formation, it will be shown that a surprising variability of patterns can be generated by minor modification of the basic process or by a coupling of a few such pattern forming reactions.

2. Primary pattern formation: Help yourself and suppress your neighbor. Pattern formation from initially more or less homogeneous situations is not restricted to living systems. The formation of sand dunes, patterns of erosion, lightning, stars and galaxies are examples for pattern formation in non-animated systems. We have proposed that pattern formation is based at local self-enhancement and long-range inhibition (Gierer and Meinhardt, 1972): small local elevations above the average grow further on the expense of the surroundings. It is easy to see that this mechanism is also the basis for the cases of inorganic pattern formation mentioned above.

A possible molecular realization of this concept could consist of a substance that has an autocatalytic feedback on its own synthesis; we have called this substance the activator $a(x)$. The production rate of the activator is slowed down by a long ranging molecule, the inhibitor $h(x)$, which is produced under the control of the activator. The following set of partial differential equations describe a possible interaction:

$$\frac{\partial a}{\partial t} = \frac{\rho a^2}{h} - \mu_a a + D_a \frac{\partial^2 a}{\partial x^2} + \sigma_a \tag{1}$$

$$\frac{\partial h}{\partial t} = \rho a^2 - \mu_h h + D_h \frac{\partial^2 h}{\partial x^2} + \sigma_h \tag{2}$$

3. Basic types of patterns. A necessary condition for the formation of a stable pattern is that the inhibitor diffuses much faster than the activator and has a shorter half life, i.e., $D_h \gg D_a$ and $\mu_h > \mu_a$ must be satisfied. Whenever the size of the field exceeds the range of the activator, a homogeneous distribution of both substances is unstable (Fig. 1a). A first maximum can appear only at the margin of the field. This is very important for biological application since the resulting graded distribution can be used as positional information (Wolpert, 1969). In other words, such a mechanism is appropriate to generate an embryonic axis. The local high concentration acts as an organizing region. The pattern can be initiated by small fluctuations or by maternally supplied asymmetries. A stable situation is reached when the activator increase is balanced by the surrounding cloud of inhibition. The resulting pattern is in a wide range independent of the mode of initiation.

In contrast, if the inhibitor has a longer half life than the activator, oscillations will occur (Fig. 1b). A non-diffusible inhibitor can lead under this condition to traveling waves (Fig. 1c). Such a behavior is well-known from waves in an epidemic. The epidemic can spread since the autocatalytic agent, the virus, can be transmitted from one individual to the next, while the antagonistic reaction – the action of the immune system – remains confined to the individuum. Oscillations and traveling waves play

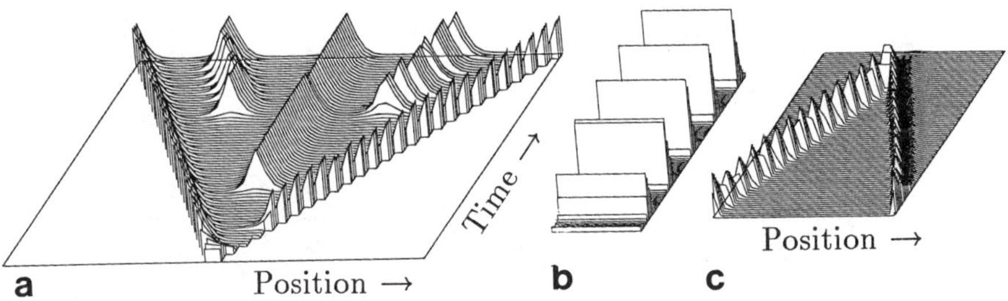

FIG. 1. *Stable patterns, oscillations and traveling waves: elementary patterns generated by self-enhancement coupled with an antagonistic reaction. (a) Stable patterns result if the inhibitor has a long range and a shorter half life than the activator. In growing field, first a monotonic gradient is formed. Insertion of further maxima in the enlarging interstices leads to periodic patterns. (b) Oscillations occur if the half life of the inhibitor is longer than that of the activator. (c) Traveling waves are possible if under these conditions the activator but not the inhibitor diffuses. Such waves annihilate each other upon collision. For initiation, they need a either a local initiation or pacemaker region.*

an important role in many biological systems, for instance in the signaling of neurons or in formation of pigment patterns on sea shells (see Fig. 4).

The two activator-independent production terms σ_a and σ_h may have important functions. σ_a is required to initiate autocatalysis when necessary, e.g. during regeneration (see Fig. 3) or during oscillations (Fig. 1b). A non-zero σ_b can suppress the spontaneous onset of activation, the system remaining asleep until an inducing trigger occurs that lifts the cells above a threshold. In the case of traveling waves, this trigger is obtained from the neighboring cells (Fig. 1c).

Stable patterns periodic in space are formed if the size of the field is larger than the range of the inhibitor. In a sheet of cells, maxima with more or less regular spacing can be formed. This is appropriate for the initiation of periodic structures such as bristles. Stripes, a pattern frequently encountered in development, can be formed if the rate of activator autocatalysis saturates at high activator concentrations (Fig. 2). This leads also to a limitation of the inhibitor production. More cells become activated at a lower level until sufficient inhibitor is produced. In other word, the activated regions have the tendency to enlarge. However, in order to become activated, a close neighborhood to the non-activated cells is essential into which the inhibitor can be dumped (or from which the necessary substrate can be obtained, see below). These requirements, of large activated patches and a direct neighborhood of non-activated cells would seem to contradict each other. However, this is not the case. In a stripe-like activation pattern, each activated cell has an activated neighbor and non-activated cells are close by. Transitions between patch- and stripe-like patterns can be frequently seen in the skin pattern of tropical fishes (Fig. 2). Kondo and Asai (1995) have observed the dynamic regulation of these stripes on growing fishes and have shown that they can be reproduced by models employing a saturating self-enhancement.

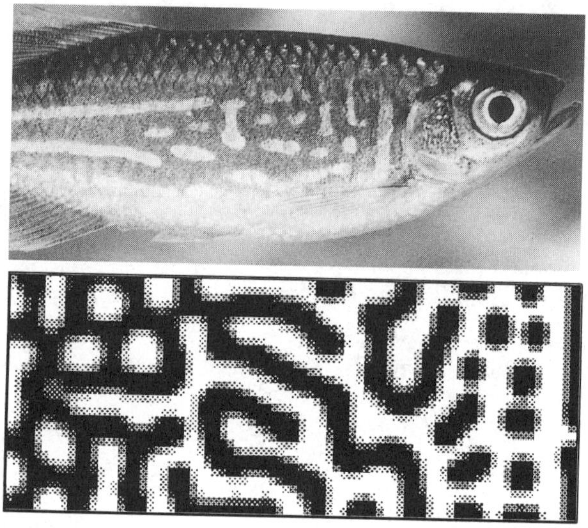

FIG. 2. *Patches and stripes. If the range of the antagonistic reaction is smaller than the field size, isolated patches with a high activator concentration emerge. If the autocatalysis has an upper bound (resulting, for instance, in Eq. (1) from a saturation term $a^2/(1+\kappa a^2)$ in the autocatalysis), stripes are the preferred pattern (Meinhardt, 1989). In the simulation, an increasing saturation κ towards the left leads to a transition from a patch- to a stripe-like pattern, as it is observed on the skin of many tropical fishes (Kondo and Asai, 1995; photograph of* D. malabaricus *courtesy of Rohan Pethiyagoda).*

4. Other molecular realizations. The activator-inhibitor reaction is, of course, only one example. In most inorganic reactions the antagonistic effect results from a depletion of ingredients that are required for the self-enhancing process. Eqs. (3) and (4) describes a possible interaction between an activator $a(x)$ and a substrate $s(x)$ that is consumed during the autocatalysis (Gierer and Meinhardt, 1972):

$$\frac{\partial a}{\partial t} = \rho s a^2 - \mu_a a + D_a \frac{\partial^2 a}{\partial x^2} + \sigma_a \quad (3)$$

$$\frac{\partial s}{\partial t} = \sigma_b - \rho s a^2 - \mu_b s + D_s \frac{\partial^2 s}{\partial x^2} \quad (4)$$

This reaction has similarities to the Brusselator reaction (Prigogine and Lefever, 1968; Lefever, 1968, Lefever et al., 1988) but is somewhat simpler. It has some properties which differ from the pure activator-inhibitor mechanism: For growing fields, new maxima are formed preferentially in an activator-substrate system by a splitting and movement of existing maxima. The size of the maxima are relatively broad. In contrast, in the activator-inhibitor system new maxima are triggered at a distance to existing maxima (see Fig. 1a). Such maxima can be a small fraction of the total field (see Fig. 3). This difference can be explained as follows: in the activator-substrate reaction the autocatalysis has an intrinsic upper bound since it comes to rest if all substrate is used up. Therefore, this reaction behaves like an activator-inhibitor system with saturation. Saturation causes an enlargement of the activated regions. The inhibition in the centers of such a maximum can become so strong that a de-activation and thus a splitting of the maximum occurs.

5. Comparison with the Turing mechanism. The possibility of generating a pattern by the interaction of two substances with different diffusion rates was discovered by Alan Turing (1952). In his paper, however, he did not mention anything like 'lateral inhibition'. It seems that he only later recognized this crucial point. In unpublished notes found after his death the following sentence occurred "The amplitude of the waves is largely controlled by the concentration V of 'poison'" (see Hodges, 1983, p. 494). Turing exemplified the mechanism he envisaged by the following set of equations (Turing, 1952, p.42).

$$\frac{dx}{dt} = 5x - 6y + 1 \tag{5}$$

$$\frac{dy}{dt} = 6x - 7y + 1 \quad (+\text{ diffusion}) \tag{6}$$

Both Eqs. (5) and (6), look very similar. It is not immediately obvious why such a reaction leads to pattern formation. It is easy to see, however, that this interaction satisfies our conditions since x has a feedback on its own production rate while the long ranging y molecule, produced under x control, acts antagonistically by destroying the x molecules. Therefore, Turing's mechanism can generate basically the same types of pattern as the lateral inhibition mechanism, i.e., graded concentration profiles and isolated maxima (Bard and Lauder, 1974; Lacalli and Harrison, 1978). Knowing that self-enhancement must be balanced by a long-ranging inhibitory reaction, however, facilitates substantially the design of appropriate reactions.

The particular mechanism proposed by Turing has an essential drawback: its molecular basis is unreasonable. According to Eq. (5), the number of x molecules disappearing per time unit is assumed to be proportional to the number of y molecules but independent of the number of x molecules

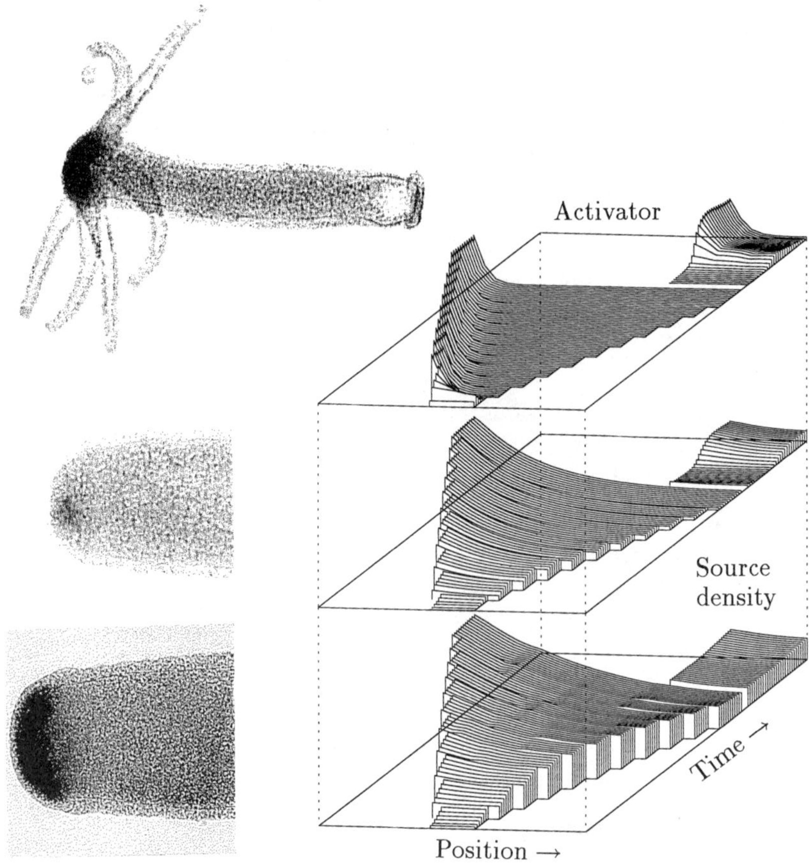

FIG. 3. *The maintenance of a polar pattern during growth: the solution of the wave length problem. (a) In the freshwater polyp* Hydra, *the activated region is presumably only a small fraction on the animal. The gene* HyBra1 *(black) has an expression region (Technau and Bode, 1999) that corresponds to the theoretical expectation for the head activation (Meinhardt, 1993). After head removal,* HyBra1 *expression reappears after 3-4h. (b) Model: Pattern formation by an activator - inhibitor system. At small field size, only a marginal maximum can be formed. A feedback of the inhibitor on the source density ρ (Eqs. 1 and 8) leads on a long time scale to a graded ρ distribution. Regions distant to the activated region (head) are unable to compete with the single existing activation. Secondary maxima are suppressed although the range of the activator is only a small fraction of the total field (compare with Fig. 1a). Nevertheless, after removal of the head and thus of the inhibitor-producing region, pattern regeneration is possible and occurs due to the ρ gradient according to the original polarity (Fig. (a) kindly supplied by U. Technau).*

present. In other words, x molecules can disappear even if no x molecules are left. This can lead to negative concentrations. Turing has seen this problem and proposed to ignore negative concentrations. One could repair this deficit by assuming a degradation also proportional to x. This requires, however, a non-linear autocatalytic activator production. A possible interaction would be similar to Eq. (1) except that the inhibitor does not lower the production but increases the destruction of the activator (the equation for h would be the same as Eq. (2)).

$$\frac{\partial a}{\partial t} = \rho a^2 - \mu_a a h + D_a \frac{\partial^2 a}{\partial x^2} + \sigma_a \tag{7}$$

This interaction has the peculiarity that the time constant of the activator changes during the formation of maxima. This may lead to a transition from a stable to an oscillating mode in the peak regions.

In summary, a system consisting of a self-enhancing and an antagonistic reaction can generate essential elementary patterns that are frequently required during development: stable gradients, periodic patterns in space, stripes, oscillating patterns and traveling waves. In the subsequent sections it will be shown that minor modifications or additions can lead to a substantial enrichment in the spectrum of patterns that can be generated.

6. How to avoid multiple maxima in growing fields. In a simple pattern-forming reaction a graded concentration profile can be maintained only over a range of about a factor two. With an increasing field size, a tendency exists to change from a monotonic into a symmetric and ultimately into a periodic distribution either by insertion of new (Fig. 1a) or by splitting of existing maxima. This is inappropriate if the graded concentration should be used in the growing embryo as positional information for the determination of the primary body axes since multiple maxima would lead, for instance, to several heads instead of one. Observations clearly demonstrate that nature was able to solve this problem. In the freshwater polyp hydra, small fragments of the body column are able to regenerate the complete animal, showing that the activated region is only a small portion of the total field. Recent gene expression studies support this view since they allow a direct visualization of the activated region (Fig. 3; Technau and Bode, 1999; Martinez et al., 1997).

The following somewhat anthropomorphic analogy should provide some intuition for the mechanism we have proposed (Meinhardt and Gierer, 1974, Meinhardt, 1993). A president (or any other local hero) usually has a strong tendency to suppress other's from becoming the president - a long range inhibition. On the other hand, he promotes individuals in his surrounding to obtain different levels in a hierarchy, becoming ministers etc.. This has two essential effects. Firstly, if the center of power were to become vacant, it is usually clear from this non-uniformity who will win the subsequent competition. The restoration of the pattern takes place in a predictable way. Due to such advantages in the competition, it does not take long to make the final decision. Secondly, the man in power does not have to inhibit everybody in the whole country since only a limited number of other individuals are able to replace him. This competence declines with increasing distance from the center of power.

In terms of our model, under the influence of the pattern, the ability of the cells to perform the pattern forming reaction has to change. This requires a feedback of either the activator or inhibitor on the source density ρ in Eqs. (1) and (2). Eq. (8) provides an example.

$$\frac{\partial \rho}{\partial t} = c\,\frac{h}{\rho} - \mu_\rho \tag{8}$$

In this case, the inhibitor with its shallow gradient leads to a similarly distributed source density ρ. Since in the system described by Eqs. (1) and (2) h increases with increasing ρ, this feedback must be slowed down at higher ρ levels to avoid an instability. This is the reason for the factor $1/\rho$. In a region of low source density, the initiation of secondary maxima becomes unlikely. The graded source density provides the long-lasting information about the polarity of the system. A small fragment regenerates a pattern according to the original polarity since the graded source density provides a systematic head start for some cells to outcompete the others. Since the source density has a much longer time constant ($\mu_a \gg \mu_\rho$), it remains essentially unchanged during pattern regeneration (Fig. 3). Although secondary maxima are successfully suppressed, the capability to regenerate is not impaired. Regeneration can be fast since no symmetry breaking and no competition over the whole field are required. In agreement, the gene *Hybra1*, presumably involved in head determination, becomes re-activated already 3h after head removal (Fig. 3; Technau and Bode, 1999). In contrast, it takes about two days to reverse the tissue polarity, for instance by transplantation of a head to a basal position (Webster, 1971). This strategy to maintain a single organizing region is presumably more generally employed. The decreasing capability of more anterior parts of the chicken blastodisk to form a primitive streak (Bachvarova et al., 1998, Spratt and Haas, 1960) may have the same reason.

7. Additional negative feedback: moving 'hot spots' and penetrating waves. As shown in the last section, a positive feedback reinforces an existing maximum. In this section, it will be shown that the opposite reaction, a negative feedback of the pattern on the ability of the cell to perform the pattern forming reaction, leads to a destabilization. Activated regions start to move over the field or they disappear while new ones arise at different positions. Three very different patterns will be discussed under this assumption: a pigment pattern on a sea shell, the regular initiation of leaves on a growing shoot (phyllotaxis) and the formation of branching filaments. This diversity suggests that such a mechanism is a general tool in development.

On sea shells, new pattern elements are added only along the growing edge. The patterns are therefore space-time plots of a one-dimensional patterning process. Oblique lines are the records of traveling waves. Crossings of such lines (Fig. 4) demonstrate that upon collision these waves can penetrate each other - a very unusual behavior of waves in excitable media. As a rule, in such a situation an annihilation takes place (see Fig. 1c). Traveling waves that penetrate each other can result by an additional negative feedback of a second antagonist. It must have opposite properties from the first: a short range and a long time constant. Due to its action, the 'hot spots', once formed, become poisoned in the course of time. This can cause their shift into a neighboring, non-poisoned region. For the simulation Fig. 4 an activator-substrate mechanism is assumed with parameters such that a cell, once activated, would remain in a steady state. A small diffusion of the activator leads to a spread of the activation. An additional diffusible inhibitor extinguishes the activation of the preceding cell. This leads to normal appearing traveling waves. However, after a collision, no new trigger in neighboring cells can occur that could cause an extinguishing of the actual activation. Therefore, cells remain activated until the refractory period of the neighboring cells is over and these cells can be re-infected again. In other words, the collision of two waves leads to the initiation of two new diverging waves. The model also accounts for a global perturbation seen on the natural shell pattern: many lines terminate at a particular growth line while others bifurcate at same instance. In the model, such a behavior occurs after a sudden general reduction of the activation since this also reduces the long range inhibition (for details and software, see Meinhardt, 1998a)

7.1. Traveling waves without a pacemaker. The generation of traveling waves usually requires a pacemaker region, a group of cells that oscillates somewhat faster than the rest. The faster oscillating cells trigger their neighbors before they would trigger spontaneously. If the position of the pace-maker is given, the resulting patterns are very reproducible since the waves have a defined origin and direction. In a system employing two antagonists as described above, an active de-synchronization takes place. Therefore, traveling waves can emerge without a pacemaker region. Turing (1952) already discovered that the spontaneous generation of traveling waves in a homogeneous field requires the interaction of three substances, a discovery that has almost been forgotten. Since the pattern is not predetermined by a pacemaker region, it can show a much higher degree of variability. This accounts for the diversity of pattern on the corresponding shells even within the same species.

8. Arrangement of leaves and staggered dots on shells - two corresponding patterns. Some shells are decorated with pigmented dots that are arranged along oblique lines (Fig. 5). This is another pattern that can be simulated by assuming two antagonists: one causing the separation of the activations along the space-, the other along the time-coordinate. In this case, the poisoning of the maxima leads to their disappearance and to their new trigger at a displaced position, in contrast to the continuous shift of the maxima mentioned above.

FIG. 4. *Collision of traveling waves without annihilation. Oblique lines on shells are records of traveling waves. Their crossings on* Tapes literatus *indicate that these waves can penetrate each other. This is very unusual for waves in excitable media. Such a pattern results from an autocatalytic component that is balanced by two antagonistic reactions. a remarkable perturbation can be recognized on the specimen shown: at a particular time many waves either terminate or bifurcate. This seemingly contradictory behavior is reproduced by a temporal lowering of the activator concentration (Meinhardt and Klingler, 1987, Meinhardt, 1998a).*

The regular initiation of leaves behind the tip of a growing shoot, called *phyllotaxis*, seems to have nothing in common with any pattern on sea shells, but this impression is misleading. On a growing shoot as in shells, the new pattern elements also appear in a narrow zone in the course of time. The tip of the shoot, the so-called meristem, consists of undifferentiated, rapidly dividing cells. Only cells just leaving this zone are able to form new leaves. According to classical models, the initiation of a new leaf is inhibited by existing leaves (Schoute, 1913). Therefore, a new leaf can be initiated only at a certain distance from the last formed leaf. In this way, a certain distance is maintained between the sites of leaf initiation.

In many plants, leaves are initiated along spirals. Seeds on fir cones have a corresponding arrangement. Such patterns result if not only the last, but also the next-to-last leaf has a repelling influence on initiation of a new leaf. This can be simulated with the basic model. Since the inhibitor has to diffuse rapidly, however, the directing cue resulting from the penultimate leaf on the positioning of a new leaf is minute. Therefore, such a mechanism is not robust against small perturbations, in contrast to the observations. In the simulations, there is a tendency to fall back to an alternating (distichous) or pair-wise, 90^0-rotated (decussate) arrangement.

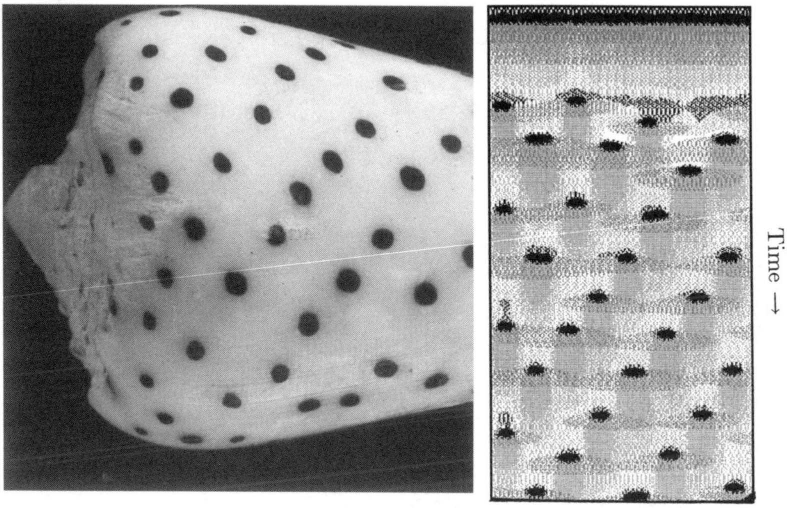

FIG. 5. *Dots on a shell arranged along oblique lines. The simulation based on an interaction of one activator and two inhibitors that act in an additive way. One has a long range (dark gray), the other a short range but a long time constant (light gray).*

Helical patterns can emerge very reliably if, as in the shifted dot model discussed above, two separate inhibitions are assumed (Fig. 6; Meinhardt et al., 1998). According to this model, the helical arrangement does not result from the inhibitory influence of the earlier leaves, but results from the long-lasting memory of cells in the leaf forming zone that a leaf has been formed at this position. Since this memory is based on a nearly non-diffusing substance, it remains localized. In Fig. 6d, the initiation of a new signal at a displacement close to the golden angle is demonstrated.

9. Formation of filament-like branching structures. Branching filaments which form net-like structures are very common pattern elements in almost all higher organisms. The venation of leaves, the tracheae of insects, the blood or lymph vessels in vertebrates and ramifying axons of neurons are examples. How can such complex networks arise?

As shown above, an additional delayed negative feedback can lead to a destabilization and to a shift of the inducing signal. Therefore, it was proposed that filaments are formed as a trace behind moving activator maxima: a local signal generated by an activator/inhibitor system provokes its own shift in space (Fig. 7; Meinhardt, 1976). The elongation of the filament can be based either on a differentiation of the cells newly exposed to the signal or by a local elongation of single cells. The elongation of the filaments can be oriented by a substrate that is produced by all cells and that is removed by the filaments. Thus, the substrate concentration is a

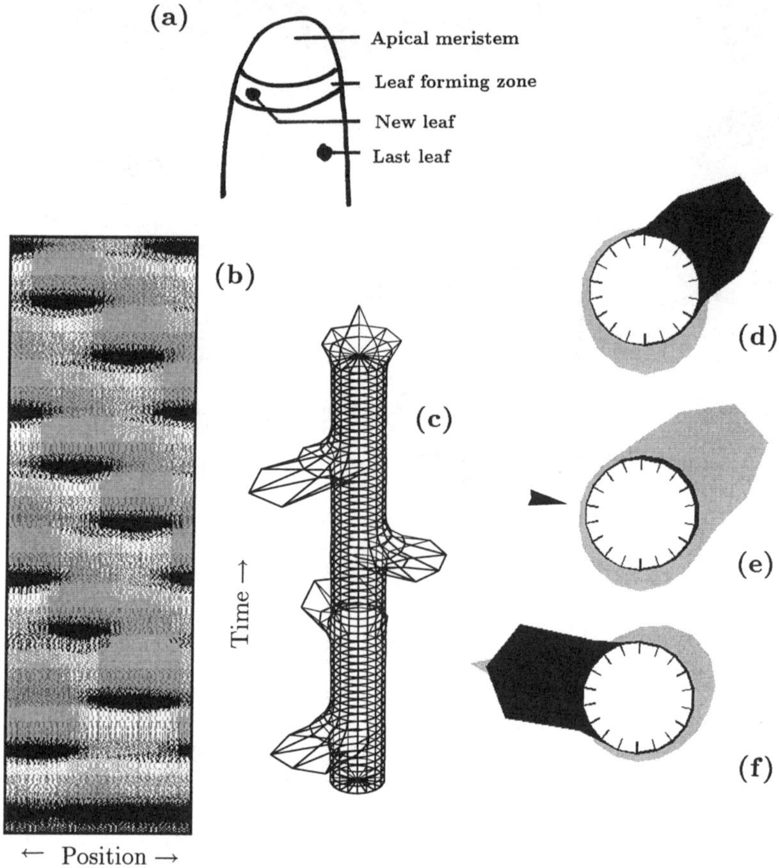

FIG. 6. *Helical initiation of leaves on a growing shoot. (a) Leaf initiation takes place only in a narrow zone below the apical meristem. (b) Simulation of an activator (black) - two inhibitor model on a ring, plotted as a function of time. The rapidly diffusing inhibitor (dark gray) leads to the spatial separation of the signals on the ring. The second, long lasting inhibitor (light gray) leads to a pulse-like activation. Its reappearance occurs at shifted positions. A helical arrangement occurs spontaneously. (c) as (b) but plotted as a cylinder. (d) Details to show the displacement of the signal; only the activator and the slowly spreading inhibitor is shown. The next signal appears at a position where this inhibition drops below a certain level (arrowhead). The displacement is close to the golden angle (Meinhardt et al., 1998).*

measure of how urgently a further ingrowth of filaments is required. If the signal depends on this substrate, the shift of the signal and thus elongation will occur toward the correct regions. Branches are formed if new activations are triggered along filaments by a baseline activator production. This happens whenever the density of the net is below a threshold and if other

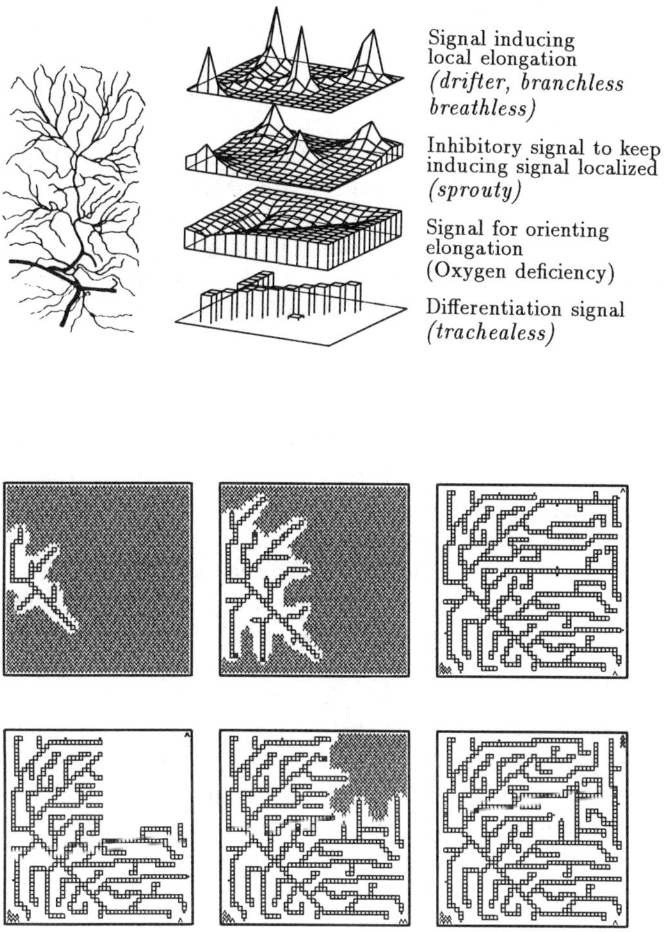

FIG. 7. *Formation of a net-like structure. (a) Tracheae of an insect (drawn after Wigglesworth, 1954). (b, c) Elements of the model and simulation in a larger field: an activator / inhibitor system generates a signal for the local elongation of a filament. This signal is used elongate the filaments (open squares). The filaments remove a substrate (or an oxygen deficiency, gray). Since the activator/inhibitor system depends on this substrate, the activator maxima are shifted to that neighboring cells that have the highest substrate concentration, i.e. the largest distance to other differentiated cells (Meinhardt, 1976, 1982, 1998a,b). After partial removal of filaments, a high substrate concentration recovers. From the unaffected regions new filaments extend into the damaged region, causing a repair. The new pattern is similar but not identical. An analogous pattern regulation has been observed after cutting a trachea in an insect (Wigglesworth, 1954).*

sites of elongation are sufficiently remote. As long as no other constraints are imposed, lateral branching occurs by 90^0. This is the fastest way to escape the trough of the substrate distribution generated by the filament itself.

The simulation in Fig. 7 shows that this model, although requiring only four components, is able to generate rather complex net-like structures. This model been supported by evidence of the interactions involved in the formation of tracheae in insects. The following is a short list of components expected from the model and genes that presumably have the corresponding functions: (i) A cell has to remember that it belongs to the network. For this, an autoregulatory gene is expected that allows an irreversible switch. A corresponding gene is *trachealess* (Wilk et al., 1996). (ii) A second autocatalytic loop is expected for the generation of the local signal accomplishing filament elongation. The transcription factor *drifter* (Anderson et al., 1996) has this property. Since the activator has to spread, but a transcription factor is restricted to the nucleus, a factor is expected that transmit the activation to neighboring cells. The corresponding molecule could be *branchless* (Sutherland et al., 1996), a member of the FGF family. The corresponding receptor is *breathless* (Glazer and Shilo, 1991; Lee et al., 1996b, Shilo et al. 1997). *Sprouty* (Hacohen et al., 1998) has the properties of the expected inhibitor that keeps the activation localized. It has a long range and blocks the FGF-signaling in a competitive fashion. The substrate removed by the filament and thus the driving force for the movement, could be the oxygen deficiency itself. Wigglesworth (1959) has observed that tracheoles (the fine endings of tracheae before their conversion into a tube) are attracted by small patches of epithelial cells. The latter are located in front of the tracheoles and make contact by extending protrusions. The tip of the tracheoles are moved by retraction of these protrusions. Afterwards, new and more distant epithelial cells repeat this process, thus causing the oriented extension, in full agreement with the model proposed: the signal for elongation shifts, causing a further elongation, causing a further shift, and so on.

It is a property of such a system that elongation occurs away from existing filaments, a necessary condition to cover a region evenly with filaments. Therefore, such a system does not have the tendency to generate closed loops. If connections between filaments are to be made, different cell types are expected: one type attracts the elongating branches of the other type (an attraction of the same type would abolish branching). Veins and arteries may build loops in this way. In the tracheal network, different cell types are involved on the one hand to accomplish the oxygen supply within a segment and, on the other, to make connections with filaments in neighboring segments (see Shilo et al., 1997).

10. Polygonal pattern: the skin of giraffes and the venation on the wing of a dragon fly. The variability and complexity of animal coat patterns have attracted several theoretical approaches. Models are available for zebra stripes (Bard, 1981; Murray, 1981a, 1989), for the coloration of butterfly wings (Nijhout, 1978, 1980; Murray, 1981b; Bard and French, 1984), for patterns on snake skins (Cocho et al., 1987; Murray and

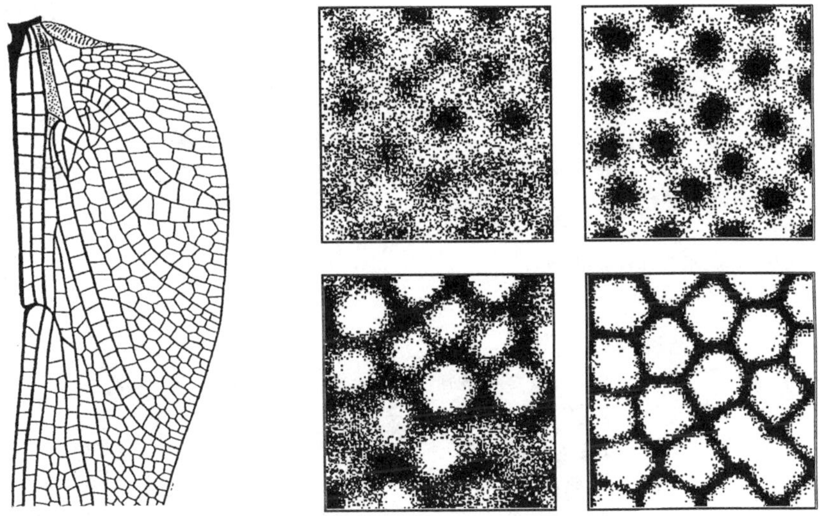

FIG. 8. *A polygonal pattern: venation on the wing of a dragonfly (after Séguy, 1973). Simulation based on a spot-forming system (top) that repels a stripe-forming system (bottom). Closed loops of polygonal shape are formed around the founding spots (Koch and Meinhardt, 1994).*

Myerscough, 1991) and for the pattern on tropical fishes (Kondo and Asai, 1995). A frequent prototype is a polygonal pattern with closed loops. We have shown that this pattern can be generated by the coupling of a system that generates local maxima with a system that generates stripes. The 'hot spots' determine where *no* stripe should be formed. The stripes, instead of being randomly oriented as in Fig. 2, emerge at the largest distance from the spots. In a mathematical term, these patterns are therefore Dirichlet domains (Koch and Meinhardt, 1994).

An example for a filamentous patterns with closed loops is the venation pattern on the wings of dragonflies (Fig. 8). In the corresponding simulation an activator-substrate system (Eqs. 3, 4) is assumed that causes local maxima. Since the substrate is depleted in the activator autocatalysis, the highest substrate concentration remains at positions with the largest distance to the maxima. The substrate has a promoting influence on the stripe-forming system. Although the substrate has a smooth profile due to its rapid diffusion, the stripes obtain a sharp delineation and a uniform appearance due to the self-shaping property of the stripe system.

The complex venation pattern of a dragon fly is presumably not produced in a single step at a particular stage. It is rather likely that a simple pattern is laid down at an early stage and in a small field. By analogy to the *Drosophila* wing venation (see Biehs et al., 1998), we assume that the positions of the main veins are genetically determined. In this process, borders between cells of different determinations are used as the new reference points to supply positional information (Meinhardt, 1983a). The finer veins are presumably added later in order to strengthen the growing wing blade and to maintain an approximately constant the mesh size. The system used for the simulation in Fig. 8 has this property: whenever the interstices become too large, the maximum at the center of a polygon splits. Thus, a new stripe will be inserted in the newly-emerging valley of the stripe-avoidance signal.

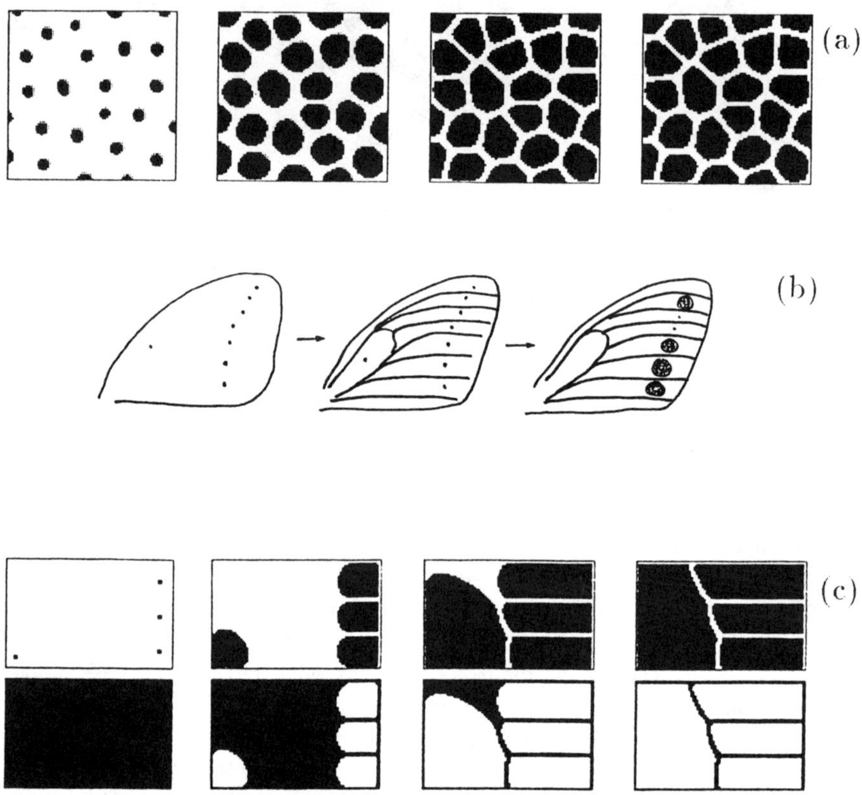

FIG. 9. *Polygon formation by non-touching waves. (a) A spot-forming system is assumed that initiates traveling waves. These come to rest if a counter wave is encountered. The resulting pattern closely resembles the skin pattern on a giraffe (Koch and Meinhardt, 1994). (b) A possible mode to specify vein formation in a complex butterfly wing. It is presumably much simpler to generate the positional information for the intervein pattern than for each element of the vein pattern itself. These founding hot spots may be used a second time to initiate the eye spots on butterfly wings at intervein positions. (c) A simulation of vein formation by non-touching waves that suppress vein formation. They spread from a simple pattern of 'hot spots'.*

10.1. Polygon formation by waves that don't touch. A similar pattern results if the hot spots are the initiation points of waves that spread until a counter wave approaches. The simulation shown in Fig. 9 demonstrates that the resulting pattern closely resembles the skin pattern on a giraffe. The narrow non-pigmented stripes result by the arrest of two approaching waves. Again, they are formed at positions that have the largest distance from the founding spots. Depending on the sensitivity to the counter waves, pattern on leopards or cheetahs can be simulated as well (Koch and Meinhardt, 1994).

Another example for a polygonal pattern can be found in the somatosensory cortex. The facial vibrissae of the mouse project on the primary cortex in domains called barrels (Steindler et al., 1989; Jacobson, 1991). The shape of the barrels can be visualized by a labeling with tenascin-specific antibodies. During the first postnatal days, the barrel pattern has dynamic properties: removal of vibrissae disrupts the formation of the associated barrels. In terms of the model, the activity of vibrissa initiates barrel formation. The enlargement of a domain comes to rest if another spreading domain is encountered.

According to this model, the centers of the polygons finally formed play an active role, either by suppressing a stripe forming system or by the initiation of non-touching waves. Evidence for such an active role of non-vein forming regions comes from early blood vessel formation in vertebrates. One way to produce the finer tube-like structures is to start with a coherent sheet of endothelial cells, the primary plexus. Some spot-like holes appear that grow (like bubbles in a Swiss cheese) until the remaining veins have the correct size (Riesau and Flamme, 1995; see also Meinhardt, 1998b).

Many wings of butterflies carry eye spots. They appear preferentially between existing veins, suggesting that veins have a repulsive effect on eye spot formation (see Nijhout, 1991; Murray, 1981b). The mechanism discussed above suggests the possibility that such hot spots originally played an active role by directing vein formation from the largest possible distance to these spots. Using the compartment boundaries of an imaginal disk as reference points (Meinhardt, 1983a), it is presumably simpler to generate the positional information for such repulsive hot spots than to specify each point of the veins directly (Fig. 9b,c). In butterflies, these hot spots at intervein positions may be used a second time to initiate eye spot formation. In this view, the hot spot between the veins could have had at least initially an active role. At present, these functions are certainly uncoupled since the eye spots pattern can be easily changed by mutations without affecting other pattern elements (Brakefield et al., 1996).

11. Conclusions. Minor extensions or combination of reactions that are based on self-enhancement and one or two antagonistic reactions can generate a remarkable variability of pattern. Thus, part of the spatial complexity of a developing organism may be generated by an appropriate combination of elements taken from a basic toolbox.

An essential property in these models is their inherent capability for self-regulation. This accounts for the fact that development is a very robust process up to the point that normal development can remain possible even after removal of some essential parts (Figs. 3, 7). By this self-corrections the propagation of errors into subsequent levels can be avoided. Thus, an understanding of the dynamics of the interactions on which pattern formation is based provides a key for the reliability that is characteristic of many developmental processes.

REFERENCES

Anderson, M.G., Certel, S.J., Certel, K., Lee, T., Montell, D.J. and Johnson, W.A. (1996). Function of the *Drosophila* POU domain transcription factor *Drifter* as an upstream regulator of *Breathless* receptor kinase expression in developing trachea. Development **122**, 4169–4178.

Bachvarova, R.F., Skromme, I. and Stern, C.D. (1998). Induction of primitive streak and Hensens's node by the posterior marginal zone in the early chick embryo. Development **125**, 3521–3534.

Bard, J.B. and Lauder, I. (1974). How well does Turings theory of morphogenesis work. J. theor. Biol. **45**, 501–531.

Bard, J.B.L. (1981). A model for generating aspects of Zebra and other mammalian coat patterns. J. theor. Biol. **93**, 363–385.

Bard, J.B.L. and French, V. (1984). Butterfly wing patterns: how good a determining mechanism is the simple diffusion of a single morphogen J. Embryol. exp. Morph. **84**, 255–274.

Biehs, B., Sturtevant, M.A. and Bier, E. (1998). Boundaries in the *Drosophila* wing imaginal disc organize vein-specific genetic programs. Development **125**, 4245–4257.

Brakefield, P.M., Gates, J., Keys, D., Kesbeke, F., Wijngaarden, P.J., Monteiro, A., French, V. and Carroll, S.B. (1996). Development, plasticity and evolution of butterfly eyespot patterns. Nature **384**, 236–242.

Culi, J. and Modolell, J. (1998). Proneural gene self-stimulation in neural precursors - an essential mechanism for sense organ development that is regulated by *Notch* signaling. Genes Dev. **12**, 2036–2047.

Gierer, A. and Meinhardt, H. (1972). A theory of biological pattern formation. Kybernetik **12**, 30–39.

Glazer, L. and Shilo, B.Z. (1991). The *Drosophila FGF-R* homolog is expressed in the embryonic tracheal system and appears to be required for directed tracheal extensions. Genes Dev. **5**, 697–705.

Gurdon, J.B., Mitchell, A. and Mahony, D. (1995). Direct and continuous assessment by cells of their position in a morphogen gradient. Nature **376**, 520–521.

Hacohen, N., Kramer, S., Sutherland, D., Hiromi, Y. and Krasnow, M.A. (1998). *Sprouty* encodes a novel antagonist of *FGF* signaling that patterns apical branching of the *Drosophila* airways. Cell **92**, 253–263.

Hodges, A. (1983). Alan Turing: the enigma. Simon and Schuster, New York.

Ingham, P.W. (1991). Segment polarity genes and cell patterning within the *Drosophila* body segment. Curr. Op. Gen and Dev. **1**, 261–267.

Jacobson, M. (1991). Developmental Neurobiology. Plenum Press, New York.

Koch, A.J. and Meinhardt, H. (1994). Biological pattern-formation - from basic mechanisms to complex structures. Reviews Modern Physics **66**, 1481–1507.

Kondo, S. and Asai, R. (1995). A viable reaction-diffusion wave on the skin of *Pomacanthus*, the marine Angelfish. Nature **376**, 765–768.

LACALLI, T.C. AND HARRISON, L.G. (1978). The regulatory capacity of Turing's model for morphogenesis, with application to slime moulds. J. theor. Biol. **70**, 273–295.

LEE, E.C., HU, X.X., YU, S.Y. AND BAKER, N.E. (1996a). The scabrous gene encodes a secreted glycoprotein dimer and regulates proneural development in *Drosophila* eyes. Molec. Cell. Biol. **16**, 1179–1188.

LEE, T., HACOHEN, N., KRASNOW, M. AND MONTELL, D.J. (1996b). Regulated *breathless* receptor tyrosine kinase-activity required to pattern cell-migration area branching in the *Drosophila* tracheal system. Genes Dev. **10**, 2912–2921.

LEPTIN, M. (1991). *twist* and *snail* as positive and negative regulators during *Drosophila* mesoderm development. Genes Dev. **5**, 1568–1576.

MARTIN, G.R. (1995). Why thumbs are up. Nature **374**, 410–411.

MARTINEZ, D.E., DIRKSEN, M.L., BODE, P.M., JAMRICH, M., STEELE, R.E. AND BODE, H.R. (1997). *Budhead*, a *forkhead hnf-3* homolog, is expressed during axis formation and head specification in hydra. Dev. Biol. **192**, 523–536.

MEINHARDT, H. (1976). Morphogenesis of lines and nets. Differentiation **6**, 117–123.

MEINHARDT, H. (1978). Space-dependent cell determination under the control of a morphogen gradient. J. theor. Biol. **74**, 307–321.

MEINHARDT, H. (1982). Models of biological pattern formation. Academic Press, London.

MEINHARDT, H. (1983a). Cell determination boundaries as organizing regions for secondary embryonic fields. Dev. Biol **96**, 375–385.

MEINHARDT, H. (1983b). A boundary model for pattern formation in vertebrate limbs. J. Embryol. exp. Morphol. **76**, 115–137.

MEINHARDT, H. (1986). Models of segmentation. In: Somites in developing embryos (R. Bellairs, D.A. Edie, J.W. Lash, Edts), Nato ASI Series A, Vol. **118**, 179–189, Plenum Press, New York.

MEINHARDT, H. (1989). Models for positional signalling with application to the dorsoventral patterning of insects and segregation into different cell types. *Development (Supplement)* 1989, 169–180.

MEINHARDT, H. (1993). A model for pattern-formation of hypostome, tentacles, and foot in hydra: how to form structures close to each other, how to form them at a distance. Dev. Biol. **157**, 321–333.

MEINHARDT, H. (1998a). The Algorithmic Beauty of Sea Shells. 2nd enlarged edition (with PC-software). **Springer**, Heidelberg, New York.

MEINHARDT, H. (1998b). Models for the formation of netlike structures. In: Vascular Morphogenesis: In Vivo, In Vitro and In Sapiente, (C.D. Little, V. Mironov and E.H. Sage, Edts), pp. 147–172. Birkhäuser, Boston, Basel, Berlin.

MEINHARDT, H. AND GIERER, A. (1974). Applications of a theory of biological pattern formation based on lateral inhibition. J. Cell Sci. **15**, 321–346.

MEINHARDT, H. AND GIERER, A. (1980). Generation and regeneration of sequences of structures during morphogenesis. J. theor. Biol. **85**, 429–450.

MEINHARDT, H. AND KLINGLER, M. (1987). A model for pattern formation on the shells of molluscs. J. theor. Biol **126**, 63–69.

MEINHARDT, H., KOCH, A.J. AND BERNASCONI, G. (1998). Models of pattern formation applied to plant development. In: Symmetry in Plants, (D. Barabe and R. V. Jean, Eds), World Scientific Publishing, Singapore , pp. 723–758.

MURRAY, J.D. (1981a). A prepattern formation mechanism for animal coat markings. J. theor. Biol. **88**, 161–199.

MURRAY, J.D. (1981b). On pattern forming mechanisms for lepidopteran wing patterns and mammalian coat markings. Phil. Trans. R. Soc. Lond. B **295**, 473–496.

NIJHOUT, H.F. (1978). Wing pattern formation in *Lepidoptera*: a model. J. exp. Zool. **206**, 119–136.

NIJHOUT, H.F. (1980). Pattern formation in lepidopteran wings: determination of an eyespot. Dev. Biol. **80**, 267–274.

NIJHOUT, H.F. (1991). Development and Evolution of Butterlfly Wing Patterns. Smithonian Inst. Press Washington.

PALMEIRIM, I., HENRIQUE, D., ISHHOROWICZ, D. AND POURQUIE, O. (1997). Avian *hairy* gene-expression identifies a molecular clock linked to vertebrate segmentation and somitogenesis. Cell **91**, 639–648.

PFEIFER, M., BEJSOVEC, A. (1992). Knowing your neighbors: Cell interactions determine intrasegmental patterning in *Drosophila*. Trends Genetics **8**, 243–248.

REGULSKI, M., DESSAIN, S., MCGINNIS, N. AND MCGINNIS, W. (1991). High-affinity binding-sites for the deformed protein are required for the function of an autoregulatory enhancer of the deformed gene. Genes Dev. **5**, 278–286.

RIESAU, W. AND FLAMME, I. (1995). Vasculogenesis. Annu. Rev. Cell Dev. Biol., **11**, 73–91.

SÉGUY, E. (1973). L'aile des insectes. In *Traité de Zoologie*, Grassé P., editor, Vol. VIII, Masson et Cie, Paris.

SCHOUTE, J.C. (1913). Beiträge zur Blattstellung. Rec. trav. bot. Neerl. **10**, 153–325.

SHILO, B.Z., GABAY, L., GLAZER, L., REICHMANFRIED, M., WAPPNER, P., WILK, R. AND ZELZER, E. (1997). Branching morphogenesis in the *Drosophila* tracheal system. Cold Spring Harbor Symposia on Quantitative Biol. **62**, 241–247.

SPRATT, N.T. AND HAAS, H. (1960). Integrative mechanisms in development of the early chick blastoderm. I. Regulative potentiality of separated parts. J. Exp. Zool. **145**, 97–137.

STEINDLER, D.A., FAISSNER, A. AND SCHACHNER, M. (1989). Brain "cordones": Transient boundaries of glia and adhesion molecules that define developing functional units. Comments Dev. Neurobiol. **1**, 29–60.

SUN, Y., JAN, L.Y. AND JAN, Y.N. (1998). Transcriptional regulation of *atonal* during development of *Drosophila* peripheral nervous system. *Development* **125**, 3731–3740.

SUTHERLAND, D., SAMAKOVLIS, C. AND KRASNOW, M.A. (1996). Branchless encodes a *Drosophila FGF* homolog that controls tracheal cell-migration and the pattern of branching. Cell **87**, 1091–1101.

TECHNAU, U. AND BODE, H.R. (1999). *HyBra1*, a *Brachyury* homolog, acts during head formation in *Hydra*. Development **126**, (in press).

TURING, A. (1952). The chemical basis of morphogenesis. Phil. Trans. B. **237**, 37–72.

VINCENT, J.P. AND LAWRENCE, P.A. (1994). It takes three to distalize. Nature **372**, 132–133.

WEBSTER, G. (1971). Morphogenesis and pattern formation in hydroids. Biol. Rev. **46**, 1–46.

WIGGLESWORTH, V.B. (1954). Growth and regeneration in the tracheal system on an insect *Rhodnius prolixus* (Hemipter) Quart. J. micr. Sci. **95**, 115–137.

WIGGLESWORTH, V.B. (1959) The role of the epidermal cells in the "migration" of tracheoles in *Rhodnius prolixus* (Hemipter). J. exp. Biol. **36**, 632–640.

WILK, R., WEIZMAN, I. AND SHILO, B.Z. (1996). *trachealess* encodes a bHLH-PAS protein that is an inducer of the tracheal cell fate in *Drosophila*. Genes Dev. **10**, 93–102.

WOLPERT, L. (1969). Positional information and the spatial pattern of cellular differentiation. J. theor. Biol. **25**, 1–47.

SPATIOTEMPORAL PATTERNING IN MODELS OF JUXTACRINE INTERCELLULAR SIGNALLING WITH FEEDBACK

NICHOLAS A.M. MONK*, JONATHAN A. SHERRATT†, AND MARKUS R. OWEN‡

Abstract. Juxtacrine signalling is the class of intercellular communication mediated by ligands and receptors that are both anchored in the cell membrane. Two particularly well documented examples of such signalling pathways are the Delta-Notch and TGFα–EGF-R interactions. In this review, we discuss mathematical models for juxtacrine signalling, focussing on these two specific examples. We discuss the various model formulations that have been used, and consider gradient, travelling front, and spatial pattern type solutions. We show that juxtacrine mechanisms can explain a wide range of observed behaviours in each of these categories, in a manner that is genuinely different from that in traditional diffusion-based models for intercellular signalling.

1. Introduction to juxtacrine signalling. Signalling between cells (intercellular signalling) is an essential process in the development and maintenance of multicellular systems. The signals employed can take a variety of forms and act over a wide range of length scales. In general, signalling depends on the production of ligand (the mediator of the signal) by signalling cells and detection of this ligand by specific receptors expressed by receiving cells. Traditionally, intercellular signalling has been classified as either autocrine or paracrine. In autocrine signalling, a cell signals specifically to itself, whereas paracrine signalling involves signalling between distinct cells that are spatially separated and depends on secreted diffusible ligands such as growth factors and hormones.

Juxtacrine signalling is a distinct class of intercellular signalling that is mediated by ligands and receptors that are both anchored to the cell membrane (Massagué, 1990; Bosenberg & Massagué, 1993). Juxtacrine ligands can be either membrane-anchored precursors of soluble forms of the ligand (for example transforming growth factor α, TGFα) or purely membrane bound (for example, Delta). In the former case, the juxtacrine and soluble forms of the ligand can trigger qualitatively distinct responses in receiving cells due either to activation of distinct receptors (as for tumour necrosis factor—Grell *et al.*, 1995) or due to differences in ligand presentation to the receptor (as for certain ephrin-A ligands—Davis *et al.*, 1994).

*Mathematical Modelling and Genetic Epidemiology Group, Division of Molecular and Genetic Medicine, University of Sheffield, Royal Hallamshire Hospital, Sheffield, S10 2JF, UK; n.monk@sheffield.ac.uk.

†Centre for Theoretical Modelling in Medicine, Department of Mathematics, Heriot-Watt University, Edinburgh EH14 4AS, UK; jas@ma.hw.ac.uk.

‡Nonlinear and Complex Systems Group, Department of Mathematical Sciences, Loughborough University, Loughborough, Leicestershire LE11 3TU, UK; m.r.owen@lboro.ac.uk.

An immediate consequence of the membrane anchoring of juxtacrine ligands is that direct signalling can occur only between cells that are in intimate contact within a tissue (Fagotto & Gumbiner, 1996). In considering spatiotemporal patterning it is thus important to determine whether or not juxtacrine activation of receptor results in changes in the levels of expression of the juxtacrine ligand and/or receptor in the activated cell itself. If this is not the case (no feedback) then signals cannot be propagated over a range of more than one cell diameter (as is the case for Boss–Sevenless signalling in the developing *Drosophila* eye—Krämer *et al.*, 1991). However, if there is feedback between receptor activation and expression levels of ligand and/or receptor, then juxtacrine signalling provides an important mechanism for the long-range propagation of localised signals and the *de novo* generation of spatiotemporal pattern. It is the latter case that we are concerned with in this review.

Focussing on two particularly well documented biological examples, Delta–Notch and TGFα–EGF-R signalling, we will show that the implications of juxtacrine signalling with feedback depend crucially on whether binding of ligand to receptor up- or downregulates the further expression of ligand and receptor on the cell surface. For the Delta–Notch system, it has been well established that activation of the Notch pathway by Delta can lead to the downregulation of Delta activity, thus establishing a negative feedback within signalling cells that can act to amplify any small differences in levels of Notch pathway activity between neighbouring cells (reviewed in Simpson, 1997). However recent evidence suggests that this feedback regulation is context dependent, with experiments on *Drosophila* wing development showing upregulation of Notch and Delta expression by binding (see below). In the case of TGFα binding EGF-R, positive feedback is well established. Auto-induction of TGFα synthesis has been demonstrated in human keratinocytes (Coffey *et al.*, 1987), and EGF has been shown to stimulate production of EGF-R (Clark *et al.*, 1985; Earp *et al.*, 1986; Earp *et al.*, 1988; Kudlow *et al.*, 1986). Similar positive feedback loops have been documented in a number of other cell types (e.g. Zigmond, 1982).

1.1. Delta–Notch signalling. The Notch/Lin-12 family of transmembrane receptor proteins[1] mediate a wide range of cell-fate decisions during the development of flies, vertebrates and nematodes (reviewed in Artavanis-Tsakonas *et al.*, 1995; Fortini & Artavanis-Tsakonas, 1993; Kimble & Simpson, 1997; Lewis, 1996; Muskavitch, 1994). The receptors are characterised by the presence of multiple epidermal growth factor (EGF)-like repeats in the extracellular domain, and by a number of other conserved domains (reviewed in Greenwald, 1994). A number of ligands have been described that bind to the Notch/Lin-12 EGF-repeats and activate the in-

[1]The *Notch* mutation was named after the phenotype of heterozygous flies, which have little notches taken out of the wing margin.

tracellular Notch signalling pathway. These include Delta and Serrate in *Drosophila*, and their homologues in nematodes and vertebrates.

To date, all the Notch-binding ligands that have been described are large transmembrane proteins with multiple EGF-like repeats and a conserved DSL motif in their extracellular domains (see Tax *et al.*, 1994; Simpson, 1995). The DSL motif is critical for binding to Notch. The transmembrane nature of Notch-binding ligands suggests that they act in a purely juxtacrine fashion. This is supported by a number of functional studies (for example, Heitzler *et al.*, 1996). However, recent studies have revealed that Delta can, in some instances, be proteolytically cleaved to yield a soluble secreted ligand (Klueg *et al.*, 1998; Qi *et al.*, 1999). There is as yet no direct evidence that secreted forms of Notch-binding ligands contribute to Notch signalling *in vivo*.

The binding of Delta-like ligands to Notch activates an intracellular signal transduction pathway that regulates the expression of tissue-specific target genes (for a recent review, see Weinmaster, 1998). If Delta-Notch signalling occurs between members of an initially equivalent group of cells, each of which can act both as a source and recipient of signalling, robust spatial patterning of cell fate can result. The process by which this happens, called lateral inhibition or lateral specification, appears to be the dominant mode of action of Delta–Notch signalling in development. Lateral inhibition has been studied most extensively in the contexts of the development of the central and peripheral nervous systems in *Drosophila* and of vulval development in *C. elegans* (Campos-Ortega, 1993; Goriley *et al.*, 1991; Heitzler & Simpson, 1991; Simpson, 1990; Wilkinson *et al.*, 1994). During the development of the central nervous system, Delta–Notch-mediated lateral inhibition acts to ensure that a reproducible proportion of a population of initially equipotential neural progenitor cells goes on to differentiate as neurons. Mutations in *Delta* and *Notch*, as well as in other genes that code for components of the signalling pathway, lead to a striking phenotype in which there is a massive overproduction of neurons at the expense of epidermis (Lehmann *et al.*, 1983); it is for this reason that genes involved in Delta–Notch-mediated lateral signalling are often referred to as neurogenic genes.

In at least some cases, Delta–Notch-mediated lateral specification depends on the inhibition of Delta activity in cells that are receiving Delta signalling from their neighbours. More specifically, it has been proposed that the level of expression of Delta in a cell is a decreasing function of the level of activity of the Notch signalling pathway in that cell (Heitzler & Simpson, 1991, 1993; Heitzler *et al.*, 1996). When this intracellular regulation operates in neighbouring cells, a feedback loop is formed between the cells; this feedback loop acts to amplify any small differences in the level of activity of the Notch pathway in neighbouring cells (Simpson, 1997; Sternberg, 1993). Since the level of activity of the Notch pathway within a cell is known to be a critical determinant of fate choice (elevated Notch activity inhibits differentiation and exit from the cell cycle in general—see, for example, Artavanis-Tsakonas *et al.*, 1995), the differences in Notch pathway activity in neighbouring cells can result in the cells adopting radically different fates. This mechanism can act within small populations of cells to single-out one cell for differentiation (such as in proneural clusters of five or six cells in the *Drosophila* neuroectoderm—Skeath & Carroll, 1992), or in large populations of cells to generate a fine-grained pattern of differentiated cells surrounded by inhibited cells (such as in the *Drosophila* endoderm—Tepass & Hartenstein, 1995).

In this review, we shall focus on the generation of spatial patterns of cell fate by juxtacrine lateral signalling. However, the Delta–Notch signalling system is undoubtedly more versatile than this. In particular, recent data on the development of the *Drosophila* wing veins and margin suggest that in some instances Notch activation can lead to an upregulation of the expression of its ligands Delta and Serrate, thus generating a positive feedback loop between neighbouring cells (Huppert *et al.*, 1997; de Celis & Bray, 1997; Micchelli *et al.*, 1997; Panin *et al.*, 1997). Other evidence suggests that Notch activity can also upregulate expression of Notch itself (Christensen *et al.*, 1996; de Celis *et al.*, 1997; Heitzler *et al.*, 1996; Wilkinson *et al.*, 1994). These data tend to suggest that the Notch signalling pathway can also play a role in the generation of boundaries between two cell types, and in the functioning of these boundaries as organising centres. Recent results have also implicated the Notch pathway in vertebrate segmentation (reviewed in Jiang *et al.*, 1998; McGrew & Pourquié, 1998); however, the mode of action of Notch signalling in these systems is currently unclear.

1.2. TGFα–EGF-R signalling. TGFα is a member of the epidermal growth factor family, and is an important regulator of epithelial cell behaviour, in particular cell division—see Kumar *et al.* (1995) for a review. The name of the growth factor derives from its original identification in cultures of transformed cells (de Larco & Todaro, 1978), and TGFα production is thought to be elevated in a number of malignancies (Cohen *et al.*, 1994; Derynck *et al.*, 1987); this has been particularly well-studied in breast cancer (Ciardiello *et al.*, 1989). TGFα is now known to play an important role in untransformed epithelia (Bates *et al.*, 1990; Derynck, 1988), by binding to EGF-R, and thus promoting cell division. The TGFα–EGF-R interaction is one of the best studied examples of a cellular control loop (van de Vijver *et al.*, 1991); it operates via an extracellular pathway, as opposed to intracellular autocrine loops such as v-*sis*-PDGF-R (Bejcek *et al.*, 1989; Keating & Williams, 1988).

TGFα is synthesised in the cell as a 160 amino acid membrane-bound precursor, pro-TGFα. The 50 amino acid soluble form of TGFα is generated by two separate cleavages of this precursor—see Brachmann *et al.*(1989) and Massagué (1990) for details. Originally it was assumed TGFα activity was due to the soluble form of the growth factor, but in the last decade it has emerged that pro-TGFα can also activate EGF-R (Brachmann *et al.*, 1989). Moreover, cleavage of pro-TGFα, which has a half-life of about 4 hours, is typically slower than the turnover rate of pro-TGFα, so that the membrane-bound precursor is in fact the dominant form of the growth factor (Massagué, 1990), and the TGFα–EGF-R control loop is now recognised as being a prime example of the juxtacrine signalling mechanism. A number of other soluble growth factors similarly derive from membrane-bound precursors that can themselves bind to receptors, making them candidates for juxtacrine receptor activation (Bosenberg & Massagué, 1993). For example, Tumour Necrosis Factor-α has a membrane bound precursor, and has been found to kill cells in a juxtacrine fashion (Perez *et al.*, 1990), and to mediate B-cell activation (Macchia *et al.*, 1993).

TGFα is of particular interest because of its role in epidermal wound healing (Martin *et al.*, 1992a; Schultz *et al.*, 1991). In adult mammals, such wounds heal by a combination of cell crawling at the wound edge, and enhanced proliferation further back—see Martin (1996) for review. Although this combined mechanism of healing was established many years ago (Winter, 1972), the underlying molecular details remain unclear. Growth factor regulation is known to be central to the wound healing process in general, with TGFα, keratinocyte growth factor, and epidermal growth factor all contributing to epidermal repair. TGFα is implicated as an important element of the process in humans, since normal human keratinocytes produce TGFα both *in vivo* and *in vitro* (Coffey *et al.*, 1987), and TGFα upregulates both migration and proliferation of keratinocytes in culture (Barrandon & Green, 1987). Moreover, Schultz *et al.* (1987) have shown that addition of exogenous TGFα accelerates epithelial wound healing. The realisation that TGFα communication is mainly juxtacrine raises a key question: Can such a nearest neighbour signalling mechanism account for the observed increase in cell proliferation many cell diameters away from from the wound edge? This question will be answered by our discussion of gradient-type solutions to juxtacrine models.

2. Mathematical modelling of juxtacrine signalling. We consider only juxtacrine signalling within either a one-dimensional line of epithelial cells or a two-dimensional epithelial sheet; these are much the most important cases in development, and also include signalling within the epidermis, such as occurs in response to wounding. The most natural way in which to model this system is to represent the cells individually, with the model variables being ligand and receptor levels for each of these cells; thus, mathematically, the model has the form of a large system of coupled ordinary differential equations.

2.1. Discrete formalism. In the case of the Delta–Notch interaction, Collier *et al.* (1996) use this approach, solving the equations

(1a) $$dN/dt = F(\langle D\rangle) - \mu N$$

$$dD/dt = G(N) - \rho D \tag{1b}$$

for each cell in a regular array. Here $N(t)$ and $D(t)$ represent the levels of activity of Notch and Delta on the cell, relative to a typical activity level, and the functions $F(.)$ and $G(.)$ represent the feedback control. Thus $F(.)$ is an increasing function, corresponding to the activation of Notch (the receptor) by binding with the ligand Delta on neighbouring cells, while $G(.)$ is decreasing, representing downregulation of Delta activity by binding. The notation $\langle . \rangle$ indicates an average over neighbouring cells. In a one dimensional line of cells, one can index the cells by a single integer j, so that $\langle D_j \rangle = (D_{j-1} + D_{j+1})/2$ (Figure 1). In the two-dimensional case, Collier *et al.* (1996) consider a hexagonal array of cells, indexed as described in Figure 1, so that

$$\langle D_{i,j} \rangle = \frac{1}{6} \left(D_{i-1,j-1} + D_{i,j-1} + D_{i-1,j} + D_{i+1,j} + D_{i,j+1} + D_{i+1,j+1} \right) .$$

It is important to stress that this type of local averaging is quite different from the more traditional diffusion mechanism of signalling. In particular, the above formula is quite different from a discrete representation of diffusion, which would involve the difference between concentrations on nearby cells, rather than their average (pattern formation in arrays of discrete cells coupled by diffusion is discussed in Othmer & Scrivens, 1971; Babloyantz, 1977).

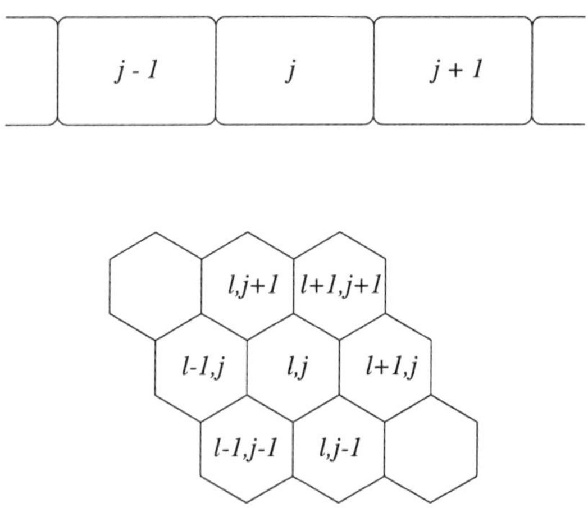

FIG. 1. *The labelling scheme used for cells in linear and two-dimensional arrays.*

For TGFα binding to EGF-R, Owen & Sherratt (1998) consider only a two-dimensional cell sheet, in which the cells are assumed to occupy a rectangular grid. In their model, free and bound receptors are included as explicit variables in the model, with binding represented by the kinetic scheme illustrated in Figure 2. Their model thus consists of a series of coupled ordinary differential equations for the numbers of ligand molecules $a_{i,j}(t)$, unoccupied receptors $f_{i,j}(t)$, and occupied receptors $b_{i,j}(t)$, on the surface of cells in column i and row j. All ligand is assumed to be membrane-anchored. As discussed above, membrane-anchored ligand can also be cleaved to give a freely diffusing form; however, we neglect this complication in order to focus on juxtacrine signalling in isolation. Using the kinetic scheme discussed above, the model equations are:

$$da/dt = -k_a a\langle f\rangle + k_d\langle b\rangle - d_a a + P_a(b) \tag{2a}$$

$$df/dt = -k_a\langle a\rangle f + k_d b - d_f f + P_f(b) \tag{2b}$$

$$db/dt = k_a\langle a\rangle f - k_d b - k_i b \tag{2c}$$

for each cell in the array.

For a regular grid of square cells, the value of $\langle a\rangle$ for cell (i,j) is

$$\langle a\rangle = \frac{1}{4}\left(a_{i,j-1} + a_{i-1,j} + a_{i+1,j} + a_{i,j+1}\right)$$

and similarly for $\langle b\rangle$ and $\langle f\rangle$. Owen & Sherratt specialise to the case where all cells in each column i are equivalent, in which case the variables can be labelled by the single index j. $\langle a\rangle$ then reduces to

$$\langle a\rangle = \frac{1}{4}\left(a_{j-1} + 2a_j + a_{j+1}\right),$$

and similarly for $\langle b\rangle$ and $\langle f\rangle$.

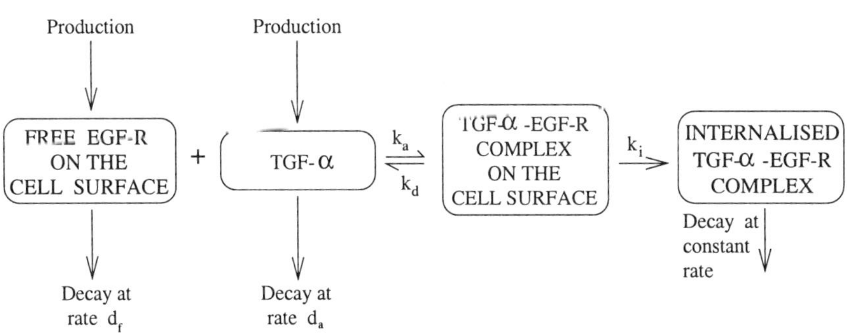

FIG. 2. *A schematic representation of the kinetic scheme used in the model for the binding of* TGFα *to* EGF-R. *The scheme is similar to that of Waters* et al. *(1990) for* EGF–EGF-R *interactions, and the parameters used are based on the values they determined from experiments on the binding of* EGF *to* EGF-R *on rat lung epithelial cells.*

The synthesis of new ligand and receptor by epithelial cells is a crucial aspect of the model. As explained above, we assume that this is controlled by a positive feedback to the level of occupied receptors on the cell surface.

Thus the production rates P_a of ligand and P_f of receptor are functions of the bound receptor number b. Our only assumption in general is that both of these production rates increase with b. In particular applications, the data available on production rates of ligand and receptors is typically extremely limited. However, the forms chosen for P_a and P_f can be specified to some extent because they must satisfy a number of conditions that relate them to quantities that are more easily measurable in experiments:

(i) In the absence of any ligand binding at the cell surface, there will be a background level of receptor expression, say r_0. This is a homogeneous steady state of the model, and so the equation for f in (2) gives

$$P_f(0) = d_f r_0. \tag{3a}$$

(ii) Normal equilibrium levels of free and bound receptors, f_e and b_e say, are often known in particular systems. Specifying f_e and b_e defines the normal steady state level of free ligand, a_e, implicitly through equation (2c), as well as the values of the feedback functions at the steady state, so that

$$a_e = \frac{(k_d + k_i)b_e}{k_a f_e}, \quad P_a(b_e) = k_i b_e + d_a a_e, \quad \text{and} \quad P_f(b_e) = k_i b_e + d_f f_e. \tag{3b}$$

(iii) In any system, there will be a maximum possible level of receptor expression, r_m say. This can be estimated experimentally by saturating cells with ligand. Such saturation means that the rate of internalisation of bound receptors must be equal to the rate of free receptor production, giving

$$P_f(r_m) = k_i r_m. \tag{3c}$$

Monk (1998) considers a less general form of juxtacrine signalling with positive feedback, in which only ligand production is explicitly enhanced by receptor activity. In addition to levels of activity of ligand and receptor, each cell has a variable competence to respond to signalling from its neighbours, which is in general a function of receptor activity in the cell. In a one-dimensional array of cells, indexed as in Figure 1, the equations for receptor, ligand and competence take the form

$$dr/dt = -\mu r + c\mathcal{R}(\langle l \rangle), \tag{4a}$$

$$dl/dt = -\rho l + \mathcal{T}(r), \tag{4b}$$

$$dc/dt = -\nu c + \mathcal{C}(r), \tag{4c}$$

where r, l and c represent receptor, ligand and cell competence, respectively. $\mathcal{R}$ and $\mathcal{T}$ are increasing functions encoding ligand–receptor binding and receptor–ligand intracellular feedback, while $\mathcal{C}$ is a decreasing (or constant) function encoding the influence of receptor activity on cell competence.

2.2. Continuous formalism. The above model frameworks involve a discrete regular representation of the cell population. In reality, cells in an epithelial sheet are not of course located in a regular grid, but rather have a range of shapes and sizes. Thus it is important to consider whether the assumption of regularity, which is implicit in the discrete models, is significant in the predictions of these models. One method of addressing this is to consider an alternative, spatially continuous framework, which does not involve this assumption. The approach of using different mathematical representations of the same phenomenon, in order to highlight the common features, is a valuable one that has been widely used in ecology (Hassell *et al.*, 1991; Savill & Hogeweg, 1997; Sherratt *et al.*, 1997). For simplicity, we restrict attention to the case of an epithelial sheet that is effectively homogeneous in one direction, so that the solutions of interest are one-dimensional. This is an important geometry for pattern formation, applying to striped patterns, and is also relevant to wound healing, where wound size is very large compared with a typical cell length, so that signal propagation away from a wound edge is effectively one-dimensional.

One benefit of using a continuous framework is that there is a very large body of previous theoretical work to call upon. In particular, a further development of our model would be to include the movement and proliferation of epidermal cells as they close the wound, and most of the previous models of wound healing have been of reaction-diffusion type. It would be relatively easy to combine a continuous model of juxtacrine signalling with such models. We have discussed previously the possibility of interactions between the epidermis and dermis, and continuous mechanochemical models including such interactions have been proposed for morphogenesis (Cruywagen *et al.*, 1992). Again, extensions of such models could include juxtacrine signalling mechanisms.

Our approach to formulating a continuous model of juxtacrine signalling is similar to that often used to model dispersal in ecological systems. In such models, the population after disperal, at a point x, is given by the sum over all y of the individuals that have moved from $(x - y)$ to x. A redistribution kernel specifies the probability of such movement as a function of y. Neubert *et al.* (1995) discuss the derivation of a number of redistribution kernels, and their pattern-forming potential in predator-prey models. In our case, we are not interested in redistribution as such, but in estimating the average contribution that spatially distributed bound receptors, which are the consequence of juxtacrine binding on the surfaces of randomly distributed cells, make to the TGFα and EGF-R production terms at each point in space. Our idea is to consider a cross section through the epidermis, parallel to the wound edge, with x representing the distance of this cross section from the edge. We expect that cells centred less than half a cell length from x have about an equal probability of contributing to the number of bound receptors at x, and cells centred between half and one cell length from x have a probability of contributing that decreases with distance from x. This idea is illustrated in Figure 3, along with the piecewise linear weighting function that we will actually use in our model.

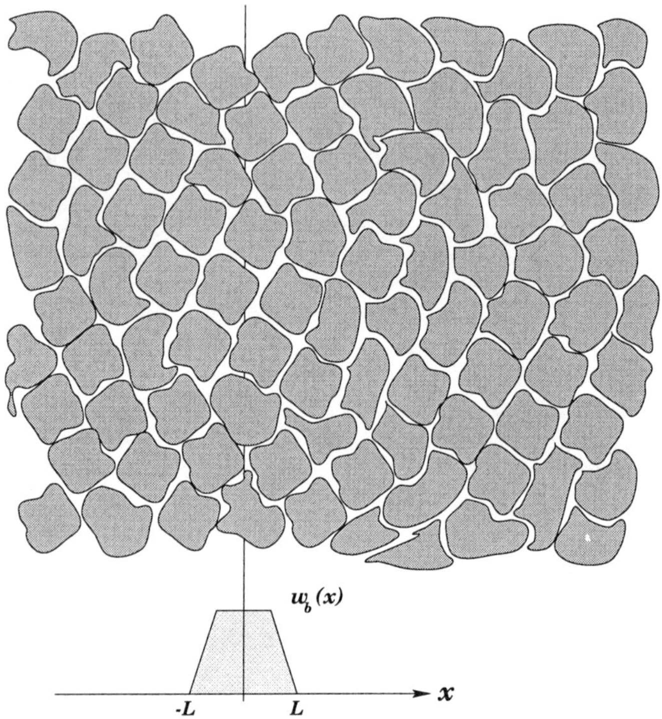

FIG. 3. *A schematic representation of the continuous representation of juxtacrine signalling. When the cells are considered to be randomly distributed with varying shapes, the model variables can be considered as averages given by some spatial weighting kernel. The particular piecewise linear kernel that we use in our numerical simulations is illustrated.*

The ideas discussed above give rise to a model in which equations (2) apply at every point in a continuous spatial domain, with local averages defined using a kernel $\omega(.)$:

$$\langle b\rangle(x,t) = \int_{-\infty}^{\infty} \omega(s)\, b(x+s,t)\, ds \tag{5}$$

(and similarly for $\langle a\rangle$ and $\langle f\rangle$). For mathematical and computational convenience, we use a piecewise linear kernel, which gives equal weights within half a cell length, and then decreases linearly to a zero weight at one cell length:

$$
(6) \qquad \omega_b(x) = \begin{cases} 0 & x < -L \\ \dfrac{4}{3L^2}(L+x) & -L < x < -\dfrac{L}{2} \\ \dfrac{2}{3L} & \text{if} \quad \dfrac{L}{2} < x < \dfrac{L}{2} \\ \dfrac{4}{3L^2}(L-x) & \dfrac{L}{2} < x < L \\ 0 & x > L \end{cases}
$$

where L is a typical cell length.

3. Types of solution. The nature of the spatiotemporal pattern that is generated by juxtacrine signalling with feedback clearly depends on the nature of the feedback between cells, on the initial and boundary conditions for ligand and receptor, and on the geometry of the cellular array. However, in general, three classes of solution can be distinguished. In all the cases considered here, the systems possess at least one spatially homogeneous steady state. If a localised perturbation is made to such a steady state (represented by a boundary condition fixed at a value different from the steady state), then stable spatial gradient and propagating front solutions can be generated. Under appropriate conditions, perturbation of the homogeneous steady state can also result in stable spatially-periodic patterns.

3.1. Spatial gradients. In the case of positive feedback of receptor activation on ligand production (models (2) and (4)), it is of interest to consider the effect of a localised disturbance on a stable homogeneous steady state. For TGFα binding to EGF-R, such a situation could represent the effect of an epidermal wound on surrounding unwounded cells (Owen & Sherratt, 1998); in a developmental context, it could represent the spread of an inductive signal within a population of cells (Monk, 1998). To investigate this situation, homogeneous steady state initial conditions are used, together with a boundary condition on one margin of the array that is fixed at a value distinct from that at the homogeneous steady state. To restrict attention to this single boundary, the boundary conditions on all other boundaries are taken to be either zero flux or fixed at the homogeneous steady state level.

Because of the anomalous boundary condition, the initial homogeneous steady state is no longer a solution of the model equations. Owen & Sherratt consider a non-zero homogeneous steady state of the model described by (2) with a zero ligand boundary condition at one margin of a two-dimensional array of cells. In the final steady solution, the levels of ligand and receptor grade smoothly from their fixed boundary values to their homogeneous steady state levels over a range of several cell diameters

(Figure 4). Since juxtacrine ligands can mediate signalling only between immediately neighbouring cells, it is not immediately obvious whether or not the spatial range of these gradient solutions is bounded above. Considering both discrete and continuous forms of the TGFα model, Owen *et al.*(1999) have shown that in principle there is not an upper bound on spatial range. As the strength of the positive feedback between activated receptor and ligand production increases, the range of the gradient solution increases without bound (Figure 5). However, as the spatial range of solutions increases, linear analysis suggests that the rate at which the solutions approach steady state decreases (Owen & Sherratt, 1998), and so there may be an upper bound to the spatial range over which a signal can be propagated in a realistic time.

Monk (1998) considers the effect of a non-zero fixed boundary condition on the zero homogeneous steady state of the model described by (4). In this case, the absence of feedback between activated receptor and receptor levels has the effect of setting an upper bound on the range over which a gradient solution can be generated (where range is defined as the maximum distance from the fixed boundary at which a given proportion of the receptors on a cell are bound by ligand). Typical solutions are shown in Figure 6. As for the TGFα model, the range is an increasing function of the strength of feedback between the cells; furthermore, for a given strength of feedback, the range is an increasing function of the magnitude of the fixed boundary condition. However, there are two effects that impose a strict upper bound on the spatial range of the gradient solutions in this case. Firstly, as the strength of feedback between the cells is increased, other non-zero homogeneous steady states can come into existence. When this occurs, the fixed boundary condition can initiate a propagating front solution which dominates over gradient solutions (see below). Secondly, in cases where gradient solutions are stable, it is not possible to increase their range indefinitely by increasing the magnitude of the fixed boundary condition. The reason for this is that each of the cells has a fixed number of receptors; once the receptors in the cell neighbouring the boundary have been saturated, there is no possibility of increasing the ligand output from this cell (see Figure 7). The differing results of models (2) and (4) demonstrate that the presence or absence of feedback control of the amount of receptor expressed by the cells can have a significant impact on the properties of gradient solutions.

3.2. Travelling fronts. As mentioned in the previous section, a localised perturbation of a homogeneous steady state can also generate travelling front solutions (Monk, 1998). For suitable forms of binding and feedback functions in (4), in addition to a stable zero homogeneous steady state, there exists a stable non-zero homogeneous steady state in which levels of bound receptor are high, together with an intermediate unstable homogeneous steady state. When this is the case, the intermediate steady

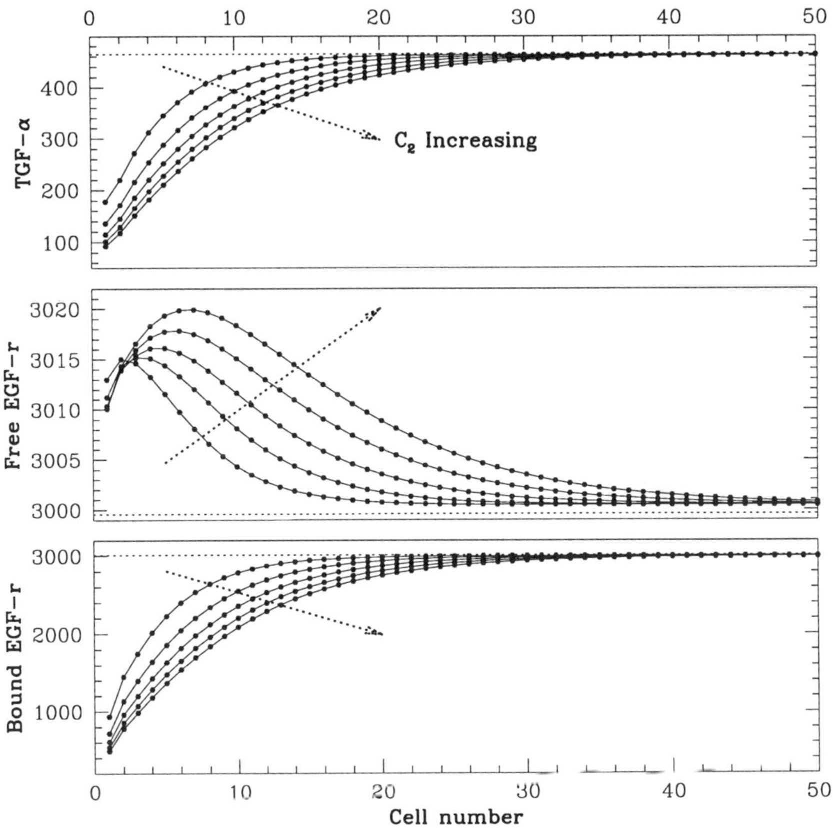

FIG. 4. *Numerically calculated solutions of the model (2). The solutions are shown after 166.7 hours (10000 minutes) of evolution with the wounded boundary condition ($a = f = b = 0$ at cell 0), for C_2 increasing from 10000 to 50000 at intervals of* 10000. *The distance of propagation of the wound-induced perturbation clearly increases as the parameter C_2, which measures the strength of feedback in* TGFα *production, increases. The other parameters are* $k_a = 0.0003\text{molecules}^{-1}\text{min}^{-1}, k_d = 0.12\text{min}^{-1}, k_i = 0.019\text{min}^{-1}, d_a = 0.006\text{min}^{-1}, d_f = 0.03\text{min}^{-1}, f_e = 3000, b_e = 3000, r_0 = 3000, r_m = 25500$. *For details, see Owen & Sherratt, 1998.*

state defines a threshold of receptor activation and the nature of the solution depends on the magnitude of the fixed boundary condition. For values that fail to raise the level of receptor activation above the threshold, spatially-graded solutions result; in contrast, if the threshold is surpassed, then a travelling front connecting the two stable steady states is initiated (Figure 8). The front propagates away from the fixed boundary at a constant speed, maintaining an invariant profile, and leaving a homogeneous level of receptor activation in its wake. For biologically-realistic parame-

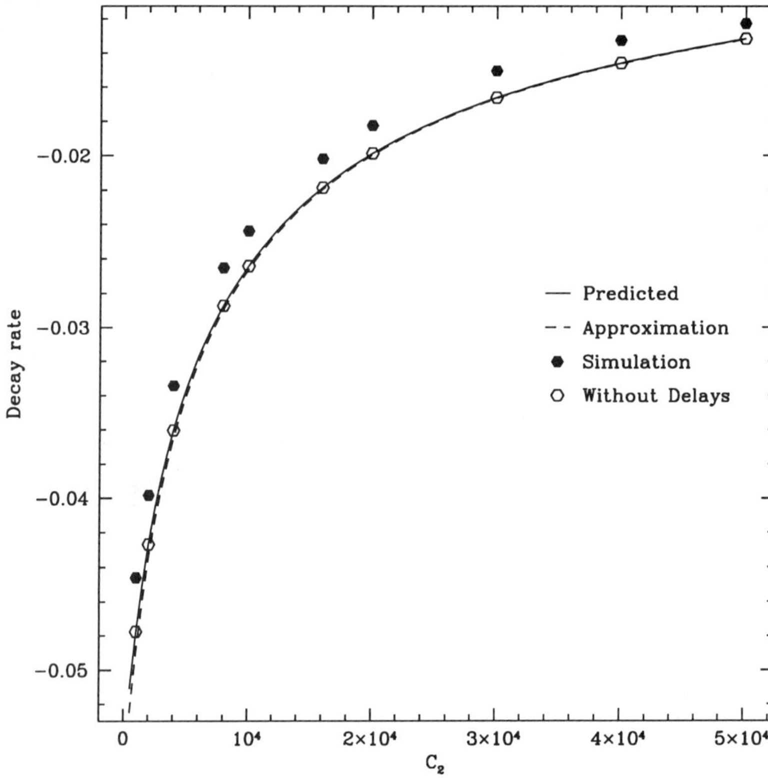

FIG. 5. *Predicted spatial decay rate of gradient solutions as C_2 varies, for the steady state of the model (2), with the wounded boundary condition $a = f = b = 0$ at cell 0. The points represent decay rates calculated from simulation data* 166.7 *hours (*10000 *minutes) after wounding. Here the spatial decay rate is calculated as* $(1/L) \cdot \ln[(a_{j+1} - a_e)/(a_j - a_e)]$, *which estimates the quantity λ in a solution of the form $a_j - a_e \propto \exp(\lambda L j)$. Thus a less negative growth rate corresponds to a greater signal range. The solid line indicates the decay rate predicted by linear analysis, and the dashed line shows the values given by a lowest order approximation to the decay rate. The other parameters are as in Figure 4.*

ters, the speed of such a front can be great enough to allow propagation over hundreds of cell diameters in the space of a few hours.

3.3. Spatial patterns. Fine-grained patterns are a ubiquitous feature of epithelia in early animal development, and juxtacrine signalling is a natural candidate for the generation of these patterns. Numerical simulations of the mathematical models described in §2 show that both the negative and positive feedback models do indeed support permanent spatial patterns for appropriate parameter values. However, the type of pattern depends crucially on feedback regulation.

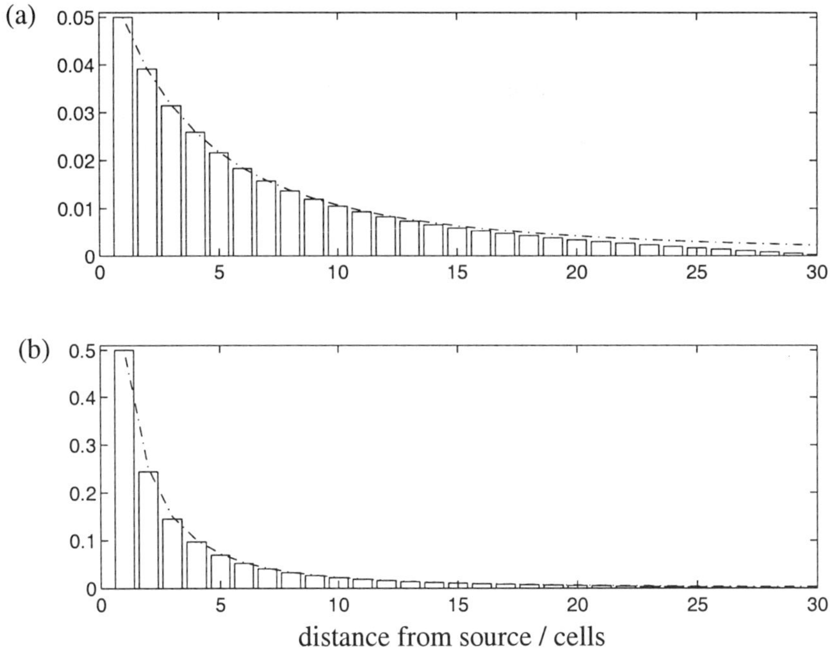

FIG. 6. *Examples of steady state spatial gradients of cell activation resulting from a fixed signal source (cell 1). The strength of signal source in (b) is 10-times greater than that in (a), illustrating the fact that the qualitative shape of the gradient is dependent on the strength of the signal source. Histograms show the results of numerical simulation of the model described by (4); continuous curves show corresponding analytical approximations to these solutions (in (b), the agreement between the two solutions is such that they are indistinguishable at distances of more than about 5 cells from the source). Responsive cells had random initial levels of activation between 0 and 0.01 and* $c_i(0) = 1$. $\mu = \rho = \nu = 1$, $\mathcal{C}(\xi) = 1$, $\mathcal{T}(\xi) = \xi$, $\mathcal{R}(\xi) = \xi/(1+\xi)$.

In the case of negative feedback, which is represented in model (1), patterns form provided that the feedback functions $F(.)$ and $G(.)$ are sufficiently strong: specifically, stability analysis shows that the condition for patterning is $F'(D^*)G'(N^*) < -\mu\rho$, where $D = D^*$, $N = N^*$ is the unique homogeneous steady state. In the case of a one-dimensional line of cells, the patterns have a characteristic and highly reproducible form, consisting of alternating high and low levels of Notch and Delta activity (Figure 9). The key property of a wavelength of two cells is independent of parameters, and arises because of the lateral inhibition intrinsic in the model. A cell with higher Notch activity than its neighbours has lower Delta activity, because of negative feedback. This in turn reduces Notch activity in the neighbouring cells, leading to increased Delta activity in these cells (negative feedback again) and hence further increase in Notch activity. Thus the feedback regulation enables a regular, alternating pattern to be sustained.

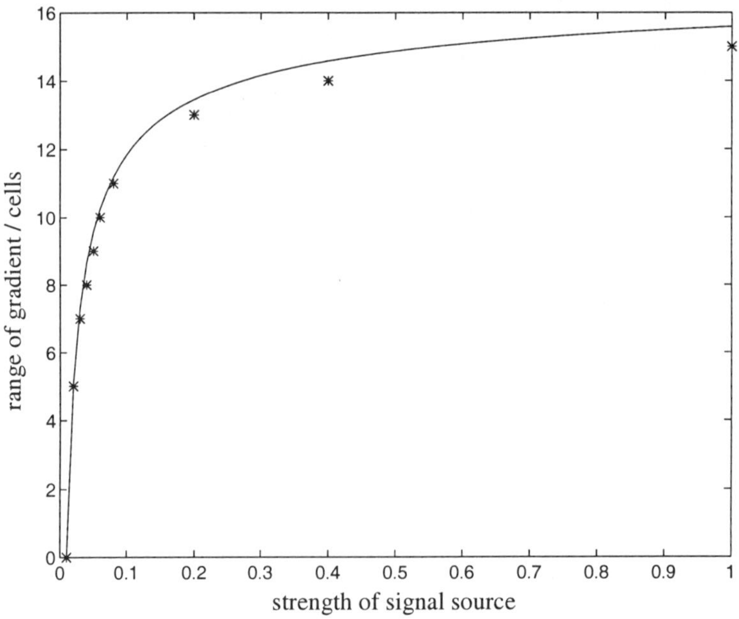

FIG. 7. *Distance over which a level of cell activation of at least 0.01 (i.e. 1% receptor saturation) can be attained at steady state as a function of the strength of signal source in the model described by (4). The continuous curve shows an analytical approximation, while stars show results of corresponding computer simulations. Parameters and functional forms as in Figure 6.*

In numerical solutions, the only significant difference between simulations is the presence of defects, where two adjacent cells have high levels of Notch activity — these are a consequence of the particular perturbations applied initially to the uniform steady state. However, solutions in which two adjacent cells have low levels of Notch activity are never seen, in agreement with experiment.

In the two-dimensional hexagonal grid of cells, the model (1) again predicts spatial pattern formation. Again, the possible patterns are all fine-grained, with scattered cells with low Notch activity surrounded by cells with high Notch activity, but a number of different pattern configurations are in principle possible (Figure 10). Numerical simulations show that the boundary conditions exert a major influence on the patterns formed, at least on relatively small domains, and thus Collier *et al.* (1996) considered patterns formed on 6×6 or 7×7 arrays of hexagonal cells with periodic boundary conditions (Figure 11), enabling consideration of potential periodic patterns on an infinite domain. On a 6×6 array, patterns with periods 1,2,3 or 6 in i and j are in principle possible. It is the period 3 pattern (with slight defects) that emerges in the simulations in Figures 11(a) and (b);

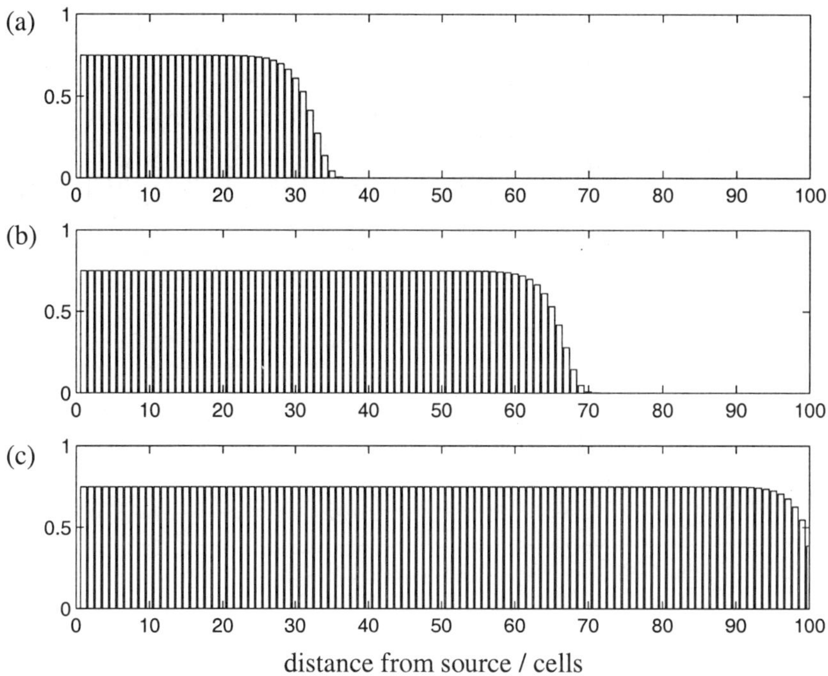

FIG. 8. *A typical travelling front of cell activation generated by (4). The responsive field is fixed at 100 cells by setting the level of activation of cell 101 to zero.*

however in the 7×7 array, a pattern with period 7 (as in Figure 10c) can be achieved by using initial conditions that are biased towards this pattern.

In the case of positive feedback in ligand and receptor production, juxtacrine signalling is again able to produce spatial patterns, as shown by numerical simulations of (2). Mathematical conditions for patterning can be derived by linear analysis (Wearing, Owen & Sherratt, 2000). Patterns form when feedback is relatively low for ligand and moderate for receptors, with alternative behaviours being a stable uniform state (when feedbacks are weak) and uncontrolled upregulation in ligand and receptors throughout the domain (when feedbacks are strong). Figure 12 illustrates such a pattern forming when the homogeneous steady state is perturbed locally at one edge of a large sheet of cells; for simplicity we assume that the perturbation is such that a one-dimensional (striped) pattern results.

Intuitively, spatial patterning in this system arises via spatially localised positive feedback. The weak feedback in ligand expression, which is a key ingredient, means that ligand levels remain relatively constant despite large variations in receptor expression. This means that both increases and decreases in receptor levels, away from homogeneous equilibrium levels, are self-reinforcing. This enables a wide range of spatial patterns, with clear division of cells into high and low receptor occupancy.

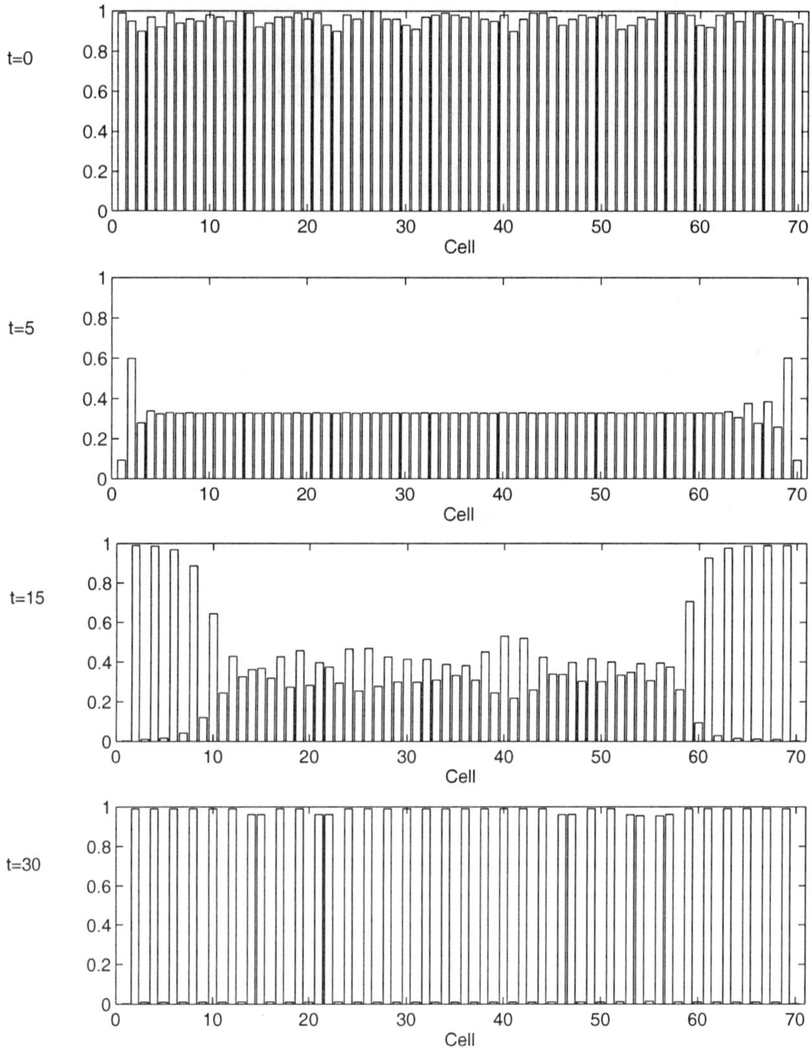

FIG. 9. *Time evolution of the level of Notch activity in a line of 70 cells starting from an initial near-homogeneous state. By* $t = 30$*, the levels of Notch and Delta activity have almost reached equilibrium. Initial conditions for* n_j *are shown in the top panel* ($t = 0$)*, and* $d_j(0) = 1$ *for* $1 \leq j \leq 70$.

4. Further development of juxtacrine models. The models we have discussed here represent a first step in the investigation of the properties of juxtacrine signalling systems. The complexity of the models has been kept to a minimum in order to exhibit most clearly the properties of this type of signalling system, and to avoid where possible including any assumptions that cannot be experimentally justified. However, there are many extensions of the models that should be considered. Although

(a)
(b)
(c)

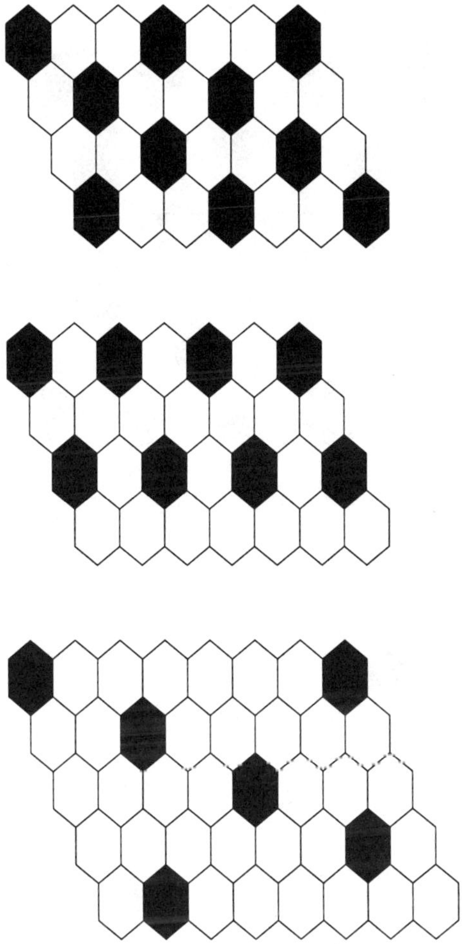

FIG. 10. *Examples of possible periodic steady-state patterns of Notch activity in (unbounded) two-dimensional arrays of hexagonal cells. Shaded cells have low Notch activity, while white cells have high Notch activity.*

the discrete and continuous models outlined above are equally suited to describing simple juxtacrine systems, they each have features that make them more or less suitable when considering more complex systems. For example, in order to investigate the effects of juxtacrine signalling in the context of cell proliferation and movement, it would be most effective to employ a continuous formalism. In contrast, a discrete formalism would be more suited to an investigation of the effects of internal cellular dynamics on patterning by juxtacrine signalling.

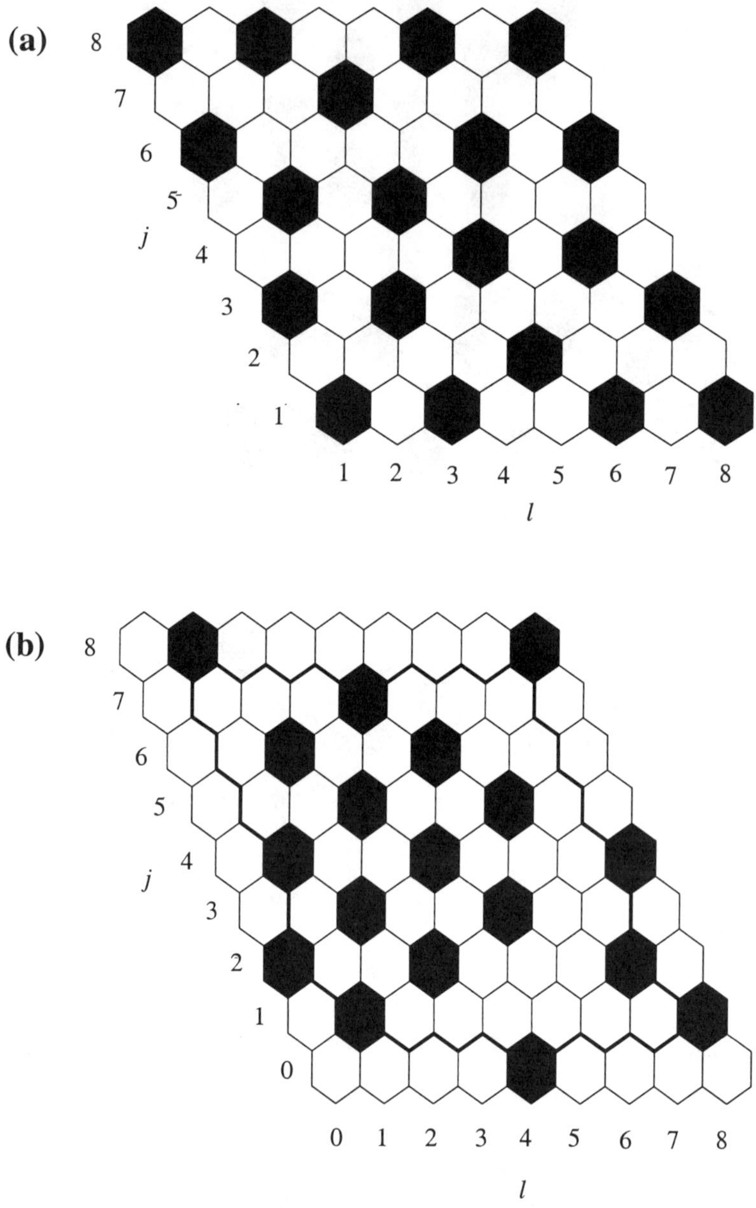

FIG. 11. *Typical steady state patterns of Notch activity in an* 8×8 *(a) and* 7×7 *(b) array of hexagonal cells starting from near-homogeneity. Shading as in Fig. 10. Boundary conditions: (a)* $d_{lj} = 0$ *for* $l, j = 0$ *or 9; (b) periodic. Initial conditions:* $n_{lj}(0)$ *and* $d_{lj}(0)$ *have arbitrary values between 0.95 and 1.0.*

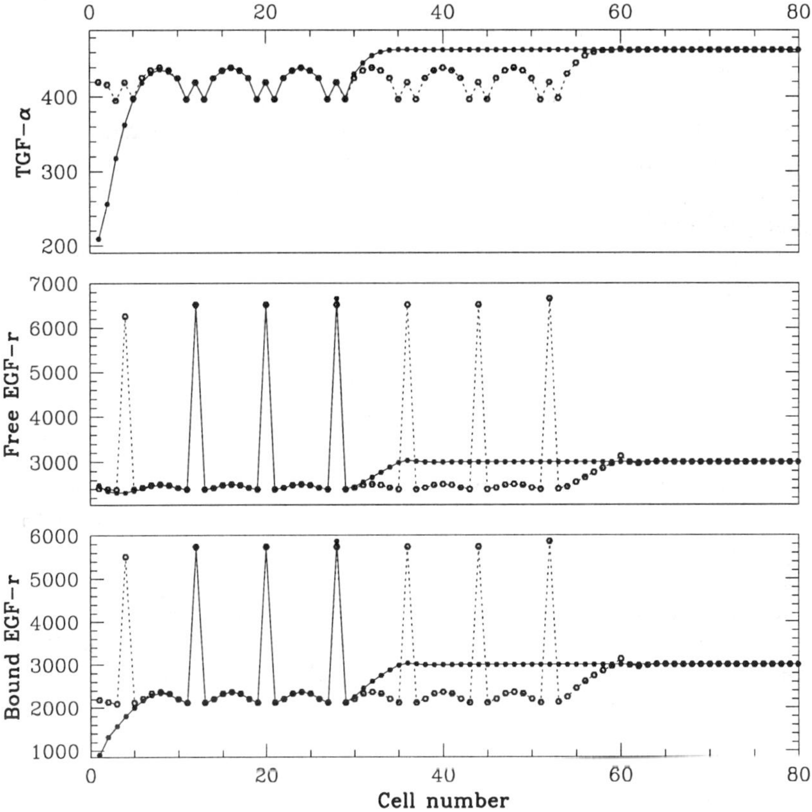

FIG. 12. *Numerical simulation of the model (2), specified with equations (1), and with* $m = 1.0$ *and* $n = 3.0$. *Linear analysis predicts pattern formation in this case. The solid points and lines indicate the solution after 500 hours of evolution with a wounded boundary condition* ($a = f = b = 0$ *at cell 0), and the open points and dotted lines indicate the solution a further 500 hours after the reintroduction of the healed boundary condition (values of* a, f *and* b *the same at cell 0 and cell 1). Interestingly the patterned solution continues to persist and spread, despite the completion of healing at the wound edge. Thus, this figure shows the early spread of a spatial pattern which would eventually spread to cover the entire domain. This simulation has no relevance to wound healing, but periodic patterns with a wavelength of a few cells are observed in a range of processes in early vertebrate development (see Lewis, 1998, for review). The other parameters are* $k_a = 0.0003\text{molecules}^{-1}\text{min}^{-1}, k_d = 0.12\text{min}^{-1}, k_i = 0.019\text{min}^{-1}, d_a = 0.006\text{min}^{-1}, d_f = 0.03\text{min}^{-1}, f_e = 3000, b_e = 3000, r_0 = 3000, r_m = 25500, C_2 = 8000$.

An important consideration that has been neglected in the models outlined here is the time taken for the operation of intracellular feedback. It is necessary to have at least an estimate of the time taken for a change in the level of receptor activity in a cell to have an effect on the levels of ligand and free receptor in that cell. For TGFα signalling, it is known that changes in the number of bound EGF-R receptors in a cell result in changes in the level of transcription of the genes coding for ligand and receptor. For Delta–Notch signalling, it is known that such transcriptional control can occur, though it is not clear to what extent it is important in lateral signalling. If feedback is mediated by control of the rates of transcription of ligand and receptor, then the delay between a change in the level of receptor activity and a concomitant change in the rate of appearance of mature ligand and receptor in the cell membrane can reasonably be assumed to depend on their molecular weight. EGF-R and TGFα have molecular weights of 170kD and 18kD respectively (Ciardiello *et al.*, 1989; Kumar *et al.*, 1995). Notch and Delta are significantly larger, having molecular weights of around 300kD and 100kD respectively (Fehon *et al.*, 1990).

For all of these proteins, with the possible exception of TGFα, transcription alone would be expected to take a significant time. For example, the region of genomic DNA corresponding to *Delta* in *Drosophila* is approximately 25kb in length (Kopczynski *et al.*, 1988; Vässin *et al.*, 1987). Assuming a transcriptional rate of 1kb per minute, production of the primary mRNA transcript for *Delta* would take around 25 minutes. When mRNA processing, translation, protein processing and transport to the membrane are taken into account, it is likely that there would be a lag of at least 30 minutes in a Notch–Delta feedback mediated by transcriptional regulation. Thus the delays in regulation of the production of ligands and in particular receptors are likely to be significant in some applications. In this review we have neglected this complication, but models in which delays are included are discussed elsewhere (Owen & Sherratt, in preparation; Monk, in preparation).

The models we have discussed here are all based on the assumption that cells signal with equal strength to each of their immediate neighbours. However, it is not uncommon for epithelial cells to be overtly polarised within the plane of the epithelium (termed planar polarisation—see Eaton, 1997). In *Drosophila*, a genetic pathway centred on the transmembrane receptor Frizzled and the intracellular protein Dishevelled has been shown to be involved in the generation and coordination of polarisation in a wide range of epithelia during development (reviewed in Shulman *et al.*, 1998). While the precise mechanisms underlying planar polarisation remain unclear, the non-uniform activation of the Frizzled signalling pathway within epithelial cells has recently been implicated as an important step in the process (Axelrod *et al.*, 1998).

In order to investigate the effects of planar polarisation on patterning mediated by juxtacrine signalling, the models described here can be generalised to take into account spatially non-uniform activity of ligand and receptors in the cell membrane. This is most easily achieved in the discrete formalism, at least for simple forms of non-uniformity. Generalisations of this type are particularly relevant to the study of spatial patterning by lateral signalling. There is growing evidence for non-trivial interactions between the Notch and Wingless pathways, mediated by Dishevelled (Axelrod *et al.*, 1996; Martinez Arias, 1998), and it has been suggested that such an interaction could modulate the periodicity of the fine-grained patterns generated by Delta–Notch signalling (for example, in the *Drosophila* stomatogastric nervous system—González-Gaitán & Jäckle, 1995). Models including spatially non-uniform signalling within the cell membrane are discussed elsewhere (Monk, in preparation).

5. Discussion. The models we have described in this review are based on the basic biological premise that cells signal to their immediate neighbours via a transmembrane signalling system. We have shown that such models can account for the generation of a range of stable and propagating patterns within populations of interacting cells. The best-studied theoretical model for spatial pattern formation in biological systems is the Turing mechanism (Turing, 1952; for a recent review, see Maini *et al.*, 1997), in which diffusion-driven instability sets up a prepattern in two or more reacting chemical morphogens; this is translated into a discrete pattern of distinct cell fates via a threshold level of one of the morphogens. More generally, models based on the Turing mechanism belong to a class of models for spatial pattern formation that are based on the idea of a localised self-activating process balanced by a longer-range inhibitory process (Oster, 1988).

The juxtacrine mechanisms that we have described here are quite different to previous models based on local activation and long-range inhibition, both in their mathematical form and in their biological interpretation. In juxtacrine models, all interactions are local: Feedback depends on biochemical events restricted to the intracellular environment, while signalling between cells is restricted to immediate neighbours. They thus constitute a new and quite distinct class of mechanism for the generation of spatial patterns in biological systems.

In their spatially discrete form, juxtacrine signalling models bear some resemblance to cellular automata models (see, for example, Ermentrout & Edelstein-Keshet, 1993). However, the simple discrete rules that dictate the behaviour of conventional cellular automata (Wolfram, 1994) are replaced in juxtacrine models by functions based on well-characterised biochemical interactions (between membrane-anchored ligand and receptors). Consequently, the patterns generated lend themselves very naturally to biological interpretation: Bound receptor activity is the clear biological determinant of cell fate. Since this activity is either very high or very low in all the patterns we have observed, there is no requirement for the imposition of a particular (and arbitrary) threshold level of the type required in local activation and long-range inhibition models.

Acknowledgements. N.A.M.M. and J.A.S. thank the IMA and British Society for Developmental Biology (N.A.M.M.) for support, enabling them to participate in the Program on Mathematical Biology. We thank Julian Lewis, Philip Maini, Simon Myers and Helen Wearing for helpful discussions.

REFERENCES

[1] S. Artavanis-Tsakonas, K. Matsuno and M. Fortini. Notch signaling. *Science*, **268**:225–232, 1995.

[2] J.D. Axelrod, K. Matsuno, S. Artavanis-Tsakonas and N. Perrimon. Interaction between *wingless* and *Notch* signalling pathways mediated by *dishevelled*. *Science*, **271**:1826–1832, 1996.

[3] J.D. Axelrod, J.R. Miller, J.M. Shulman, R.T. Moon and N. Perrimon. Differential recruitment of Dishevelled provides signaling specificity in the planar cell polarity and Wingless signaling pathways. *Genes Dev.*, **12**:2610–2622, 1998.

[4] A. Babloyantz. Self-organization phenomena resulting from cell-cell contact. *J. Theor. Biol.*, **68**:551–561, 1977.

[5] Y. Barrandon and H. Green. Cell migration is essential for sustained growth of keratinocyte colonies: The roles of Transforming Growth Factor α and epidermal growth factor. *Cell*, **50**:1131–1137, 1987.

[6] S.E. Bates, E.M. Valverius, B.W. Ennis, D.A. Bronzert, J.P. Sheridan, M.R. Stampfer, J. Mendelsohn, M.E. Lippman and R.B. Dickson. Expression of the transforming growth factor-alpha/epidermal growth factor receptor pathway in normal human breast epithelial cells. *Endocrinology*, **129**:596–607, 1990.

[7] B.E. Bejcek, D.Y. Li and T.F. Deuel. Transformation of v-*sis* occurs by an internal autoactivation mechanism. *Science*, **245**:1496–1499, 1989.

[8] M.W. Bosenberg and J. Massagué. Juxtacrine cell signalling molecules. *Curr. Opin. Cell Biol.*, **5**:832–838, 1993.

[9] R. Brachmann, P.B. Lindquist, M. Nagashima, W. Kohr, T. Lipari, M. Napier and R. Derynck. Transmembrane TGF-α precursors activate EGF/TGF-α receptors. *Cell*, **56**:691–700, 1989.

[10] J.A. Campos-Ortega. Early neurogenesis in *Drosophila melanogaster*. In: *The Development of Drosophila melanogaster* (M Bate and A Martinez Arias, eds) Vol. **2**, pp. 1091–1129. New York: Cold Spring Harbor Laboratory Press, 1993.

[11] S. Christensen, V. Kodoyianni, M. Bosenberg, L. Friedman and J. Kimble. *lag-1*, a gene required for *lin-12* and *glp-1* signaling in *Caenorhabditis elegans*, is homologous to human CBF1 and *Drosophila* Su(H). *Development*, **122**:1373–1383, 1996.

[12] F. Ciardiello, N. Kim, D.S. Liscia, C. Bianco, R. Lidereau, G. Merlo, R. Callahan, J. Greiner, C. Szpak, W. Kidwell, J. Schlom and D.S. Salomon. Messenger-RNA expression of transforming growth factor α in human breast carcinomas and its activity in effusions of breast cancer patients. *J. Natl. Cancer Inst.*, **81**:1165–1171, 1989.

[13] D.W. Cohen, R. Simak, W.R. Fair, J. Melamed, H.I. Scher, and C. Cordon-cardo. Expression of transforming growth factor α and the epidermal growth factor receptor in human prostate tissues. *J. Urol.*, **152**:2120–2124, 1994.

[14] J.F. de Celis and S. Bray. Feedback mechanisms affecting Notch activation at the dorsoventral boundary in the *Drosophila* wing. *Development*, **124**:3241–3251, 1997.

[15] J.F. de Celis, S. Bray and A. Garcia-Bellido. Notch signalling regulates *veinlet* expression and establishes boundaries between veins and interveins in the *Drosophila* wing. *Development*, **124**:1919–1928, 1997

[16] J.R. Collier, N.A.M. Monk, P.K. Maini and J.H. Lewis. Pattern formation by lateral inhibition with feedback: A mathematical model of Delta-Notch intercellular signalling. *J. theor. Biol.*, **183**:429–446, 1996.

[17] A.J.L. Clark, S. Ishii, N. Richert, G.T. Merlino and I. Pastan. Epidermal growth factor regulates the expression of its own receptor. *Proc. Natl. Acad. Sci. USA*, **82**:8374–8378, 1985.

[18] R.J. Coffey, R. Derynck, J.N. Wilcox, T.S. Bringman, A.S. Goustin, H.L. Moses and M.R. Pittelkow. Production and auto-induction of transforming growth Factor-α in human keratinocytes. *Nature*, **328**:817–820, 1987.

[19] G.C. Cruywagen, P.K. Maini and J.D. Murray. Sequential pattern formation in a model for skin morphogenesis. *IMA J. Math. Appl. Med. Biol.*, **9**:227–248, 1992.

[20] S. Davis, N.W. Gale, T.H. Aldrich, P.C. Maisonpierre, V. Lhotak, T. Pawson, M. Goldfarb and G.D. Yancopoulos. Ligands for EPH-related receptor tyrosine kinases that require membrane attachment or clustering for activity. *Science*, **266**:816–819, 1994.

[21] J.E. de Larco and G.J. Todaro. Growth factors from murine sarcoma virus transformed cells. *Proc. Natl. Acad. Sci. USA*, **75**:4001–4005, 1978.

[22] R. Derynck, D.V. Goeddel, A. Ullrich, J.U. Gutterman, R.D. Williams, T.S. Bringman and W.H. Berger. Synthesis of messenger RNAs for transforming growth factors α and β and the epidermal growth factor receptor by human tumors. *Cancer Res.*, **47**:707–712, 1987.

[23] R. Derynck. Transforming growth factor α. *Cell*, **54**:593–595, 1988.

[24] H.S. Earp, K.S. Austin, J. Blaisdell, R.A. Rubin, K.G. Nelson, L.W. Lee and J.W. Grisham. Epidermal growth factor (EGF) stimulates egf receptor synthesis. *J. Biol. Chem.*, **261**:4777–4780, 1986.

[25] H.S. Earp, J.R. Hepler, L.A. Petch, A. Miller, A.R. Berry, J. Harris, V.W Raymond, B.K. McCune, L.W. Lee, J.W. Grisham and T.K. Harden. Epidermal growth factor (EGF) and hormones stimulate phosphoinositide hydrolysis and increase EGF receptor protein synthesis and mRNA levels in rat liver epithelial cells. *J. Biol. Chem.*, **263**:13868–13874, 1988.

[26] S. Eaton. Planar polarization of *Drosophila* and vertebrate epithelia. *Curr. Opin. Cell Biol.*, **9**:860–866, 1997.

[27] G.B. Ermentrout and L. Edelstein-Keshet. Cellular automata approaches to biological modeling. *J. Theor. Biol.*, **160**:97–133, 1993.

[28] F. Fagotto and B. Gumbiner. Cell contact-dependent signaling. *Dev. Biol.*, **180**:445–454, 1996.

[29] R.G. Fehon, P.J. Kooh, I. Rebay, C.L. Regan, T. Xu, M.A.T. Muskavitch and S. Artavanis-Tsakonas. Molecular interactions between the protein products of the neurogenic loci *Notch* and *Delta*, two EGF-homologous genes in *Drosophila*. *Cell*, **61**:523–534, 1990.

[30] M.E. Fortini and S. Artavanis-Tsakonas. *Notch*: Neurogenesis is only part of the picture. *Cell*, **75**:1245–1247, 1993.

[31] M. González-Gaitán and H. Jäckle. Invagination centers within the *Drosophila* stomatogastric nervous system anlage are positioned by *Notch*-mediated signaling which is spatially controlled through *wingless*. *Development*, **121**:2313–2325, 1995.

[32] A. Goriley, N. Dumont, C. Dambly-Chaudière and A. Ghysen. The determination of sense organs in *Drosophila*: Effect of neurogenic mutations in the embryo. *Development*, **113**:1395–1404, 1991.

[33] I. Greenwald. Structure/function studies of lin-12/Notch proteins. *Curr. Opin. Genetics Dev.*, **4**:556–562, 1994.

[34] M. Grell, E. Douni, H. Wajant, M. Lohden, M. Clauss, W. Maxeiner, S. Georgopoulos, W. Lesslauer, G. Kollias, K. Pfizenmaier and P. Scheurich. The transmembrane form of tumor necrosis factor is the prime activating ligand of the 80kDa tumor necrosis factor receptor. *Cell*, **83**:793–802, 1995.

[35] M.P. Hassell, H.N. Comins and R.M. May. Spatial structure and chaos in insect population dynamics. *Nature*, **353**:255–258, 1991.

[36] P. Heitzler and P. Simpson. The choice of cell fate in the epidermis of *Drosophila*. *Cell*, **64**:1083–1092, 1991.

[37] P. Heitzler and P. Simpson. Altered epidermal growth factor-like sequences provide evidence for a role of Notch as a receptor in cell fate decisions. *Development*, **117**: 1113–1123, 1993.

[38] P. Heitzler, M. Bourouis, L. Ruel, C. Carteret and P. Simpson. Genes of the *Enhancer of split* and *achaete-scute* complexes are required for a regulatory loop between *Notch* and *Delta* during lateral signalling in *Drosophila*. *Development*, **122**:161–171, 1996.

[39] S.S. Huppert, T.L. Jacobson and M.A.T. Muskavitch. Feedback regulation is central to Delta-Notch signalling required for *Drosophila* wing vein morphogenesis. *Development*, **124**:3283–3291, 1997.

[40] T.L. Jacobsen, K. Brennan, A. Martinez Arias and M.A.T. Muskavitch. *Cis*-interactions between Delta and Notch modulate neurogenic signalling in *Drosophila*. *Development*, **125**:4531–4540, 1998.

[41] Y.J. Jiang, L. Smithers and J. Lewis. Vertebrate segmentation: The clock is linked to Notch. *Curr. Biol.*, **8**:R868–R871, 1998.

[42] M.T. Keating and L.T. Williams. Autocrine stimulation of intracellular pdgf receptors in v-*sis*-transformed cells. *Science*, **239**:914–916, 1988.

[43] J. Kimble and P. Simpson. The LIN-12/Notch signaling pathway and its regulation. *Annu. Rev. Cell Dev. Biol.*, **13**:333–361, 1997.

[44] K.M. Klueg, T.R. Parody and M.A.T. Muskavitch. Complex proteolytic processing acts on Delta, a transmembrane ligand for Notch, during *Drosophila* development. *Mol. Biol. Cell*, **9**:1709–1723, 1998.

[45] C.C. Kopczynski, A.K. Alton, K. Fechtel, P.J. Kooh and M.A.T. Muskavitch. *Delta*, a *Drosophila* neurogenic gene, is transcriptionally complex and encodes a protein related to blood coagulation factors and epidermal growth factor of vertebrates. *Genes Dev.*, **2**:1723–1735, 1988.

[46] H. Krämer, R.L. Cagan and S.L. Zipursky. Interaction of *bride of sevenless* membrane-bound ligand and the *sevenless* tyrosine-kinase receptor. *Nature*, **352**:207–212, 1991.

[47] J.E. Kudlow, C-Y M. Cheung and J.D. Bjorge. Epidermal growth factor stimulates the synthesis of its own receptor in a human breast cancer cell line. *J. Biol. Chem.*, **261**:4134–4138, 1986.

[48] V. Kumar, S.A. Bustin and I.A. McKay. Transforming growth factor alpha. *Cell Biol. Intl.*, **19**:373–388, 1995.

[49] R. Lehmann, F. Jimenez, U. Dietrich and J.A. Campos-Ortega. On the phenotype and development of mutants of early neurogenesis in *Drosophila melanogaster*. *Roux's Arch. Dev. Biol.*, **192**:62–74, 1983.

[50] J. Lewis. Neurogenic genes and vertebrate neurogenesis. *Curr. Opin. Neurobiol.*, **6**:3–10, 1996.

[51] J. Lewis. Notch signalling and the control of cell fate choices in vertebrates. *Sem. Cell Dev. Biol.*, **9**:583–589, 1998.

[52] P.K. Maini, K.J. Painter and H.N.P. Chau. Spatial pattern formation in chemical and biological systems. *J. Chem. Soc., Faraday Trans.*, **93**:3601–3610, 1997.

[53] M.J. McGrew and O. Pourquié. Somitogenesis: segmenting a vertebrate. *Curr. Opin. Genetics Dev.*, **8**:487–493, 1998.

[54] D. Macchia, F. Almerigogna, P. Parronchi, A. Ravina, E. Maggi and S. Romagnani. Membrane tumour necrosis Factor-α is involved in the polyclonal B-cell activation induced by HIV-infected human T-cells. *Nature*, **363**:464–466, 1993.

[55] P. Martin, J. Hopkinson-Woolley and J. McCluskey. Growth factors and cutaneous wound repair. *Prog. Growth Factor Res.*, **4**:25–44, 1992.

[56] P. Martin. Mechanisms of wound healing in the embryo and fetus. *Curr. Topics Dev. Biol.*, **32**:175–203, 1996.

[57] A. Martinez Arias. Interactions between Wingless and Notch during the assignation of cell fates in *Drosophila*. *Int. J. Dev. Biol.*, **42**:325–333, 1998.

[58] J Massagué. Transforming growth Factor-α: A model for membrane-anchored growth factors. *J. Biol. Chem.*, **265**:21393–21396, 1990.

[59] C.A. Micchelli, E.J. Rulifson and S.S. Blair. The function and regulation of *cut* expression on the wing margin of *Drosophila*: Notch, Wingless and a dominant negative role for Delta and Serrate. *Development*, **124**:1485–1495, 1997.

[60] N.A.M. Monk. Restricted–range gradients and travelling fronts in a model of juxtacrine cell relay. *Bull. math. Biol.*, **60**:901–918, 1998.

[61] M.A. Muskavitch. Delta–Notch signalling and *Drosophila* cell fate. *Dev. Biol.*, **166**:415–430, 1994.

[62] M.G. Neubert, M. Kot and M.A. Lewis. Dispersal and pattern formation in a discrete-time predator-prey model. *Theor. Pop. Biol.*, **48**:7–43, 1995.

[63] G. Oster. Lateral inhibition models of developmental processes. *Math. Biosci.*, **90**:265–286, 1988.

[64] H.G. Othmer and L.E. Scrivens. Instability and dynamic pattern in cellular networks. *J. theor. Biol.*, **32**:507–537, 1971.

[65] M.R. Owen and J.A. Sherratt. Mathematical modelling of juxtacrine cell signalling. *Math. Biosci.* **153**:125–150, 1998.

[66] M.R. Owen, J.A. Sherratt and S.R. Myers. How far can a juxtacrine signal travel? *Proc. R. Soc. Lond. B* **266**:579–585, 1999.

[67] V.M. Panin, V. Papayannopoulos, R. Wilson and K.D. Irvine. Fringe modulates Notch-ligand interactions. *Nature*, **387**:908–912, 1997.

[68] C. Perez, I. Albert, K. DeFay, N. Zachariades, L. Gooding and M. Kriegler. A nonsecretable cell-surface mutant of tumour necrosis factor (TNF) kills by cell-to-cell contact. *Cell*, **63**:251–258, 1990.

[69] H.L. Qi, M.D. Rand, X.H. Wu, N. Sestan, W.Y. Wang, P. Rakic, T. Xu and S. Artavanis-Tsakonas. Processing of the Notch ligand Delta by the metalloprotease Kuzbanian. *Science*, **283**:91–94, 1999.

[70] N.J. Savill and P. Hogeweg. Competition and dispersal in predator-prey waves. Submitted, 1997.

[71] G.S. Schultz, M. White, R. Mitchell, G. Brown, J. Lynch, D.R. Twardzik and G.J. Todardo. Epithelial wound healing enhanced by transforming growth Factor-α and vaccinia growth factor. *Science*, **235**:350–352, 1987.

[72] G. Schultz, D.S. Rotatori and W. Clark. EGF and TGFα in wound healing and repair. *J. Cell Biochem.*, **45**:346–352, 1991.

[73] J.A. Sherratt, B.T. Eagan and M.A. Lewis. Oscillations and chaos behind predator-prey invasion: Mathematical artefact or ecological reality? *Phil. Trans. R. Soc. Lond. B*, **352**:21–38, 1997.

[74] J.M. Shulman, N. Perrimon and J.D. Axelrod. Frizzled signalling and the developmental control of cell polarity. *Trends in Genetics*, **14**:452–458, 1998.

[75] P. Simpson. Lateral inhibition and the development of the sensory bristles of the adult peripheral nervous system of *Drosophila*. *Development*, **109**:509–519, 1990.

[76] P. Simpson. The Notch connection. *Nature*, **375**:736–737, 1995.

[77] P. Simpson. Notch signalling in development on equivalence groups and asymmetric developmental potential. *Curr. Opin. Genetics Dev.*, **7**:537–542, 1997.

[78] J.B. Skeath and S.B. Carroll. Regulation of proneural gene expression and cell fate during neuroblast segregation in the *Drosophila* embryo. *Development*, **114**:939–946, 1992.

[79] P.W. Sternberg. Falling off the knife edge. *Curr. Biol.*, **3**:763–765, 1993.

[80] F.E. Tax, J.J. Yeargers and J.H. Thomas. Sequence of *C. elegans* lag-2 reveals a cell-signaling domain shared with Delta and Serrate of *Drosophila*. *Nature*, **368**:150–154, 1994.

[81] U. Tepass and V. Hartenstein. Neurogenic and proneural genes control cell fate specification in the *Drosophila* endoderm. *Development*, **121**:393–405, 1995.

[82] A.M. Turing. The chemical basis of morphogenesis. *Phil. Trans. R. Soc.* **237**:37–72, 1952.

[83] M.J. van de Vijver, R. Kumar and J. Mendelsohn. Ligand-induced activation of A431 cell epidermal growth factor receptors occurs primarily by an autocrine pathway that acts upon receptors on the surface rather than internally. *J. Biol. Chem.*, **266**:7503–7508, 1991.

[84] H. Vässin, K.A. Bremer, E. Knust and J. Campos-Ortega. The neurogenic gene *Delta* of *Drosophila melanogaster* is expressed in neurogenic territories and encodes a putative transmembrane protein with EGF-like repeats. *EMBO J.*, **6**:3433–3440, 1987.

[85] C.M. Waters, K.C. Oberg, G. Carpenter and K.A. Overholser. Rate constants for binding, dissociation, and internalization of EGF: Effect of receptor occupancy and ligand concentration. *Biochemistry*, **29**:3563–3569, 1990.

[86] H.J. Wearing, M.R. Owen and J.A. Sherratt. Mathematical modelling of juxtacrine patterning. *Bull. Math. Biol.* In press.

[87] G. Weinmaster. Notch signaling: direct or what? *Curr. Opin. Genetics Dev.*, **8**:436–442, 1998.

[88] H.A. Wilkinson, K. Fitzgerald and I. Greenwald. Reciprocal changes in expression of the receptor lin-12 and its ligand lag-2 prior to commitment in a *C. elegans* cell fate decision. *Cell*, **79**:1187–1198, 1994.

[89] G.D. Winter. Epidermal regeneration studied in the domestic pig. In: *Epidermal Wound Healing* (H.I. Maibach and D.T. Rovee, eds.) Chicago: Year Book Med. Publ., 1972.

[90] S. Wolfram. *Cellular Automata and Complexity* Reading, MA: Addison-Wesley, 1994.

[91] S.H. Zigmond, S.J. Sullivan, and D.A. Lauffenburger. Kinetic analysis of chemotactic peptide receptor modulation. *J. Cell. Biol.*, **92**:34–43, 1982.

MODELLING *DICTYOSTELIUM DISCOIDEUM* MORPHOGENESIS

BAKHTIER VASIEV* AND CORNELIS J. WEIJER*

Key words. chemotaxis, cell movement, excitable medium, computational fluid dynamics, hydrodynamics.

Abstract. Morphogenesis of the social amoebae *Dictyostelium discoideum* results from the aggregation of individual cells to form a multicellular hemispherical cell mass, the mound. In the mound the cells differentiate into several cell types. These cell types arise initially in random location in the mound, but then sort out from one another to form a slug. In the slug these cell types are arranged in a simple axial pattern. The slug can migrate and under suitable conditions transforms into a fruiting body consisting of a stalk supporting a mass of spores. It is well established that cells aggregate in response to propagating waves of the chemoattractant cAMP. There is increasingly good experimental evidence that the later stages of morphogenesis are also controlled by cAMP wave propagation and chemotaxis. Here we present a hydrodynamic model to describe *Dictyostelium* development from early aggregation up to migrating slug. We consider the population of cells as an excitable medium, which supports propagation of waves of the chemoattractant cAMP. To model the chemotactic cell movement we consider the masses of moving cells as a fluid flow. The morphogenesis of this multicellular organism is basically modelled as shape changes occurring in a drop of liquid with a free surface. At the mound stage this liquid consists of two randomly mixed component fluids corresponding to two cell types. Cell sorting can be effectively modelled as the separation of the component fluids driven by differential chemotaxis. Finally, our model calculations show that migration of the slug can result from chemotactic flows inside the slug.

1. Introduction. Morphogenesis, i.e. the generation of form is central to biology. Form is most often generated during the embryonic development of organisms. In higher organisms development starts from a fertilised egg that goes through a great number of cell divisions to generate more cells. These cells differentiate into many different cell types. Sometimes they differentiate *in-situ*, however in many cases they also form in one place, start to differentiate, and then move to their final destination. Besides cell division, programmed cell death plays an important role in the shaping of the embryo. All these processes have to be precisely co-ordinated in space and time to reproducibly result in a functional adult, with its characteristic shape. These processes and their co-ordination are clearly very complex and in most cases not well understood. We have therefore concentrated on understanding the cellular principles governing morphogenesis of a relatively simple organism, the social amoebae *Dictyostelium discoideum.* This organism forms by the chemotactic aggregation of single cells, which are generated during a unicellular growth phase. In *Dictyostelium*, growth and development occur in separate parts of the life cycle and once development is initiated there is no significant cell death until terminal differentiation

*Department of Anatomy and Physiology, WTB/MSI Complex, University of Dundee, Dundee, DD1 5EH, UK; c.j.weijer@dundee.ac.uk.

of the stalk cells. Furthermore the cells only differentiate into a limited number of cell types. This implies that development is the result of the co-ordinated movement of individual differentiating cells. All these characteristics make *Dictyostelium* a prime object for the study of the principles controlling simple multicellular morphogenesis (Maeda et al., 1997).

Dictyostelium morphogenesis is initiated by chemotactic aggregation of free living single amoebae (Fig. 1) (Loomis, 1982). During the initial phase of aggregation some cells start to produce and secrete cAMP in a periodic fashion. This cAMP diffuses away to excite neighbouring cells. These cells detect the cAMP via specific transmembrane cAMP receptors and upon stimulation start to produce and secrete cAMP themselves and thereby in turn excite their neighbours. This is the so-called cAMP relay response. cAMP binding to the surface receptor also induces an adaptation process which results in a shutting down of cAMP production. This adaptation process ensures the outward propagation of cAMP waves, since cells which have just relayed are refractory to further stimulation. The cells also secrete an extracellular phosphodiesterase which breaks down cAMP and enables the cells to de-adapt and regain sensitivity to further stimulation. These processes result in the repeated outward propagation of cAMP waves from the place of initiation. Since the cells are also chemotactically sensitive to cAMP and move up gradients as long as the concentration is increasing in time, they move in the direction of the signal source and accumulate at the site of wave initiation. After a while the aggregating cells form patterns of bifurcating streams, in which the cells move in a directed fashion towards the aggregation centre. In the aggregates the cells start to differentiate into prestalk and prespore cells. Differentiated cells sort out so that the aggregate (mound) transforms into a polarised cylindrical shaped structure, the slug. During slug formation the precursor cells of the later stalk, the prestalk cells, sort out to the anterior end of the slug and a sub population of prestalk cells forms the tip which guides all further morphogenesis. The slug falls over and migrates away, guided by signals from its environment such as light and temperature gradients. Under the influence of the right environmental signals (light and low humidity) the slug transforms into a fruiting body consisting of stalk and spore cells. The stalk cells are dead and vacuolated while the spores survive and await favourable conditions to germinate and release single amoebae again.

There have been a number of models describing different stages of *Dictyostelium* development. The streaming phenomenon in aggregation fields has puzzled many researchers for over 30 years (Höfer and Maini, 1997; Höfer et al., 1995; Keller and Segel, 1970; Levine and Reynolds, 1991; Mackay, 1978; Nanjundiah, 1973; Novak and Seelig, 1976). Models devoted to this phenomenon have been studied numerically and analytically and theoretical mechanisms responsible for stream formation have been suggested. It was shown that streams can occur due to an instability in cell distribution due to a dependence of the velocity of the chemoattractant

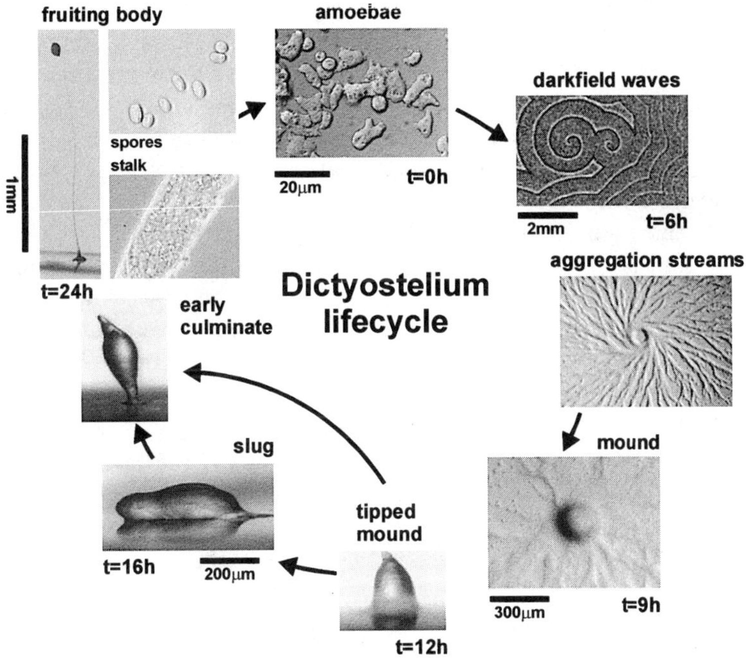

FIG. 1. *The Dictyostelium discoideum life cycle. Shown are in a clockwise order starting at the top, vegetative amoebae, darkfield waves, as observed during aggregation (they reflect the cells in different phases of the movement cycle in response to cAMP waves), aggregation streams, a top view of a mound with incoming streams, a side view of a tipped mound, a side view of a migrating slug and an early culminate and a fruiting body with on its side high magnification images of the stalk cells and spores. This developmental cycle is starvation induced and takes 24 hours at room temperature.*

waves on the density of cells (Vasiev et al., 1994, van Oss et al., 1996;). With the formation of streams and the mound, i.e. when cells get closely packed, mechanical interactions between cells become as important as chemical signalling. Different ways to model these interactions have been proposed in a number of models describing the formation of the mound or migration of the slug (Odell and Bonner, 1986; Bretschneider et al., 1997; Levine et al., 1997; Savill and Hogeweg, 1997). Experimental observations of the mode of cell movement have shown them to be periodic at the aggregation stage when the cells are still single and more continuous at the mound stage (Rietdorf et al., 1996; Siegert et al., 1994; Varnum et al., 1986; Varnum-Finney et al., 1988). These observations suggest that a good way to describe cell movement in the mound is by considering the mound as a drop of liquid, the cells as fluids, and their motion as a flow, which is initiated by chemotactic forces and affected by pressure and viscosity.

We show here that such a hydrodynamic approach can be used to model aggregation, mound formation, cell sorting, slug formation and slug migration (Vasiev et al., 1997; Vasiev and Weijer, 1999). Our model describes the cAMP relay response and resulting cAMP wave propagation as the propagation of a chemical signal in a generic excitable medium and the chemotactic cell movement of the amoebae in response to the signal as a flow of a fluid. We begin from randomly distributed cells on a plane, which in the course of aggregation form bifurcating aggregation streams and then collect into a three dimensional hemispherical mound. We do not take into account the signals responsible for the differentiation of the cells but we assume that the mound consists of two mixed liquids, corresponding to the two cell types, prestalk and prespore cells. Both liquids are chemotactically responsive to cAMP. They respond with rotational movement to a counter rotating scroll wave of cAMP in the mound. We show that sorting of prestalk cells to the top of the mound (while the prespore cells occupy the rest of the mound) takes place when the excitability of prestalk cells and their chemotactic movement is higher than that of prespore cells. In our model there is a natural transformation from the mound into a cylindrical slug. The slug once fallen over migrates. Migration is driven by internal cell flows, which gain traction from the substrate.

2. Model. To model propagation of cAMP waves during *Dictyostelium* development we use the FitzHugh-Nagumo equations, which are widely known as describing a prototype excitable medium:

$$\frac{\partial g}{\partial t} = D\Delta g + \rho\Big(k_g g(g-0.05)(g-1) - k_r r\Big) \tag{1}$$

$$\frac{\partial r}{\partial t} = \frac{(g-r)}{\tau} . \tag{2}$$

Here g is assumed to define the level of extra-cellular cAMP, and r is the fraction of active cAMP receptors (Martiel and Goldbeter, 1987) or activated α subunits of the inhibitory G-proteins (Tang and Othmer, 1995; Tang and Othmer, 1994). D is the diffusion coefficient for cAMP; τ is a time scaling factor for the variables r and g. k_g and k_r define the rate of cAMP production and hydrolysis respectively. Locally, the rate of cAMP production and decay is proportional to the density of cells ρ (Levine and Reynolds, 1991; Vasiev et al., 1994).

Cell movement is described by the Navier-Stokes equation:

$$\begin{aligned}\rho\left[\frac{\partial \mathbf{V}}{\partial t} + (\mathbf{V}\,\mathrm{div}\,)\mathbf{V}\right] \\ = \mathbf{F}_{ch} + \mathbf{F}_{fr} + \eta\Delta\mathbf{V} + \left(\xi + \frac{\eta}{3}\right)\mathbf{grad}\,\mathrm{div}\,\mathbf{V} + \mathbf{F}_{ad} - \mathbf{grad}\,p\end{aligned} \tag{3}$$

This equation defines the acceleration of the cells under the influence of various forces given in its right hand side. $\mathbf{V}$ is the velocity of the cells. $\mathbf{F}_{ch}$ is the chemotactic force, which is active on the rising front of the cAMP waves, $\mathbf{F}_{fr}$ is a friction force responsible for slowing down cell movement,

The third and fourth terms on the right hand side describe cell-cell friction: η and ξ are, respectively, the first and second viscosity coefficients. $\mathbf{F}_{ad}$ takes into account cell-cell and cell-substrate adhesion forces; ρ is the pressure due to the chemotactic accumulation of the cells. Equation (3) is given here in the most common, full notation. While modelling different stages of *Dictyostelium* morphogenesis we will use different modifications (reductions in complexity) of the right hand side.

We assume that chemotactic force is proportional to gradient of cAMP:

$$\mathbf{F}_{ch} = K_{ch}\left(\frac{\partial g}{\partial t}\right)\mathbf{grad}\, g \tag{4}$$

where K_{ch} is equal to zero when $\frac{\partial g}{\partial t} \leq 0$ and to a positive constant when $\frac{\partial g}{\partial t} \geq 0$. This step-wise function allows to distinguish a front of chemoattractant wave where cells are chemotactically active from its back where cells do not exhibit chemotactic response (Futrelle, 1982; Futrelle et al., 1982). The friction force is assumed to be proportional to velocity:

$$\mathbf{F}_{fr} = K_{fr}\mathbf{V} \tag{5}$$

where K_{fr} is a negative constant. In some computations adhesion was needed to keep the aggregate/slug attached to the substrate. It was treated as a force directed towards the substrate. This force should be considered as adhesion and is not a gravitational force. When plates with aggregates or slugs are turned upside down they keep developing/migrating in a manner indistinguishable from normal.

The last term in the right hand side of (3) is a force generated by a pressure field in the mound. This force is responsible for limiting the increase in cell density caused by chemotaxis in case of a compressible liquid. Here the pressure is assumed to be proportional to cell density. In the case of an incompressible liquid, it keeps the density constant. Pressure allows cells to reorient the direction of their motion so that they not only move towards the source of chemotactic signal, resulting in mound formation and its shape changes.

While dealing with a compressible liquid we calculate density of cells using the equation of conservation of mass:

$$\frac{\partial \rho}{\partial t} = D_\rho \Delta\rho - \operatorname{div}(\rho\mathbf{V}) \tag{6}$$

The first term on the right hand side of the equation describes the random motion of the cells, while the second term describes co-ordinated (chemotactic) cell movement.

Cell sorting and slug migration can be best described as flows in a heterogeneous incompressible liquid, consisting of two kinds of fluids. These fluids correspond to prestalk and prespore cells and each fluid is characterised by its volume fraction, α_1 and α_2:

$$\alpha_1 + \alpha_2 = 1 \text{ inside the mound}; \qquad \alpha_1 + \alpha_2 = 0 \text{ outside the mound} \tag{7}$$

To model differences in excitability between prestalk and prespore cells we assume that they differ in their rate of cAMP production:

$$k_g = k_1\alpha_1 + k_2\alpha_2 \tag{8}$$

where k_1 and k_2 define the rate of cAMP production by each cell type.

We model differential chemotactic movement by introducing parameters, K_1 and K_2, which define the chemotactic force developed by prestalk and prespore cells:

$$K_{ch} = K_1\alpha_1 + K_2\alpha_2 \tag{9}$$

Consequently the velocities of prestalk, $\mathbf{V}_1$, and prespore, $\mathbf{V}_2$, cells are different. They are obtained from the momentum balance equation for each component-liquid:

$$\rho\alpha_i\left(\frac{\partial \mathbf{V}_i}{\partial t} + (\mathbf{V}_i \,\mathrm{div}\,)\mathbf{V}_i\right) = \mathbf{F}_i + \eta\alpha_i\Delta\mathbf{V}_i - \alpha_i\mathbf{grad}\,p \tag{10}$$

where the index $i = 1,2$ denotes prestalk or prespore cells, $\mathbf{F}_i$ – corresponds to chemotactic forces $\mathbf{F}_i = \alpha_i K_i\left(\frac{\partial g}{\partial t}\right)\mathbf{grad}\,g$. The effects of viscosity and pressure are proportional to their volume fractions. Volume fractions for prestalk and prespore cells are found using the equation for the conservation of mass:

$$\frac{\partial \alpha_i}{\partial t} = -\,\mathrm{div}\,(\alpha_i\mathbf{V}_i) \qquad \text{where } i = 1,2 \tag{11}$$

We do not include a diffusion term in (11) since random motion at this stage of development is small compared to chemotactic movement. In addition in (6) it was necessary for the stability of the computations but we find that (11) is stable without this term.

In principle the velocity obtained from (3) and the velocities from (10) and volume fractions from (11) should satisfy the following equation:

$$V = \alpha_1 V_1 + \alpha_2 V_2\ . \tag{12}$$

Our calculations showed that this is true during the early stages of the computations but that the difference between the left and right side of the equation increased over the course of the simulations up to 20% for Figure 3 and up to 10% for simulations shown in Figure 4 and 5. The inaccuracy stems from (10), which is a simplified version of the full equation from which we removed all terms involving derivatives of volume fractions, which result in instability of the computations.

All calculations were performed in three-dimensional domains using the finite volume method. Equations (1), (2) were integrated by the Euler explicit method using a forward time centred space method for the diffusion term. Equation (3) was integrated explicitly using forward time centred space method for diffusion term and the upwind method for the convective term (Press et al., 1988). In the case of an incompressible liquid it was integrated by the two-step projection method (Kothe et al., 1991) using a simultaneous over-relaxation scheme (SOR) for the pressure Poisson equation (PPE). Equations (10), (11) were integrated explicitly, using the upwind method for the convection terms and taking values for pressure, p, from solution of equation (3). The location of the free surface was determined by the level $\rho = 0.5$ in the compressible liquid model or detected by tracking massless particles distributed in the volume of the mound (MAC method (Harlow and Welch, 1965)) in the incompressible liquid model.

For the cAMP concentration (1), density (6) and volume fraction (11) fields we have used Neumann "no flux" boundary conditions at the boundary of the medium as well as at the free boundary of the aggregate. For the velocity fields (3), (10) we used both no flux (Neumann) and no slip (zero value) boundary conditions (on the free boundary of the aggregate and free-slip (zero value for the normal component and Neumann condition for the tangential components) conditions on the boundaries of the medium. For pressure in (3) we used zero value boundary conditions on the free boundary of the aggregate.

3. Results.

Aggregation streams and mound formation. We simulate *Dictyostelium* morphogenesis up to the mound stage treating the population of cells as an inviscid compressible liquid (Vasiev et al., 1997). At this stage of development the cells move towards each other and thereby increase in density. Therefore compressibility is essential. Since most of cells are separated from each other and do not interact mechanically, we neglect viscosity in a first approximation. We initiate a spiral cAMP wave in the two- dimensional field of randomly distributed cells (Fig. 2). This wave causes periodic changes in cell movement and results in the formation of aggregation streams. Bifurcating aggregation streams form due to the dependence of wave propagation speed on the cell density (Höfer and Maini, 1997; Levine and Reynolds, 1991; Vasiev et al., 1994). As more cells move towards the centre, a hemispherical mound forms. The pressure p between the cells is responsible for the mound formation. It increases during aggregation and forces the cells up into the third dimension. The aggregation patterns observed in the simulations are remarkably similar to those from real experiments (compare aggregation patterns and mound in Figs. 1 and 2). Experiments have shown that during aggregation there is a decrease in the period and propagation speed of the cAMP waves which results in a decrease in the wavelength of the spiral wave (Gross et al., 1976; Siegert and Weijer, 1989). This behaviour is also observed in the model calculations (wavelength in Fig. 2 decreases from A to D). Including a viscous term in the computations basically does not alter the overall phenomena but results in wider aggregation streams that look less similar to the experimentally observed aggregation patterns.

Different phenotypes of aggregation patterns observed in experiments can be simulated by variation of model parameters. For example, decreasing the excitability of the medium in the computer simulations by varying the rate of cAMP production leads to a large cell free region in the centre of the aggregates. This is very similar to the effect of caffeine seen in experimental conditions, which is known to decrease the excitability of the cells by inhibiting their cAMP production (Brenner and Thoms, 1984; Siegert and Weijer, 1989). To explain this phenomenon we have to take into account that the cell free region in the centre of the aggregate represents the core of spiral wave of chemoattractant. Decrease in the excitability of the medium results in an increase of the core so that the latter can become enormously large. Simulations show that, in addition, excitability effects the behaviour of the mound. A mound made of a very low excitability is not stable and exhibits oscillatory motion (meandering), which is also observed in experiments (Vasiev et al., 1997).

Cell sorting in the mound. Contrary to early aggregation, cells in mounds and slugs constitute a compact body, therefore we treat them as an incompressible liquid in which viscosity plays an important role. To study cell sorting we have performed computations starting with a hemispherical mound (drop of incompressible viscid liquid) consisting of two cell types that are initially randomly distributed. The most realistic cell sorting patterns are obtained when the cell types differ in velocity and excitability, i.e. such that prestalk cells are faster and more excitable compared to prespore cells (Vasiev and Weijer, 1999). An example of cell sorting using these conditions is shown in Fig. 3. A scroll wave of cAMP rotating in a hemispherical mound causes cell movement in the mound. Cells tend to move inward towards the core of the scroll. There is competition for the space in the middle of the mound between cells of different type. Faster cells, which are able to move more effectively win this competition and accumulate in the middle of the mound. In the middle of the mound there is an upward flow and most of the faster cells move further up and finally form a plume-like structure pointing to the top surrounded by slower cells. If the cell types only differ in their velocity, cell sorting stops at this stage. If the cell types differ, in addition, in excitability, the structure formed by highly excitable cells deforms the shape of the scroll wave. The plume-like structure formed by prestalk cells results in an anisotropy in the mound, i.e. the top of the mound becomes more excitable than its bottom. As a result the scroll wave becomes twisted and gets a new downward component. Waves propagate from the top down, which leads to further accumulation of the fast moving cells on top and elongation of the mound upwards causing further upward cell flows in the mound. There is again the competition

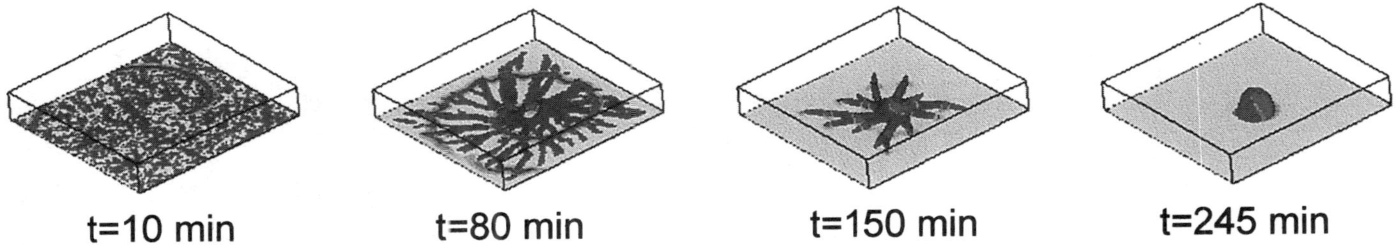

FIG. 2. Development of *Dictyostelium* from single cells to the mound stage. *Successive images of aggregation as calculated by the model. The cell density is shown as an iso-surface* ($\rho = 0.5$) *and the cAMP concentrations are mapped on this surface from low cAMP (blue) to high cAMP (red). The initial density of cells was zero everywhere in 3d-space except for the bottom plane. A random number varying between 0 and 1 represented the cell density in each grid of this plane so that average density in this plane was equal to 0.5. In response to cAMP spiral waves cells move and form aggregation streams* ($t =$ *40–200 min) and then mound* ($t =$ *250 min) which represents a stable solution of the system. Model parameters:* $\tau = 5$; $k_g = 10.5$; $k_r = 3; \eta = 1$; $D = 1$; $\xi = 0.5$; $D_\rho = 0.005$; $K_{ch} = 0.2$; $\mathbf{F}_{fr} = -\mathbf{V}$; $\mathbf{grad}\, p = 0.01\mathbf{grad}\, \rho$; *space and time steps* $h_x = 1.2$; $h_t = 0.01$; *size of the medium:* $100 \times 100 \times 20$ *volume elements. Scaling to real time is made assuming that a time unit corresponds to 15 sec.*

for the space on the top of the mound. Finally all the faster cells collect at the top of the mound and form a tip.

During this process the period of the scroll wave decreases from 48 to 21 time units (or from 4.8 to 2.1 min according to our scaling). The mound's shape also changes over time: the hemispherical mound elongates and gradually transforms into a cylindrical slug.

Slug migration. To study slug migration we have performed computations starting with a cylindrical slug, which consists of two cell types: 20% of prestalk cells located at the anterior end of a cylinder and 80% of prespore cells occupying the more posterior positions. We have checked the modes of slug migration driven by a pacemaker located at leading edge of a slug and a twisted scroll wave rotating inside the slug.

Migration of a slug controlled by pacemaker at its anterior end. Let us assume that the movement of the slug is controlled by waves of a chemoattractant, which are initiated by pacemaker located in the tip of the slug. For simplicity we will first consider a case where there is no difference between cell types or, in other words, a slug consisting of only one cell type. The pacemaker is simulated by the repeated external stimulation of a small area in the tip of the slug. The behaviour of such a slug is shown in Fig. 4. Waves of chemoattractant originate in the tip and propagate along the slug axis backward, while the slug migrates forward in the direction of the pacemaker. The shape of the slug is gradually changing: it remains more- or-less cylindrical, however it gets narrower at the anterior and posterior ends and becomes more similar in shape to experimentally observed slugs.

Simulations where differences in the excitable properties of the cell types are taken into account show that all the results described above do not change. Variations of excitability as described in the model section do not affect the motive forces. However, when motive force generated by prestalk and prespore cells differ from each other, the behaviour of the slug changes dramatically. The tip and the tail of the slug are moving at different speeds. The slug is gradually elongating until the prestalk and prespore zones are completely separated. After this occurs, only the prestalk zone is moving since it is the only part containing a pacemaker. Such a phenotype has also been observed experimentally, when slugs are placed on cAMP containing agar. The prestalk zone keeps moving, while the prespore zone is immobilised (Weijer et al., unpublished observations). This effect is most likely due to the quantitative differences in phosphodiesterase produced by prestalk (more) and prespore cells (less), the higher amounts of adenylate cyclase, the enzyme that produces cAMP, and the lower affinity of the cAMP receptors in prestalk cells (Firtel, 1996; Parent and Devreotes, 1996). This could result in higher amplitude cAMP oscillations in the prestalk zone of the slug compared to the prespore zone and make the cAMP signalling in the prespore zone more sensitive to inhibition by external cAMP.

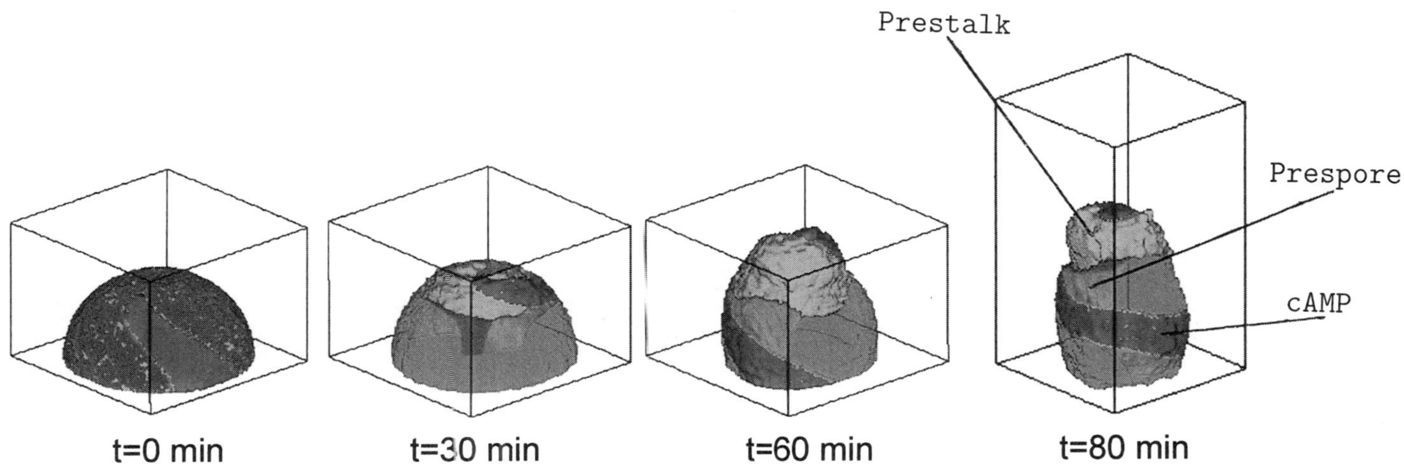

FIG. 3. *Cell sorting in the mound. The mound consists of 20% of prestalk cells (yellow) and 80% prespore cells (blue). The cell types differ in chemotactic velocity ($K_1 = 2$ and $K_2 = 1$ in Equation (9)) and in excitability $k_1 = 6.0$ and $k_2 = 5.4$ in Equation (1)). Initially the mound is a hemisphere in which a cAMP scroll wave (purple) is initiated and rotates clockwise, and both cell types are randomly mixed. Affected by the cAMP waves the cells move and sort, so that the prestalk cells collect at the top of the mound and form a tip. In the course of time hemispherical mound elongates and transforms into a cylindrical standing slug. The model parameters: $\tau = 4$; $k_r = 1.5; K_{ch} = 0.1$; $D = 1$; $\rho = 1$; $\eta = 1$; $\mathbf{F}_{fr} = 0$; (friction is delivered by zero value boundary conditions) space and time steps $h_x = 0.6$; $h_t = 0.06$. Size of the medium is $70 \times 70 \times 50$ volume elements. Scaling to real time is made assuming that a time unit is equal to 6 sec.*

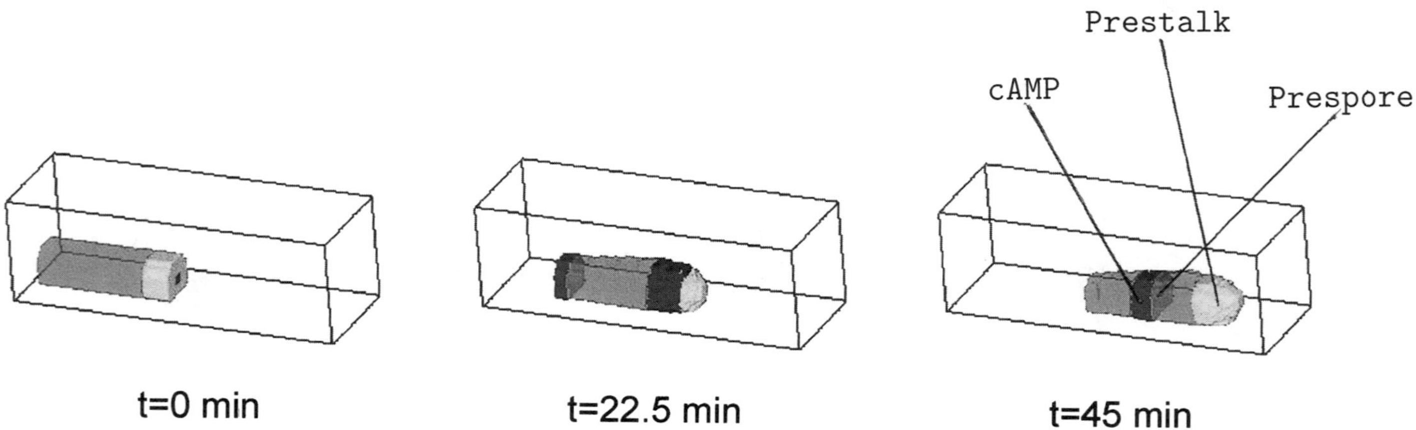

FIG. 4. *Slug organised by a pacemaker located in its tip. The slug is homogeneous, i.e. there are no differences between prestalk (yellow) and prespore (blue) cells:* $k_1 = k_2 = 5.4$ *in (1) and* $K_1 = K_2 = 2$ *in (9). Initially the slug is represented by a cylinder with diameter* $180\mu m$ *(20 volume elements) and length* $540\mu m$. *Waves (represented by a purple isosurface) are initiated by external stimulation of a small group of cells in the slug tip. The medium is* $40 \times 40 \times 150$ *volume elements. Model parameters are the same as in Fig. f3, except* $\mathbf{F}_{fr} = 0.1$ *and Neumann boundary conditions are used for velocity on the free surface of the slug.*

Migration of a slug organised by a scroll wave of cAMP. There are many indications that the cell flows in a mound result from a rotating scroll wave of cAMP. As it was shown above (Fig. 3) such a scroll wave twists during the course of slug formation, so that it has a component of velocity directed from the tip to the back. This twisted scroll can persist in a migrating slug since a slug is axially asymmetric with respect to its excitability: anterior prestalk cells are more excitable than posterior prespore cells. Such a twisted scroll wave can also organise the motion of a slug. Results of simulations with a twisted scroll wave initiated in a slug are shown in Fig. 5. Since the scroll originates in the tip, the slug migrates in the direction pointed at by the tip. The slug's shape changes over time in a way similar to that observed for a slug organised by a point source. The tip of the slug lifts off the substrate, similar to what is often observed in experiments. In these simulations the chemotactic force for prestalk and prespore cells is the same. In further computations we found that a decrease in chemotactic force (50%) for prespore cells resulted in an elongation of the slug but does not result in it breaking. This suggest a mechanism for the regulation of the slug shape which can be vastly different under different experimental conditions and between different strains.

How do cells get traction ? We will now address the following problem: how do cells inside mounds and slugs gain traction to move in a way described in the model section. When a cell has a contact with a substrate there are no difficulties with traction, it is derived from the substrate. However in mounds or in slugs most cells have no direct contact with the substrate. Odell and Bonner many years ago made the appealing assumption that cells gain traction locally from their direct neighbours (Odell and Bonner, 1986). Since the cells were all motile, a cell moving forward pushed back other cells. This would result in no net forward movement. Therefore they introduced a second factor produced by all cells, which modulated their motility. This resulted in cells in the centre moving slower than in the periphery. This assumption led to circulating cell flows in the slug, which are not in agreement with our successive experimental work in which we showed that these flows do not occur (Abe et al., 1994; Siegert and Weijer, 1992). Furthermore a fountain-like motion would require continuous cell sorting to keep the axial distribution of cell types in a slug upright, which is also not in agreement with experimental observations.

One way to avoid these problems is to assume that cells can gain traction also from distant neighbours. For example, to assume that long ranging cell-cell interactions exist. This links up cells over larger distances and introduces solid-like properties in the slug. Via such a mechanism accelerating cells can develop traction from a significantly larger area. As an extreme case we can assume that each cell gains traction from the whole slug, which delivers the reaction forces to the substrate. This assumption would perfectly agree with our formal description of chemotaxis. In addi-

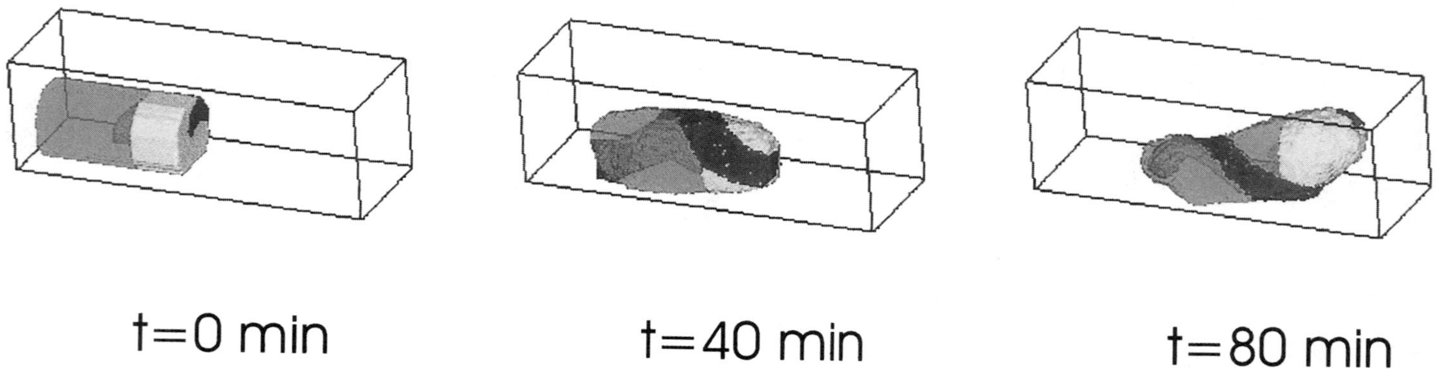

FIG. 5. *Slug organised by a twisted scroll wave of chemoattractant. Model parameters are the same as in Fig. 4 except:* $k_1 = 6.0$ *and* $k_2 = 5.4$ *in (1) the same values used in Fig. 3. The diameter of the slug is 270μm. A twisted scroll wave is initiated in the slug, it twists because there is a difference in excitability between the prestalk and prespore cells. The adhesive force* $\mathbf{F}_{ad}$ *is directed towards the substrate and equal to 0.01.*

tion, it would mean that no matter whether a cell is isolated or in contact with other cells, it exerts the same force in response to the same chemotactic signal. In reality we think that cells are made up of a "stiff" internal actin cytoskeleton which links to the substrate and neighbouring cells via specific adhesion molecules. A cell moves by locally extending its internal actin cytoskeleton in the direction of a chemotactic signal. This extended part now becomes stiff and new contacts are made with neighbouring cells. Extension obviously has to be co-ordinated with actin depolymerisation, release of adhesive contacts and retraction at the back end of the cell. These processes are co-ordinated over many cells by the propagating waves of cAMP and therefore result in waves of co-ordinated motion.

4. Conclusion. Slime mould morphogenesis results from propagation of cAMP waves, which control the chemotactic movement of individual amoebae. The main assumption made in this paper is that we can consider cell movement as the flow of a liquid. We have shown that this approach can be used successfully to describe formation of aggregation pattern and mound, cell sorting, transformation of mound into the slug, and migration of slug. A hydrodynamic description of the cell-cell interactions by pressure and viscosity terms seems to work as a good approximation. Unfortunately a quantitative comparison of our model parameters with experimental values is not yet possible since it has not yet been possible to measure chemotactic and adhesive forces produced by the cells quantitatively. Likewise viscosity of slug tissue and pressure inside the slug are unknown. Performing quantitative measurements of these parameters clearly has to be a prime experimental research objective.

REFERENCES

ABE, T., EARLY, A., SIEGERT, F., WEIJER, C., AND WILLIAMS, J. (1994). *Patterns of cell movement within the Dictyostelium slug revealed by cell type-specific, surface labeling of living cells.* Cell **77**, 687–699.

BRENNER, M. AND THOMS, S.D. (1984). *Caffeine blocks activation of cyclic AMP synthesis in Dictyostelium discoideum.* Dev. Biol. **101**, 136–146.

BRETSCHNEIDER, T., VASIEV, B., AND WEIJER, C. J. (1997). *A model for cell movement during Dictyostelium mound formation. Journal of Theoretical Biology* **189**, pp. 41.

FIRTEL, R.A. (1996). *Interacting Signaling Pathways Controlling Multicellular Development in Dictyostelium.* Current Opinion in Genetics & Development **6**, 545–554.

FUTRELLE, R.P. (1982). *Dictyostelium chemotactic response to spatial and temporal gradients. Theories of the limits of chemotactic sensitivity and of pseudochemotaxis.* J. Cell. Biochem. **18**, 197–212.

FUTRELLE, R.P., TRAUT, J., AND MCKEE, W.G. (1982). *Cell behavior in Dictyostelium discoideum preaggregation response to localized cAMP pulses.* J. Cell Biol. **92**, 807–821.

HARLOW, F.H. AND WELCH., J.E. (1965). *Numerical calculation of time dependent viscous incompressible flow of fluid with a free surface.* Phys. Fluids **8**, 2182–2189.

Höfer, T. and Maini, P.K. (1997). *Streaming instability of slime mold amoebae: An analytical model.* Physical Review E **56,** 2074–2080.

Höfer, T., Sherratt, J.A., and Maini, P.K. (1995). *Dictyostelium-Discoideum - Cellular Self-Organization in an Excitable Biological Medium.* Proceedings of the Royal Society of London Series B-Biological Sciences **259,** 249–257.

Keller, E.F. and Segel, L.A. (1970). *Initiation of slime mold aggregation viewed as an instability.* J. Theor. Biol. **26**, 399–415.

Kothe, D.B., R.C. Mjolsness, and Torrey, M.D. (1991). *RIPPLE a Computer program for incompressible flows with free surfaces.* Los Alamos Natl.Lab.

Levine, H. and Reynolds, W. (1991). *Streaming instability of aggregating slime mold amoebae.* Phys. rev. lett. **66,** 2400–2403.

Levine, H., Tsimring, L., and Kessler, D. (1997). *Computational modeling of mound development in Dictyostelium.* Physica D **106**, 375–388.

Loomis, W.F. (1982). *The development of Dictyostelium discoideum.* (New York: Ac. Press).

Mackay, S.A. (1978). *Computer simulation of aggregation in Dictyostelium discoideum.* J. Cell Sci. **33**, 1–16.

Maeda, Y., Inouye, K., and Takeuchi, I. (1997). *Dictyostelium; A Model System for Cell and Developmental Biology,* 1st Edition (Tokyo: Universal Academy Press).

Martiel, J.-L. and Goldbeter, A. (1987). *A model based on receptor desensitization for cyclic AMP signaling in Dictyostelium cells.* Biophys. J. **52**, 807–828.

Nanjundiah, V. (1973). *Chemotaxis, signal relaying and aggregation morphology.* J. Theor. Biol. **42**, 63–105.

Novak, B. and Seelig, F.F. (1976). *Phase-shift model for the aggregation of amoebae: A computer study.* J. Theor. Biol. **56**, 301–327.

Odell, G.M. and Bonner, J.T. (1986). *How the Dictyostelium discoideum grex crawls.* Phil. Trans. R. Soc. Lond. B **312**, 487–525.

Parent, C.A. and Devreotes, P.N. (1996). *Molecular genetics of signal transduction in Dictyostelium.* Annu. Rev. Biochem. **65**, 411–440.

Press, W.H., Flannely B.P., Teukovsky S.A., and Wetterling W.T. (1988). *Numerical Recipes in C* (Cambridge: Cambridge University Press).

Rietdorf, J., Siegert, F., and Weijer, C.J. (1996). *Analysis of Optical-Density Wave-Propagation and Cell-Movement During Mound Formation in Dictyostelium-Discoideum.* Developmental Biology **177**, 427–438.

Savill, N.J. and Hogeweg, P. (1997). *Modelling morphogenesis: From single cells to crawling slugs.* Journal of Theoretical Biology **184**, 229–235.

Siegert, F. and Weijer, C. (1989). *Digital image processing of optical density wave propagation in Dictyostelium discoideum and analysis of the effects of caffeine and ammonia.* J. Cell Sci. **93**, 325–335.

Siegert, F. and Weijer, C.J. (1992). *Three-dimensional scroll waves organize Dictyostelium slugs.* Proc. Natl. Acad. Sci. USA **89**, 6433–6437.

Siegert, F., Weijer, C.J., Nomura, A., and Miike, H. (1994). *A gradient method for the quantitative analysis of cell movement and tissue flow and its application to the analysis of multicellular Dictyostelium development.* J. Cell Sci. **107**, 97–104.

Tang, Y.H. and Othmer, H.G. (1995). *Excitation, oscillations and wave propagation in a G- protein-based model of signal transduction in Dictyostelium discoideum.* Phil. Trans. R. Soc. Lond. B **349**, 179–195.

Tang, Y.H. and Othmer, H.G. (1994). *A G protein-based model of adaptation in Dictyostelium discoideum.* Math. Biosci. **120**, 25–76.

van Oss, C., Panfilov, A.V., Hogeweg, P., Siegert, F., and Weijer, C.J. (1996). *Spatial pattern formation during aggregation of the slime mould Dictyostelium discoideum.* J. Theor. Biol. **181**, 203–13.

Varnum, B., Edwards, K.B., and Soll, D.R. (1986). *The developmental regulation of single-cell motility in Dictyostelium discoideum.* Dev. Biol. **113**, 218–227.

Varnum-Finney, B., Schroeder, N.A., and Soll, D.R. (1988). *Adaptation in the motility response to cAMP in Dictyostelium discoideum.* Cell Motil. Cytoskel. **9**, 9–16.

VASIEV, B., SIEGERT, F., AND WEIJER, C.J. (1997). *A hydrodynamic model for Dictyostelium discoideum mound formation.* Journal of Theoretical Biology **184**, pp. 441.

VASIEV, B. AND WEIJER, C. (1999). *Modeling Chemotactic Cell Sorting During Dictyostelium Discoideum Mound Formation.* Biophysical J. **76**, 595–605.

VASIEV, B.N., HOGEWEG, P., AND PANFILOV, A.V. (1994). *Simulation of Dictyostelium-Discoideum Aggregation Via Reaction-Diffusion Model.* Physical Review Letters **73**, 3173–3176.

MODELING BRANCHING AND CHIRAL COLONIAL PATTERNING OF LUBRICATING BACTERIA

ESHEL BEN-JACOB*, INON COHEN*, IDO GOLDING*, AND YONATHAN KOZLOVSKY*

Abstract. In nature, microorganisms must often cope with hostile environmental conditions. To do so they have developed sophisticated cooperative behavior and intricate communication capabilities, such as: direct cell-cell physical interactions via extra-membrane polymers, collective production of extracellular "wetting" fluid for movement on hard surfaces, long range chemical signaling such as quorum sensing and chemotactic (bias of movement according to gradient of chemical agent) signaling, collective activation and deactivation of genes and even exchange of genetic material. Utilizing these capabilities, the colonies develop complex spatio-temporal patterns in response to adverse growth conditions. We present a wealth of branching and chiral patterns formed during colonial development of lubricating, swimming bacteria (bacteria that produce a wetting layer of fluid so they can swim in it). Invoking ideas from pattern formation in non-living systems and using "generic" modeling we are able to reveal novel survival strategies which account for the salient features of the evolved patterns. Using the models, we demonstrate how communication leads to self-organization via cooperative behavior of the cells. In this regard, pattern formation in microorganisms can be viewed as the result of the exchange of information between the micro-level (the individual cells) and the macro-level (the colony). We mainly review known results, but include a new model of chiral growth, which enables us to study the effect of chemotactic signaling on the chiral growth. We also introduce a measure for weak chirality and use this measure to compare the results of model simulations with experimental observations.

1. Introduction. Among non-equilibrium dynamical systems, living organisms are the most challenging ones that scientists can study. A biological system constantly exchanges material, energy and information with the environment as it regulates its growth and survival. The energy and chemical balances at the cellular level involve an intricate interplay between the microscopic dynamics and the macroscopic environment, through which life at the intermediate mesoscopic scale is maintained [3]. The development of a multicellular structure requires non-equilibrium dynamics, as microscopic imbalances are translated into the macroscopic gradients that control collective action and growth [77].

Much effort is devoted to the search for basic principles of organization (growth, communication, regulation and control) on the cellular and multicellular levels [12, 32, 33, 40, 50, 52, 75, 82, 85, 86, 100]. Our approach is to use the successful conceptual framework for pattern formation in non-living systems as a tool to unravel their significantly more complex biological counterparts. Of critical importance is the choice of starting point, i.e. the choice of which phenomena to study: it has to be simple enough to allow progress, but also well motivated by the significance of the results. Cooperative microbial behavior is well suited for these require-

*School of Physics and Astronomy, Raymond and Beverly Sackler Faculty of Exact Sciences, Tel Aviv University, Tel Aviv 69978, ISRAEL.

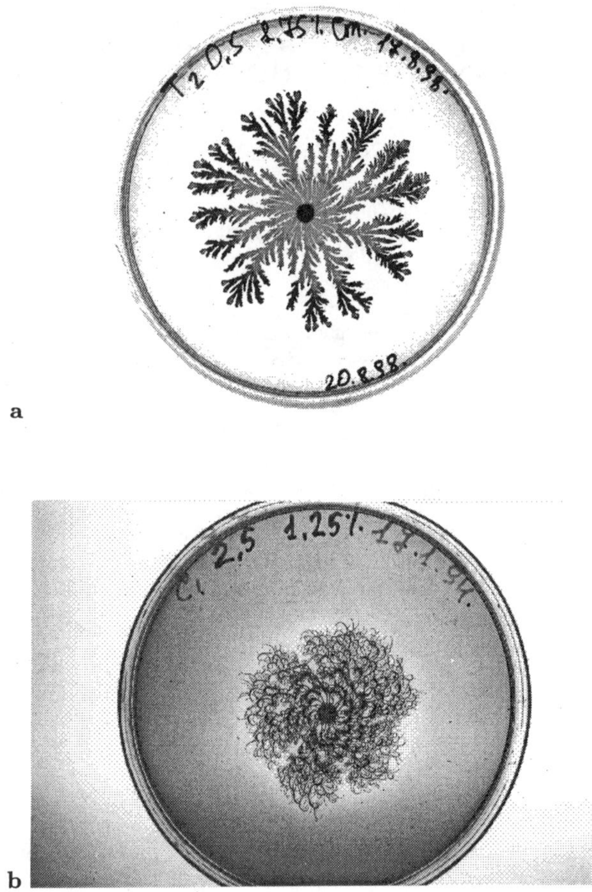

FIG. 1. *(a) Typical example of branching growth of the Paenibacillus dendritiformis var. dendron (referred to as $\mathcal{T}$ morphotype [11]) for 0.5g/l peptone and 1.75 % (17.5 g/l) agar concentration. (b) Chiral growth of the Paenibacillus dendritiformis var. chiralis (referred to as $\mathcal{C}$ morphotype [11]) for 2.5g/l peptone and 1.25 % agar concentration.*

ments, as we explain below. We focus on two examples: the branching growth in bacterial colonies of *Paenibacillus dendritiformis* var. *dendron*, and the chiral growth of *P. dendritiformis* var. *chiralis*. Examples of these patterns are shown in Fig. 1. These bacteria, which can propagate on a solid agar surface by swimming in a layer of a self-produced liquid, were originally isolated from colonies of *Bacillus subtilis* 168, which is non-motile on a solid agar surface [16, 21]. These isolates were recently identified as *P. dendritiformis* [97], whose closest known relative is *P. thiaminolyticus* (previously "Bacillus thiaminolyticus" [74]).

Traditionally, bacterial colonies are grown on substrates with a high nutrient level and intermediate agar concentration [44]. Such "friendly" conditions yield colonies of simple compact patterns, which fit well the traditional view of colonies of most bacteria as a collection of independent unicellular organisms (non-interacting "particles"). However, bacterial colonies in nature must regularly cope with hostile environmental conditions [79, 85]. When hostile conditions are created in a petri dish by using a very low level of nutrients, a hard surface (high concentration of agar), or both, very complex patterns are observed. (It should be noted that hard surfaces do not pose a problem for some bacterial species, but *P. dendritiformis* cannot propagate over an agar of concentration higher than 4%).

Drawing on the analogy with diffusive patterning in non-living systems [6, 7, 15, 49, 57], we can state that complex patterns are expected. The cellular reproduction rate that determines the growth rate of the colony is limited by the level of nutrients available for the cells. The latter is limited by the diffusion of nutrients towards the colony (for low nutrient substrate). Hence colony growth under certain conditions should be similar to diffusion limited growth in non-living systems as mentioned above [6, 7, 15]. The study of diffusive patterning in non-living systems teaches us that the diffusion field drives the system towards decorated (on many length scales) irregular fractal shapes [36, 59, 60, 83, 99]. Indeed, bacterial colonies can develop patterns reminiscent of those observed during growth in non-living systems [16–18, 21, 37, 38, 61, 63–65]. But, this is certainly not the end of the story. In fact, the colonies exhibit far richer behavior. This is ultimately a reflection of the additional levels of complexity involved [10, 13, 14, 16, 19–21, 28]. The building blocks of the colonies are themselves living systems, each having its own autonomous self-interest and internal degrees of freedom. Yet, efficient adaptation of the colony to adverse growth conditions requires self-organization on all levels which can only be achieved via cooperative behavior of these individual cells. Thus, pattern formation at the colony level may be viewed as the outcome of a dynamical interplay between the micro-level (the individual cell) and the macro-level (the colony) [16, 17, 21]. For this interplay to work, the effects of changes at the micro-level must make themselves felt at the macro-level. This is why the notion of a singular perturbation, discovered to be a key for understanding pattern selection in non-living systems, will be of even more importance here.

This manuscript is mainly a review of our modeling studies. Yet we include some important new results:

1. Explanations and modeling of chemotaxis during chiral growth.
2. Modeling and characterization of weak chirality (global weak twist) during branching growth.

For completeness, Section 2 includes a brief description of the necessary biological background needed to justify our models of the growth.

How should one approach the modeling of the complex bacterial patterning? With present computational power it is natural to use computer models as a main tool in the study of complex systems. However, one must be careful not to be trapped in the "reminiscence syndrome", described by J.D. Cowan [46], as the tendency to devise a set of rules which will mimic some aspect of the observed phenomena and then, to quote J.D. Cowan "They say: 'Look, isn't this reminiscent of a biological or physical phenomenon!' They jump in right away as if it's a decent model for the phenomenon, and usually of course it's just got some accidental features that make it look like something." Yet the reminiscence modeling approach has some indirect value. True, doing so does not reveal (directly) the biological functions and behavior. However, it does reflect understanding of geometrical and temporal features of the patterns, which indirectly might help in revealing the underlying biological principles. Another extreme is the "realistic modeling" approach, where one constructs a model that includes in detail all the known biological facts about the system. Such an approach sets a trajectory of ever including more and more details (vs. generalized features). The model keeps evolving to include so many details that it loses any predictive power.

Here we try to promote another approach – the "generic modeling" one [5, 20, 50, 53]. We seek to elicit, from the experimental observations and the biological knowledge, the generic features and basic principles needed to explain the biological behavior and to include these features in the model. We will demonstrate that such modeling, with close comparison to experimental observations, can be used as a research tool to reveal new understanding of the biological systems.

Generic modeling is not about using a sophisticated mathematical description to dress pre-existing understanding of complex biological behavior. Rather, it means a cooperative approach, using existing biological knowledge together with mathematical tools and a synergetic point of view for complex systems to reach a new understanding (which is reflected in the constructed model) of the observed complex phenomena.

The generic models can be grouped into two main categories:
1. Discrete models such as the Communicating Walkers models of Ben-Jacob *et al.* [10, 14, 20] and the Bions model of Kessler and Levine [50, 51]. In this approach, the microorganisms (bacteria in the first model and amoebae in second) are represented by discrete, random walking entities (walkers and bions, respectively) which can consume nutrients, reproduce, perform random or biased movement, and produce or respond to chemicals. The time evolution of the chemicals is described by reaction-diffusion equations.
2. Continuous or reaction-diffusion models [58, 76]. In these models the microorganisms are represented via their 2D density, and a reaction-diffusion equation of this density describes their time evolution. This equation is coupled to the other reaction-diffusion equations for the chemical fields. In the context of branching growth, this idea has been pursued recently by

Mimura and Matsushita *et al.* [62, 72], Kawasaki *et al.* [48], Kitsunezaki [54] and Kozlovsky *et al.* [55]. A summary and critique of this approach can be found in [80] and [39].

In Section 3 we present the continuous modeling of the branching growth. In Section 4 the chiral growth is modeled using the "atomistic" Communicating Spinors model, which enables us to model chemotaxis response. It is the first time that the chemotaxis effect on chiral growth has been studied. The actual study of chemotaxis, both in the chiral growth and the branching growth is done in Section 5.

Section 6 is devoted to the studies of weak chirality. The phenomenon is modeled using both continuous and discrete models. We introduce a measure for weak chirality which enables a more crucial comparison between the models' results and the observed patterns. Good agreement was found.

Conclusions are presented in Section 7. We explain that the weak chirality phenomenon is general, and show examples of weak chirality during growth of the chiral morphotype and the vortex morphotype. In the latter it results from a different mechanism, and indeed the twist is not linear with the radius of growth.

2. Observations and biological background. Following the experimental observations which are explained in this manuscript, we will describe the most relevant information for the understanding and modeling of the observed colonial patterning. We base relevancy on our previous experience and we concentrate on bacterial movement.

We work with the bacteria *Paenibacillus dendritiformis*. These bacteria were worked with as early as 1992 [16], but they were only recently identified [97]. Cells of *P. dendritiformis* are motile, sporulating, Gram-negative rods. Their closest known relative is *P. thiaminolyticus* (formally "Bacillus thiaminolyticus" [74]).

2.1. Experimental observations: branching growth of bacterial colonies.

2.1.1. Macroscopic observations. Some additional examples of the patterns exhibited by colonies of the $\mathcal{T}$ morphotype are shown in Figures 2, 3 and 4. For intermediate agar concentrations (about 1.5% – 1.5g in 100ml), at very high peptone levels (above 10g/l) the patterns are compact (Fig. 3(a)). At somewhat lower but still high peptone levels (about 5-10g/l) the patterns exhibit quite pronounced radial symmetry and may be characterized as dense fingers (Fig. 3(b)), each finger being much wider than the distance between fingers. For intermediate peptone levels, branching patterns with lower fractal dimension (reminiscent of electro-chemical deposition) are observed (Fig. 3(c)). The patterns are "bushy", with branch width smaller than the distance between branches. As the peptone level is lowered, the patterns become more ramified and fractal–like. Surprisingly,

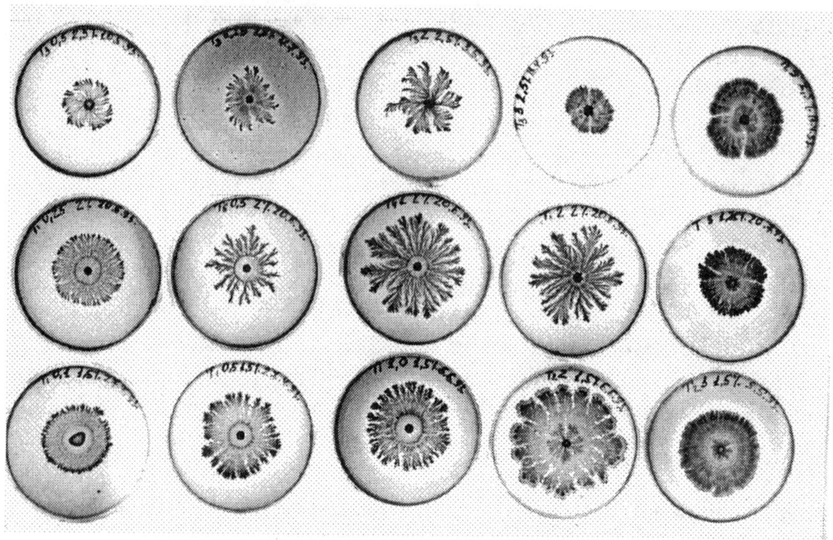

FIG. 2. *Patterns exhibited by the $\mathcal{T}$ morphotype as function of peptone level (increasing from left to right) and agar concentration (1.5% bottom row, 2% middle row, 2.5% top row).*

at even lower peptone levels (below $0.25g/l$ for 2% agar concentration) the colonies revert to organized structures: fine branches forming a well defined global envelope. We characterize these patterns as fine radial branches (Fig. 3(d)). For extremely low peptone levels (below $0.1g/l$), the colonies lose the fine radial structure and again exhibit fractal patterns (Fig. 4). For high agar concentration the branches are very thin (Fig. 4(b)).

At high agar concentration and very high peptone levels the colonies display a structure of concentric rings (Fig. 5). At high agar concentrations the branches also exhibit a global twist with the same handedness (weak chirality), as shown in Fig. 6. Similar observations during growth of other bacterial strains have been reported by Matsuyama *et al.* [64, 65]. We refer to such growth patterns as having weak chirality, as opposed to the strong chirality exhibited by the $\mathcal{C}$ morphotype.

A closer look at an individual branch (Fig. 7) reveals the phenomenon of density variations within the branches. These 3-dimensional structures arise from accumulation of cells in layers. The aggregates can form spots and ridges which are either scattered randomly, ordered in rows, or organized in a leaf vein-like structure. The aggregates are not frozen; the cells in them are motile and the aggregates are dynamically maintained.

2.1.2. Microscopic observations. Under the microscope, cells are seen to perform a random-walk-like movement in a fluid. This fluid is, we assume, excreted by the cells and/or drawn by the cells from the agar

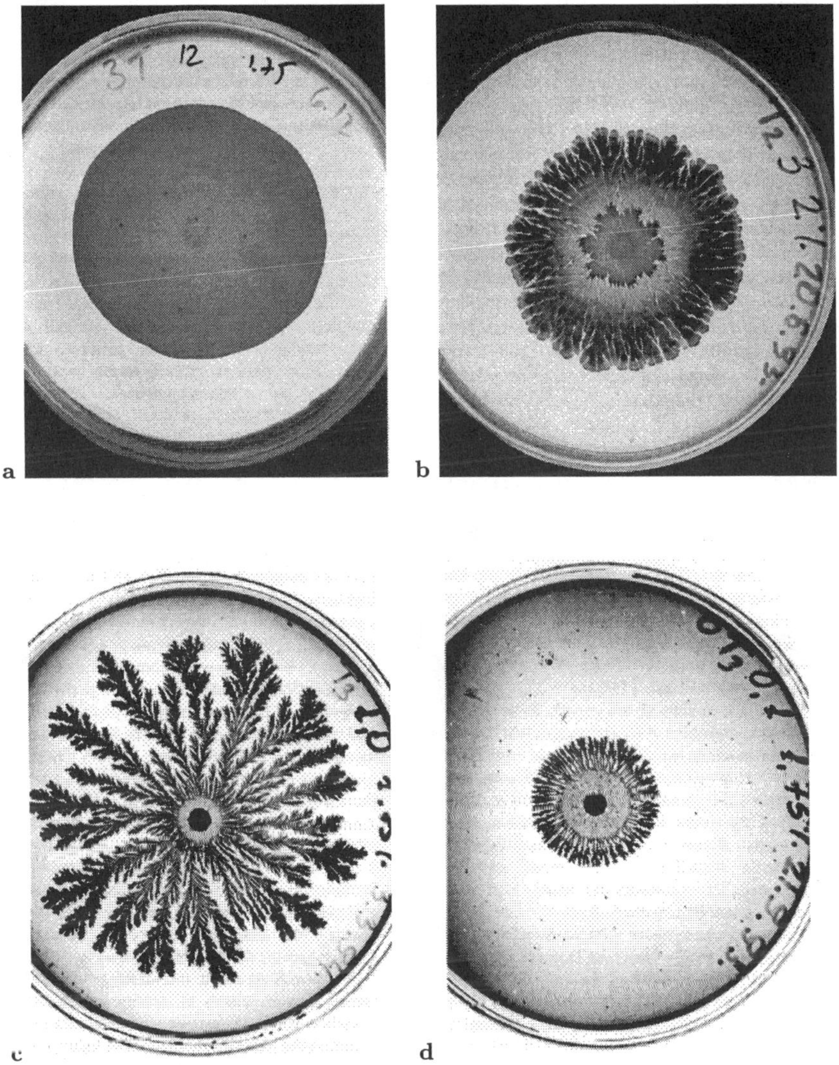

FIG. 3. *Examples of typical patterns of $\mathcal{T}$ morphotype for intermediate agar concentration. (a) Compact growth for 12g/l peptone level and 1.75% agar concentration. (b) Dense fingers for* 3g/l *peptone and 2% agar. (c) Branching fractal pattern for* 1g/l *peptone and 1.75% agar. (d) A pattern of fine radial branches for* 0.1g/l *peptone and 1.75% agar.*

[19, 20]. The cellular movement is confined to this fluid; isolated cells spotted on the agar surface do not move. The boundary of the fluid thus defines a local boundary for the branch (Fig. 8). Whenever the cells are active, the boundary propagates slowly as a result of the cellular movement pushing the envelope forward and production of additional wetting fluid.

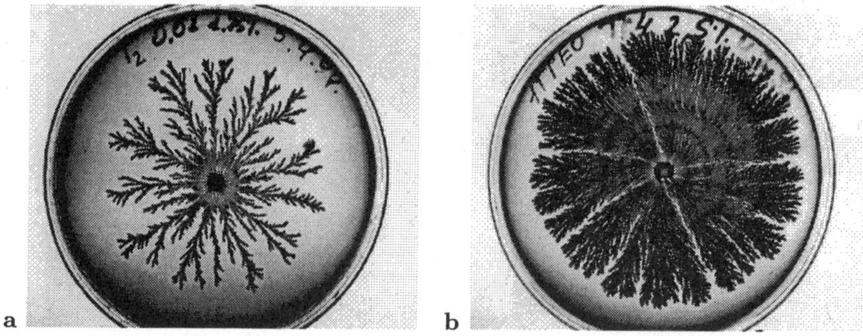

FIG. 4. *Colonial patterns of $\mathcal{T}$ morphotype . (a) Fractal pattern for* 0.01*g/l peptone level and 1.75% agar concentration. (b) Dense branching pattern for* 4*g/l peptone and 2.5% agar. Note that the branches are much thinner than those in Fig. 3(b), i.e. the branches are thinner for higher agar concentrations.*

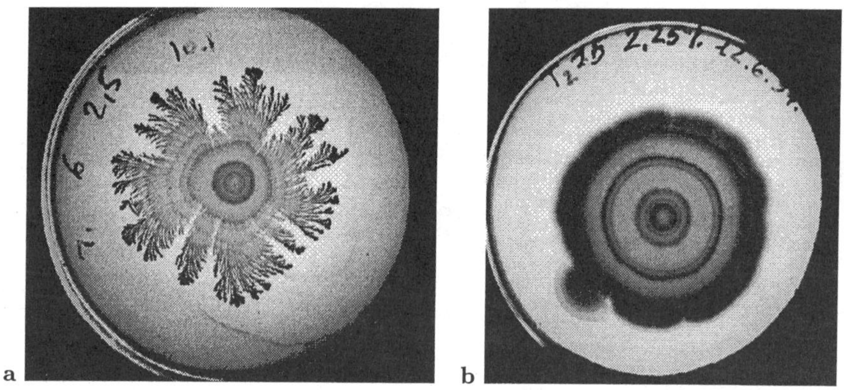

FIG. 5. *(a) Pattern of concentric rings superimposed on a branched colony for 2.5g/l peptone level and 2.5% agar concentration. (b) Concentric rings in a compact growth for* 15*g/l peptone level and 2.25% agar concentration.*

Electron microscope observations reveal that these bacteria have flagella for swimming.

The observations reveal also that the cells are active at the outer parts of the colony, while closer to the center the cells are stationary and some of them sporulate (form spores) (Fig. 9). It is known that certain bacteria respond to adverse growth conditions by entering a spore stage until more favorable growth conditions return. Such spores are metabolically inert and exhibit a marked resistance to the lethal effects of heat, drying, freezing, deleterious chemicals, and radiation.

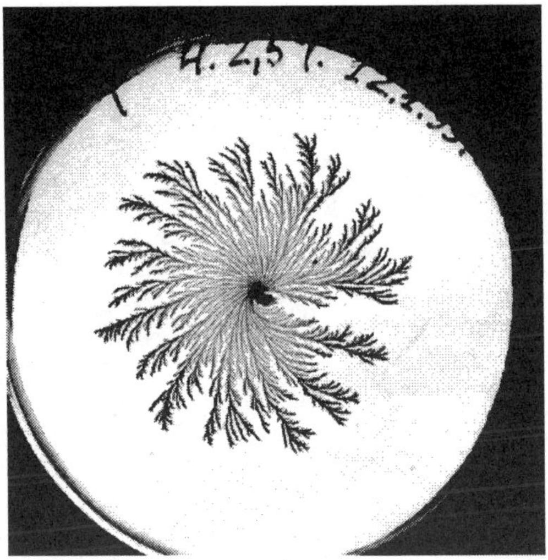

FIG. 6. *Weak chirality exhibited by the* $\mathcal{T}$ *morphotype during growth on 4g/l peptone and 2.5% agar concentration.*

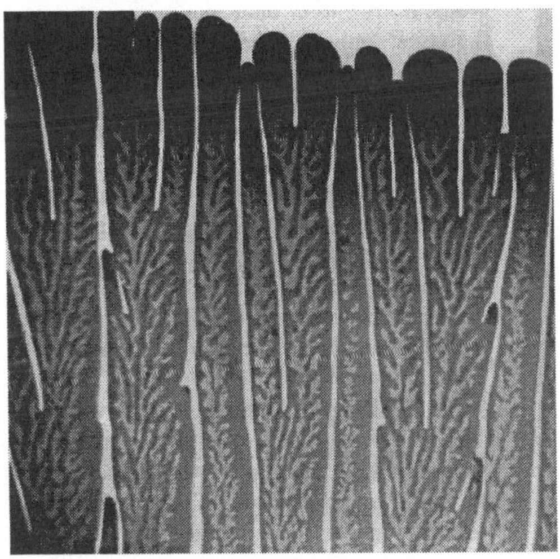

FIG. 7. *Density variations within branches of a colony of* $\mathcal{T}$ *morphotype. Optical microscope observations* ×50.

At very low agar concentrations (below 0.5%) the bacteria swim inside the agar and not on its surface. Between 0.5% and 1% agar concentration some of the bacteria move on the surface and some inside the agar.

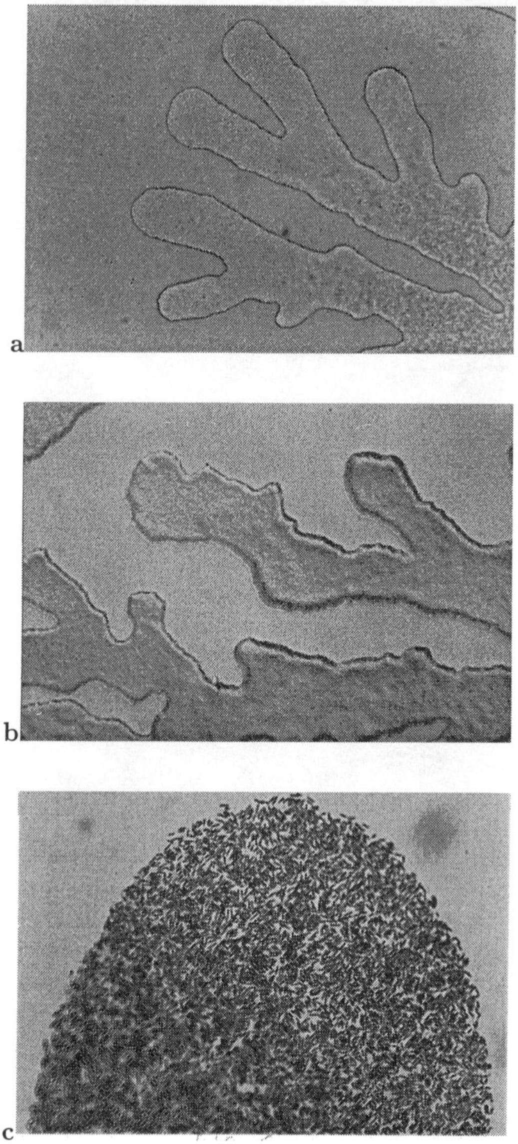

FIG. 8. *Closer look on branches of a colony. a) ×50 magnification shows the sharp boundaries of the branches. The width of the boundary is in the order of micron. b) Nomarsky (polarized light) microscopy shows the height of the branches and their envelope. What is actually seen is the layer of lubrication fluid, not the bacteria. If the colony is left to dry, the same observation technique reveals a layer of bacteria – thinner layer with more rugged surface. c) ×500 magnification shows the bacteria inside a branch. Each bar is a single bacterium. There are no bacteria outside the branch.*

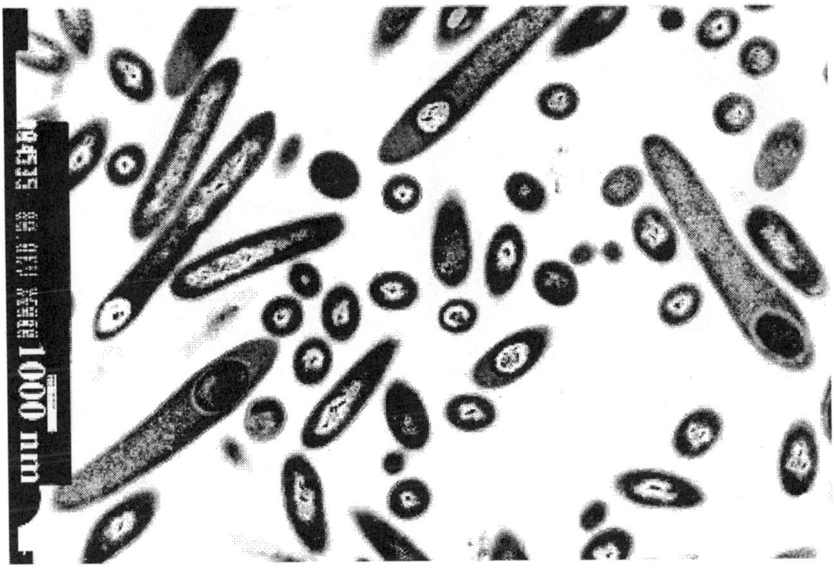

FIG. 9. *Electron microscope observation of $\mathcal{T}$ bacteria. Round or oval shapes with a bright center are spores. Elongated shapes are living cells. The cells engulfing oval shapes are pre-spores.*

2.2. Chiral patterns. Chiral asymmetry (first discovered by Louis Pasteur) exists on a whole range of scales, from subatomic particles through human beings to galaxies, and seems to have played an important role in the evolution of living systems [4, 42]. Bacteria display various chiral properties. Mendelson *et al.* [67, 68, 70, 71] showed that long cells of *B. subtilis* can grow in helices, in which the cells form long strings that twist around each other. They have shown also that the chiral characteristics affect the structure of the colony. Ben-Jacob *et al.* [14, 16, 21] have found yet another chiral property — the strong chirality exhibited by the $\mathcal{C}$ morphotype. Here, the flagella handedness acts as a microscopic perturbation which is amplified by the diffusive instability, leading to the observed macroscopic chirality. This appears to be analogous to the manner in which crystalline anisotropy leads to the observed symmetry of snowflakes [6], as discussed below.

2.2.1. A closer look at the patterns. $\mathcal{C}$ morphotype exhibits a wealth of different patterns according to the growth conditions (Fig. 10). As for $\mathcal{T}$ morphotype, the patterns are generally compact at high peptone levels and become ramified (fractal) at low peptone levels. At very high peptone levels and high agar concentration, $\mathcal{C}$ morphotype conceals its chiral nature and exhibits branching growth similar to that of $\mathcal{T}$ morphotype.

Below 0.5% agar concentration the $\mathcal{C}$ morphotype exhibits compact growth with density variations. These patterns are almost indistinguishable

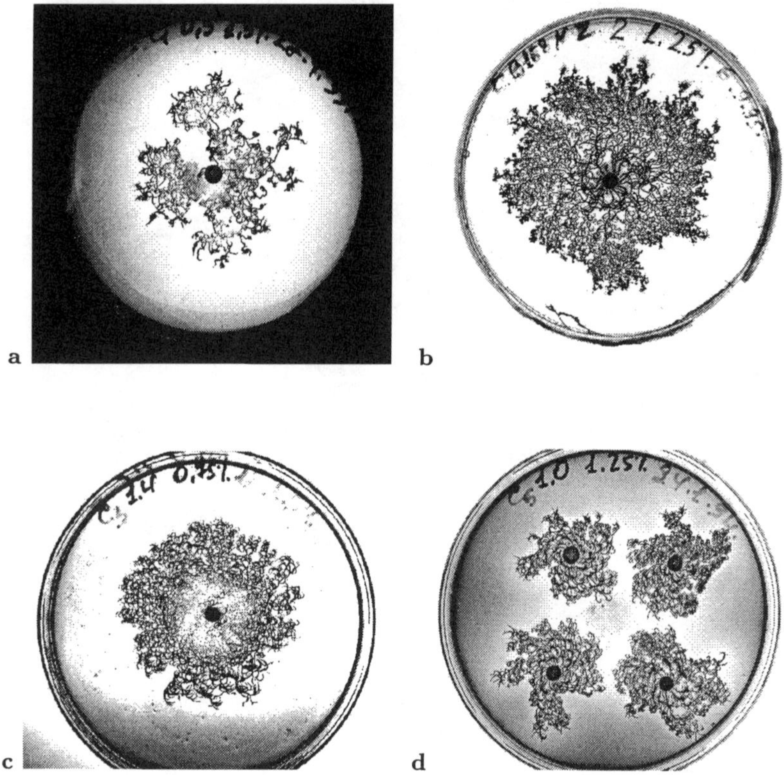

FIG. 10. *Patterns exhibited by the $\mathcal{C}$ morphotype for different growth conditions. a) Thin disordered twisted branches at 0.5g/l peptone level and 1.5% agar concentration. b) Thin branches, all twisted with the same handedness. at 2g/l peptone level and 1.25% agar concentration. c) Pattern similar to (b) but on softer agar: 1.4 g/l peptone level and 0.75% agar concentration. d) Four inocula on the same plate, conditions of 1g/l peptone level and 1.25% agar concentration.*

from those developed by the $\mathcal{T}$ morphotype. In the range of 0.4%–0.6% agar concentration the $\mathcal{C}$ morphotype exhibits its most complex patterns (Fig. 11). Surprisingly, these patterns are composed of chiral branches of both left and right handedness. Microscopic observations reveal that part of the growth is on top of the agar surface while in other parts the growth is in the agar. Our model of the chiral growth explains that indeed growth on top of the surface and in the agar should lead to opposite handedness.

Optical microscope observations indicate that during growth of strong chirality the cells move within a well defined envelope. The cells are long relative to those of $\mathcal{T}$ morphotype, and the movement appears correlated in orientation (Fig. 12). Each branch tip maintains its shape, and at the same time the tips keep twisting with specific handedness while propagating.

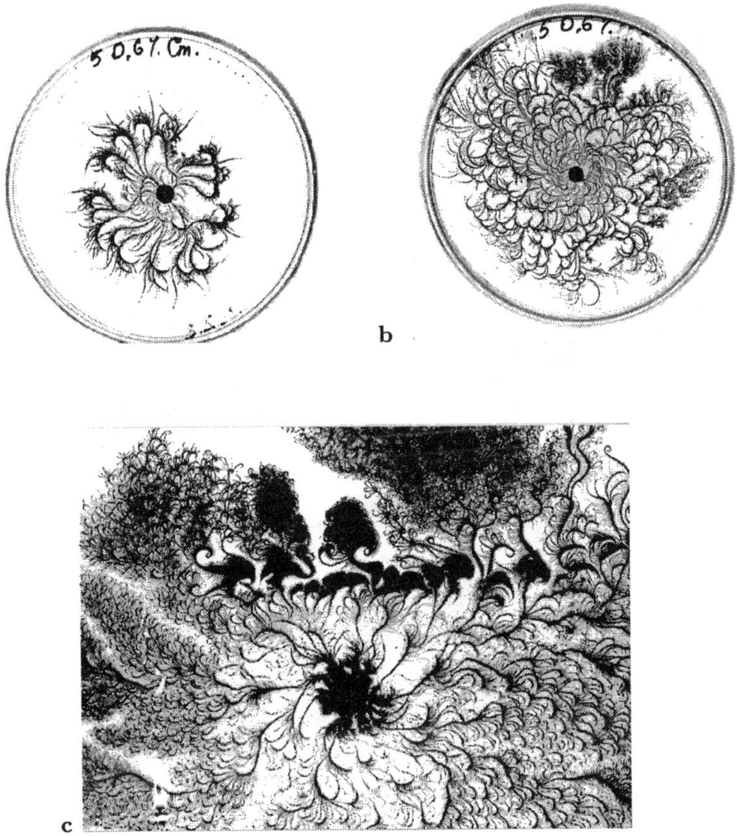

a b c

FIG. 11. *In agar soft enough for the bacteria to swim in, the branches lose the one-side handedness they have on harder agar. The two colonies of (a) and (b) are of 5g/l peptone level and 0.6% agar concentration. The two patterns are of two strains of the C morphotype, strains whose patterns are indistinguishable on harder agar. c) Closer look (×10 magnification) on a colony grown at 8g/l peptone level and 0.6% agar concentration.*

Electron microscope observations do not reveal any chiral structure on the cellular membrane [19].

2.3. Biological background.

2.3.1. Bacterial movement. In the course of evolution, bacteria have developed ingenious ways of moving on surfaces [43]. Motility on surfaces is often associated with swarming or gliding [2, 41, 45, 81, 93, 101], but $\mathcal{C}$ and $\mathcal{T}$ morphotype swim in a thin layer of liquid on top of the agar [20, 43, 69]. Swimming is a solitary movement done in liquid. A swimming bacterium moves in nearly straight runs, interrupted by short periods of tumbling. The tumbling event is a random rotation in one location. The

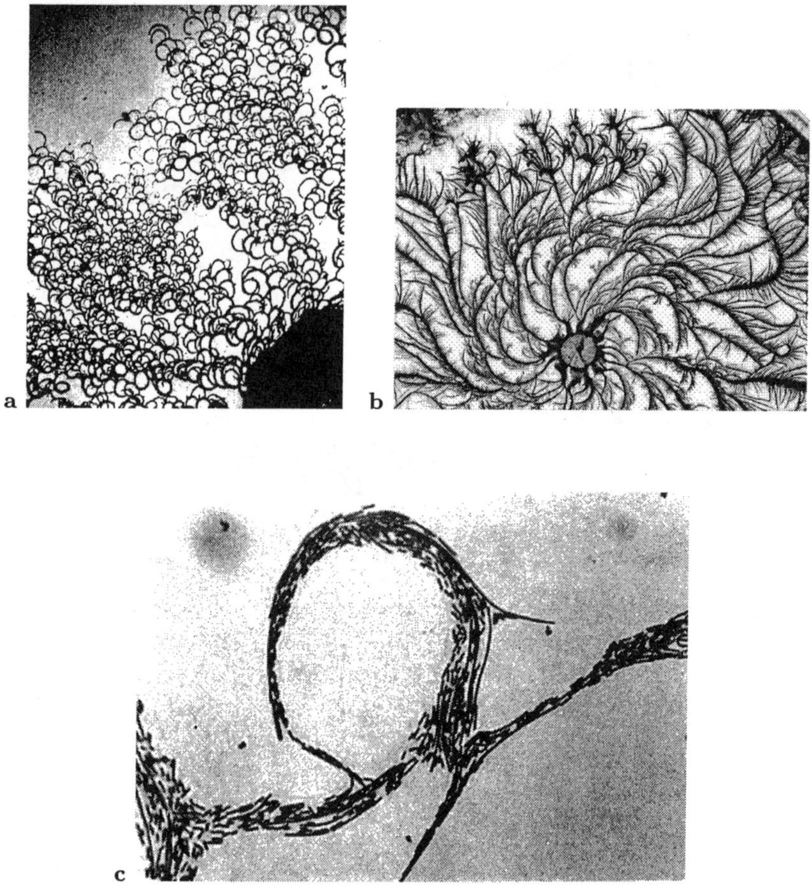

FIG. 12. *Optical microscope observations of branches of C morphotype colony. a) ×20 magnification of a colony at 1.6g/l peptone level and 0.75% agar concentration, the anti-clockwise twist of the thin branches is apparent. The curvature of the branches is almost constant throughout the growth. b) ×10 magnification of a colony at 4g/l peptone level and 0.6% agar concentration the branches are not thin, but have a feathery structure. The curvature of the branches varies, but it seems that at any given stage of growth the curvature is similar in all branches. That is, the curvature is a function of colonial growth. c) ×500 magnification of a colony at 1.6g/l peptone level and 0.75% agar concentration. Each line is a bacterium. the bacteria are long (5-50μm) and mostly ordered.*

direction of the next run is dictated by the final orientation of the tumbling bacterium.

A swimming bacterium propagates itself by rotating a bundle of flagella [30, 34]. Each flagellum is a helical protein filament which is hooked to a molecular engine transversing the bacterial membrane. The engines can rotate the flagella clockwise or counterclockwise. When the bacterium turns most of them counterclockwise, the flagella form an aligned bundle and push the bacterium forward. When they turn clockwise, the flagella disjoin and the bacterium tumbles.

In our experiments, the bacteria do not swim in a pre-supplied liquid, but in a layer of fluid on the surface of the agar. The bacterial cells move individually and at random in the same manner as flagellated bacteria move in wet mounts (i.e., nearly straight runs separated by brief tumbling). Swimming takes place only in sufficiently thick surface fluid. Microscope observations reveal no organized flow-field pattern.

Based on microscope observations of movement and electron microscope observations of flagella we identify the movement of $\mathcal{C}$ and $\mathcal{T}$ morphotype as swimming. Cells tumble about every $\tau_T \approx 1 - 5\ sec$ depending on external conditions. The speed of the bacterium between tumbling events is very sensitive to conditions such as the liquid viscosity, temperature and pH level. Typically, it is of the order of 1 - $10 \mu m/sec$.

Swimming can be approximated by a random walk with variable step size [22]. At low bacterial densities the random walk can be described by a diffusion equation with a diffusion coefficient $D_b \equiv v^2 \tau_T = 10^{-8} - 10^{-5} cm^2/sec$. Low bacterial densities mean that the mean free path between bacterial collisions l_c is longer than the tumbling length $l_T \equiv v\tau_T$, thus collisions between the bacteria can be neglected. The mean free path (or collision length) is

$$l_c \propto \begin{cases} \rho^{-\frac{1}{3}} & \text{in 3 dimensions} \\ \sigma^{-\frac{1}{2}} & \text{in 2 dimensions} \end{cases} \tag{2.1}$$

where ρ is the 3D bacterial density and σ is the 2D density — the projection of ρ on the surface.

At high densities ($l_c < l_T$), the collisions cannot be neglected. In attempting to approximate the dynamics in those conditions, one may want to consider the time of straight motion to be l_c/v instead of τ_T. Hence D_b depends on the bacterial density to yield

$$D_b \propto \begin{cases} v\rho^{-\frac{1}{3}} & \text{in 3D} \\ v\sigma^{-\frac{1}{2}} & \text{in 2D} \end{cases} . \tag{2.2}$$

This approximation is valid under the assumptions that a collision event is identical to a tumbling event (abrupt uncorrelated change in direction of motion), that a tumbling event is independent of the collisions, and that the speed between such events is not affected by their frequency.

The assumption that a collision event is like a tumbling event poses many problems. Even if the bacteria do not activate a special response to collision it is unrealistic to assume that collisions are elastic, or that the flagella adopt immediately the new orientation which changes during collisions. Thus it is reasonable to assume strong correlation between the cell's orientation before collision and the cell's orientation after collision. In addition, the orientation after the collision should be biased according to the average direction of motion of the surrounding bacteria, as they carry the liquid with them. The important parameter is not the collision length l_c but re-orientation time τ_r. The re-orientation time is the time it takes a bacterium to loose memory of its initial orientation, i.e. the time span on which the final orientation has effectively no correlation with the initial orientation. At low densities the re-orientation time τ_r is equal to the tumbling time τ_T. As the density rises and the collisions become more frequent, τ_r decreases. τ_r defines the densities above which the constant diffusion coefficient $D_b \equiv v^2 \tau_T$ is not a good approximation. It is quite possible that these densities are high enough so as to make the velocity and even the type of motion dependent on bacterial density, making relation (2.2) irrelevant. In any case, high cellular densities do mean an effective decrease in the diffusion coefficient related to the bacterial movement.

When swimming in an unstirred liquid, very low cellular densities also effect the movement. The bacteria secrete various materials into the media and some of them, e.g. enzymes and other polymers, significantly change the physical properties of the liquid making it more suitable for bacterial swimming. The secretion of these materials depends on cellular density, thus at not-too-high densities the speed of swimming rises with the cellular density. Hence the diffusion coefficient related to the bacterial movement should be a non-monotonic function of the bacterial density. Moreover, the specific functional form might depend on the specific bacterial strain.

In other conditions there is a similar but more pronounced effect. On semi-solid surface the bacteria cannot swim at all inside the agar and they have to produce their own layer of liquid to swim in. To produce such fluid the bacteria secrete lubricants (wetting agents). Other bacterial species produce known extracellular lubricants (such as surfactants, see [96, 78, 31, 66] and references therein, or the extracellular slime produced by *Proteus mirabilis* [94]). There are various materials (various cyclic lipopeptides have been identified) which can draw water from the agar. The composition and properties of the lubricant of *P. dendritiformis* is not known, but we will assume that higher concentration of lubricant is needed to extract water from a drier agar, and that the lubricant is slowly absorbed into the agar (or decomposes). It is possible that the mechanism by which *P. dendritiformis* produces the layer of liquid is different from what we postulate here, but as long as the liquid is self-produced and facilitates the bacterial movement (as revealed by the observations) the bio-chemical details of the liquid do not change the validity of the models we present here. A single bacterium on the agar surface cannot produce enough fluid to swim in, thus the bacteria cannot break out of the fluid layer and the branches of a $\mathcal{T}$ or $\mathcal{C}$ colony can be defined by this fluid. Whenever bacteria enter the shallower parts of the layer, at the edge of the branch, they become sluggish, indicating that the depth of the layer effects the bacterial movement. It can be argued (see

Section 3.2) that in such cases the bacterial speed is related to the bacterial density by a power law (at least in low densities). The diffusion coefficient related to the bacterial movement is not only a non-monotonic function of the bacterial density (as in a liquid agar), but it also vanishes for extremely low densities. In this case it is clear that the specific functional form depends on the specific bacterial strain (*B. subtilis*, for example, cannot move at all under such conditions).

2.3.2. Chemotaxis in swimming bacteria. Chemotaxis means changes in the movement of the bacteria in response to a gradient of a certain chemical field [1, 23, 56, 22, 35, 92]. The movement is biased along the gradient either in the gradient direction or in the opposite direction. Usually chemotactic response means a response to an externally produced field, as in the case of chemotaxis towards food. However, the chemotactic response can also be to a field produced directly or indirectly by the bacterial cells. We will refer to this case as chemotactic signaling. The bacteria sense the local concentration R of a chemical by membrane receptors binding to molecules of the chemical [1, 56]. It is crucial to note that when estimating gradients of chemicals, the bacterial cells actually measure changes in receptors occupancy and not in the concentration itself. When put in continuous equations [73, 39], this indirect measurement translates to measuring the gradient

$$\frac{\partial}{\partial x}\frac{R}{(K+R)} = \frac{K}{(K+R)^2}\frac{\partial R}{\partial x}. \tag{2.3}$$

where K is a constant whose value depends on receptor affinity, the speed in which the bacterium processes the signal from the receptors, etc. This means that the chemical gradient times a factor $K/(K+R)^2$ is measured, and it is known as the "receptor law" [73].

In a continuous model, we incorporate the effect of chemotaxis by introducing a chemotactic flux $\vec{J}_{chem}$:

$$\vec{J}_{chem} \equiv \zeta(\sigma)\chi(R)\nabla R\ , \tag{2.4}$$

where $\chi(R)\nabla R$ is the gradient sensed by the bacteria (with $\chi(R)$ having the units of 1 over chemical concentration). $\chi(R)$ is usually taken to be either constant or the "receptor law". $\zeta(\sigma)$ is the bacterial response to the sensed gradient (having the same units as a diffusion coefficient times the units of the bacterial density σ). It is positive for attractive chemotaxis and negative for repulsive chemotaxis.

Ben-Jacob *et al.* argued [9, 25, 8, 7] that for the colonial adaptive self-organization the bacteria employ three kinds of chemotactic responses, each dominant in a different regime of the morphology diagram (the claim was made for $\mathcal{T}$ morphotype, but the same holds for their relatives $\mathcal{C}$ morphotype). One response is the food chemotaxis mentioned above. It is expected to be dominant for only a range of nutrient levels (due to the "receptor law"). The two other kinds of chemotactic responses are chemotactic signaling. One is repulsive chemotactic signaling, a long-range signal. The repelling chemical is secreted by starved bacteria at the inner parts of the colony. The second signal is a short-range attracting signal. The length scale of each signal is determined by the diffusion constant of the chemical agent and the rate of its spontaneous decomposition.

Amplification of diffusive Instability Due to Nutrient Chemotaxis: In non-living systems, more ramified patterns (lower fractal dimension) are observed for lower growth velocity. Based on the growth dynamics and considering growth velocity as a function of nutrient level, Ben-Jacob *et al.* [20] concluded that in the case of bacterial colonies there is a need for a mechanism that can both increase the growth velocity and maintain, or even decrease, the fractal dimension. They suggested food chemotaxis to be the required mechanism. It provides an outward drift to the cellular movements; thus, it should increase the rate of envelope propagation. At the same time, being a response to an external field it should also amplify the basic diffusion instability of the nutrient field. Hence, it can support faster growth velocity together with a ramified pattern of low fractal dimension.

Repulsive chemotactic signaling: We focus now on the formation of the fine radial branching patterns at low nutrient levels. From the study of non-living systems, it is known that in the same manner that an external diffusion field leads to diffusion instability, an internal diffusion field can stabilize growth. It is natural to assume that some sort of chemotactic agent produces such a field. To regulate the organization of the branches, it must be a long-range signal. To result in radial branches it must be a repulsive chemical produced by bacteria at the inner parts of the colony. The most probable candidates are the bacteria entering a pre-spore stage.

If nutrient is deficient for a long enough time, bacterial cells may enter a special stationary state – a state of a spore – which enables them to survive much longer without food. While the spores themselves do not emit any chemicals (as they have no metabolism), the pre-spores (sporulating cells) do not move and emit a very wide range of waste materials, some of them is unique to the sporulation process [98, 88, 29] cell. These emitted chemicals might be used by other bacteria as a signal carrying information about the conditions at the location of the pre-spores. Ben-Jacob *et al.* [20, 19, 25] suggested that such materials are repelling the bacteria ('repulsive chemotactic signaling') so that they can escape from a dangerous location.

2.3.3. Food consumption, reproduction and starvation. *P. dendritiformis*, like most bacteria, reproduce by fission of the cell into two daughter cells which are practically identical to the mother cell. The crucial step in the cell division is the replication of the genetic material and its sharing between the daughter cells. Hasty replication of DNA might lead to many errors – most organisms limit the rate of replication to about 1000 bases per second. Thus reproduction must take at least a minimal time τ_R. This reproduction time τ_R is about 25 min in Bacilli.

For reproduction, as well as for movement and other metabolic processes, bacteria and all other organisms need an influx of energy. Any organism which does not get its energy directly from sunlight (by photosynthesis) needs an external supply of food. In the patterning experiments the bacteria eat nutrient from the agar. As long as there is enough nutrient and no significant amount of toxic materials, food is consumed (for cell replication and internal processes) at a maximal rate Ω_c. To estimate Ω_c we assume that a bacterium needs to consume an amount of food C_R of about $3 \times 10^{-12}g$. It is 3 times its weight – one quanta for doubling body mass, one quanta used for movement and all other metabolic processes during the reproduction time τ_R, and one quanta is for the reduced entropy of making an organized cell out of food. Hence Ω_c is about $2fg/sec$ ($1fg = 10^{-15}$gram).

If nutrient is deficient for a long enough period of time, the bacterial cells may enter a state of a spore which enables them to survive much longer without food. The bacterial cells employ very complex mechanisms tailored to the process of sporulation. They stop normal activity – like movement – and use all their internal reserves to metamorphose from an active volatile cell to a sedentary durable 'seed'. While the spores themselves do not emit any chemicals (as they have no metabolism), the pre-spores (sporulating cells, see Fig. 9) do not move and emit a very wide range of waste materials, some of which are unique to the sporulating cell [98, 88, 29]. These emitted chemicals might be used by other cells as signals carrying information about the conditions at the location of the pre-spores or as chemorpellents.

When bacteria are grown in a petri dish, nutrients are usually provided by adding peptone, a mixture including all the amino acids and sugars as source of carbon. Bacteria which are not defective in synthesis of any amino acid can grow also on a minimal agar in which a single source of carbon and no amino acids are provided. Such growth might seem to be easier to model as the growth is limited by the diffusion of a single chemical. However, during growth on minimal agar there is usually a higher rate of waste product accumulation, introducing other complications into the model. Moreover many of our strains are auxotrophic i.e. they are defective in synthesis of some amino acids and they need an external supply. Providing the bacteria with these amino acids and only a single carbon source poses the question as to what is the limiting factor in the growth of the bacteria. For all those reasons we prefer to use peptone as a nutrient source.

We said that if there is ample supply of food, bacteria reproduce at a maximal rate of one division in τ_R. If the available amount of food is limited, bacteria consume the maximum amount of food they can. In the limit of low bacterial density, the available amount of food over the tumbling time τ_T is the food contained in the area $\tau_T\sqrt{D_b D_n}$, where D_b and D_n are the diffusion coefficients of the bacteria and the food, respectively. Hence the rate of food consumption is given by $n\sqrt{D_b D_n}$ (whether D_b is constant or not).

In a continuous model, reproduction of bacteria translates to a growth term of the bacterial density which is σ times the food consumption rate per bacteria. In the limit of high nutrient it is σ/τ_R, and in the limit of low nutrient it is proportional to $n\sigma$. This brings to mind the Michaelis-Menten law [73] of $\frac{K}{1+\gamma n}n\sigma$ with K, γ constants. Many authors take only the low nutrient limit of this expression, $Kn\sigma$, although it is not biologically established that the bacteria in the experiments are limited by the availability of food and not by their maximal consumption rate.

3. Continuous models for the branching growth of $\mathcal{T}$ morphotype.

3.1. The Lubricating Bacteria model. The Lubricating Bacteria model is a reaction-diffusion model for the bacterial colonies of the $\mathcal{T}$ morphotype [39, 55]. This model includes four coupled fields. One field describes the bacterial density $b(\vec{x}, t)$, the second describes the height of the lubrication layer in which the bacteria swim $l(\vec{x}, t)$, a third field describes the nutrients $n(\vec{x}, t)$ and the fourth field is the stationary bacteria that "freeze" and begin to sporulate $s(\vec{x}, t)$ (see Section 2.3.2).

We first describe the dynamics of the bacteria and the nutrient. The two reaction-diffusion equations governing those fields are:

$$\begin{aligned} \frac{\partial b}{\partial t} &= movement + \Gamma_b(b, n) \\ \frac{\partial n}{\partial t} &= D_n \nabla^2 n - g(n, b) \end{aligned} \tag{3.1}$$

where $\Gamma_b(b, n)$ is the bacterial reproduction term. The nutrient diffusion is a simple diffusion process with a constant diffusion coefficient D_n. The bacteria consume nutrients at the rate $g(n, b)$ which is taken to be:

$$g(n, b) = nb \tag{3.2}$$

This approximate term is correct at the limit of low nutrient level and low bacterial density.

The nutrient consumed by bacteria serves as an energy source and as a precursor for synthesis of macromolecules. There is probably a minimum amount of energy necessary to maintain cell structure and integrity, known as maintenance energy, and nutrients used to supply the maintenance energy are not available for cell growth. We assume that those nutrients are required at a constant rate μ. Bacteria then cannot utilize all the nutrients for reproduction:

$$\Gamma_b(b, n) = g(n, b) - \mu b \tag{3.3}$$

There is no explicit term for sporulation in (3.1). Instead the reproduction term Γ_b can be used also to model sporulation. Sporulation is initiated by starvation so it is complementary to reproduction. When

$\Gamma_b > 0$ bacteria reproduce and so sporulation is excluded. When $\Gamma_b < 0$ bacteria reduce in number. This should represent sporulation. Note that $\Gamma_b < 0$ when the nutrient level is low so indeed we can claim that bacteria are starved. This simple sporulation scheme has its limitations. Its rate is set by the nutrient consumption and effects such as density dependence are neglected. We represent the sporulating bacteria by the field $s(\vec{x}, t)$, whose time evolution is:

$$\frac{\partial s}{\partial t} = \begin{cases} 0 & \text{if } \Gamma_b > 0 \\ -\Gamma_b & \text{if } \Gamma_b < 0 \end{cases} \tag{3.4}$$

In other continuous models, bacterial reproduction and the sporulation process are modeled differently [54, 48, 62, 39]: Bacterial reproduction is proportional to the nutrient consumption rate and sporulation is an independent process that proceeds at a rate μ which could depend on other variables such as the bacterial density or the nutrient concentration. Note that if these modifications are applied to this model, the functional form of the bacterial reaction terms (3.1), (3.3) will not change, but the equation for the field s (3.4) will change to:

$$\frac{\partial s}{\partial t} = \mu b \tag{3.5}$$

Since the dynamics of the other variables are separated from s, its different dynamics are not significant. Moreover, the modification of the equation for s is minor. The difference is only in the biological interpretation of the terms.

We now turn to bacterial movement. In a uniform layer of liquid, bacterial swimming is a random walk with a variable step length and can be approximated by diffusion. The layer of lubricant is not uniform, and its height affects the bacterial movement. An increase in the amount of lubricant decreases the friction between the bacteria and the agar surface. We suggest that the bacterial movement depends on the local lubricant height through a power law with exponent $\gamma > 0$:

$$\textit{movement} = \nabla \cdot (D_b l^\gamma \nabla b) \tag{3.6}$$

where D_b is a constant with dimensions of a diffusion coefficient. D_b is related to the fluid's viscosity and the dryness of the agar might affect this viscosity. Gathering the various terms gives the partial model:

$$\begin{aligned} \frac{\partial b}{\partial t} &= \nabla \cdot (D_b l^\gamma \nabla b) + nb - \mu b \\ \frac{\partial n}{\partial t} &= D_n \nabla^2 n - nb \\ \frac{\partial s}{\partial t} &= -\min(nb - \mu b, 0) \end{aligned} \tag{3.7}$$

It is possible to define dimensionless time and space variables $t' = t\mu$ and $\vec{x'} = \vec{x}\sqrt{\mu/D_n}$. In those units the parameters D_n and μ are equal to 1.

We model the dynamics of the lubricating fluid also by a reaction diffusion equation. There are two reaction terms: production by the bacteria and absorption into the agar. The dynamics of the field are:

$$\frac{\partial l}{\partial t} = -\nabla \vec{J_l} + f_l(b,n,l) - \lambda l \tag{3.8}$$

where $\vec{J_l}$ is the fluid flux, $f_l(b,n,l)$ is the fluid production term and λ is the absorption rate of the fluid into the agar.

We assume that the fluid production depends on the bacterial density. As the production of lubricant probably demands substantial metabolic efforts, it should also depend on the nutrient level (the model gives the same qualitative results if the lubricant production does not depend directly on food consumption). We take a simple form where the production depends linearly on the concentrations of both the bacteria and the nutrient. The exact relation should depend on the synthetic pathway of the active agents composing the lubricant. If they are, for example, secondary metabolites, then their production does not depend on the current nutrient level, but on the prior accumulation of primary metabolites. However, the model is not sensitive to the exact dependence of the lubricant production on the nutrient level since lubricant production is important at the front of the expanding colony, where the nutrient level is close to the initial level. It is reasonable to assume that the bacteria produce lubricant up to a height, denoted as l_M, which is sufficient for their swimming motion. We therefore take the production term to be:

$$f_l(b,n,l) = \Gamma bn(l_M - l) \tag{3.9}$$

where Γ is the production rate.

We turn to the flow of the lubricating fluid. The physical problem is very complicated. However a simplified model will suffice. We model the lubricant flux as a non-linear diffusion process:

$$\vec{J_l} = -D_l l^\nu \nabla l \tag{3.10}$$

where D_l is a constant with dimensions of a diffusion coefficient. The diffusion term of the fluid depends on the height of the fluid to the power $\nu > 0$. The nonlinearity causes the fluid to have a sharp boundary at the front of the colony, as is observed in bacterial colonies. The equation for the lubricant field is:

$$\frac{\partial l}{\partial t} = \nabla \cdot (D_l l^\nu \nabla l) + \Gamma bn(l_M - l) - \lambda l \tag{3.11}$$

The functional form of the terms that we introduced are simple and plausible, but they are not derived from basic physical principles. Therefore we do not have quantitative relations between the parameters of these terms and the physical properties of the agar substrate. However we can suggest some relations. In the experiments, the agar concentration is controlled. Higher agar concentration gives a drier and more solid substrate. We shall try to find what are the effects on the lubricant layer. We recall that the lubricant fluid is composed of water and active components such as surfactants. A drier agar can increase the absorption rate λ. Alternatively it can diminish the amount of water extracted by the active components. Then either the lubricant layer will be thinner or the bacteria will have to produce more of the active components. The former case should decrease D_b while the latter should decrease the production rate Γ. In both cases the composition of the lubricant fluid will change as the concentration of the active components will increase. The lubricant fluid should become more viscous with the effect of D_b and D_l decreasing.

Equation (3.11) together with equations (3.7) form the Lubricating Bacteria model. For the initial conditions, we set n to have a uniform distribution of level n_0, b to be zero everywhere but in the center, and the other fields to be zero everywhere. We use no-flux boundary conditions. We solve the equations on a 400×400 tridiagonal lattice using an explicit scheme. The simulations were stopped when the simulated colonies reached a specific radius (much as is done with the bacterial colonies).

Our results show that the model can reproduce branching patterns, similar to the bacterial colonies. In the experiments there are two control parameters: the agar concentration and the initial nutrient concentration. First we examine the effect of changing the initial nutrient concentration n_0. As Fig. 13 shows, the model produced a dense circular colony when n_0 was large. The pattern became more branched and ramified as n_0 decreased until n_0 was close to 1, which is the minimal value of n_0 to support growth (since n is never larger than n_0, $n_0 \leq 1$ means that the growth term is always non-positive and no growth is possible).

Changing the agar concentration affects the dynamics of the lubricant fluid. Previously we demonstrated that a higher agar concentration relates to a larger absorption rate λ and to lower production rate Γ and lower diffusion coefficients D_l and D_b. In Fig. 14 we show patterns obtained with different values of the parameters Γ and λ. As we expected, increasing λ or decreasing Γ produced a more ramified pattern, similar to the effect of a higher agar concentration on the patterns of bacterial colonies. Similar effects are obtained by decreasing D_b. Some more details about the results of the model (morphology diagram, growth velocity and fractal dimension as function of parameters, etc.) can be found in Refs. [55, 27].

3.2. The Non-Linear Diffusion model. Under certain assumptions, the Lubricating Bacteria model can be reduced to the Non-Linear Diffusion model of Kitsunezaki [54] and Cohen [24]. The additional assumptions needed are about the dynamics near zero bacterial density:

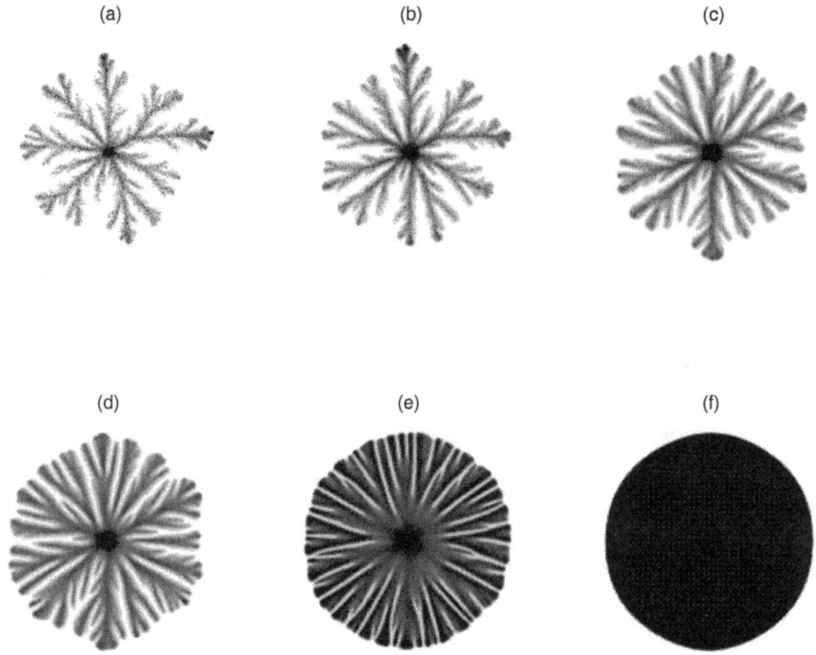

FIG. 13. *Growth patterns of the Lubricating Bacteria model, for different values of initial nutrient level* n_0. n_0 *increases from left to right: (a) 1.2, (b) 1.4, (c) 1.7, (d) 2, (e) 3, (f) 6. The minimal value of* n_0 *to support growth is* 1.

1) The lubricant height l is much smaller than l_{max}, so that the production of the lubricant can be assumed to be independent of its height.

2) The production of lubricant is proportional to the bacterial density to the power $\alpha > 0$ (in the simplest case taken above $\alpha = 1$).

3) The absorption of the lubricant is proportional to the lubricant height to the power $\beta > 0$ (in the simplest case taken above $\beta = 1$).

4) Over the bacterial length scale, the two above processes are much faster than the diffusion process, so the lubricant height is proportional to the bacterial density to the power of β/α.

5) The friction is proportional to the lubricant height to the power $\gamma < 0$.

Given these assumptions, the lubricant field can be removed from the dynamics and be replaced by a density dependent diffusion coefficient. This diffusion coefficient is proportional to the bacterial density to the power $k \equiv -2\gamma\beta/\alpha > 0$.

The resulting model is:

$$\frac{\partial b}{\partial t} = \nabla(D_0 b^k \nabla b) + nb - \mu b \tag{3.12}$$

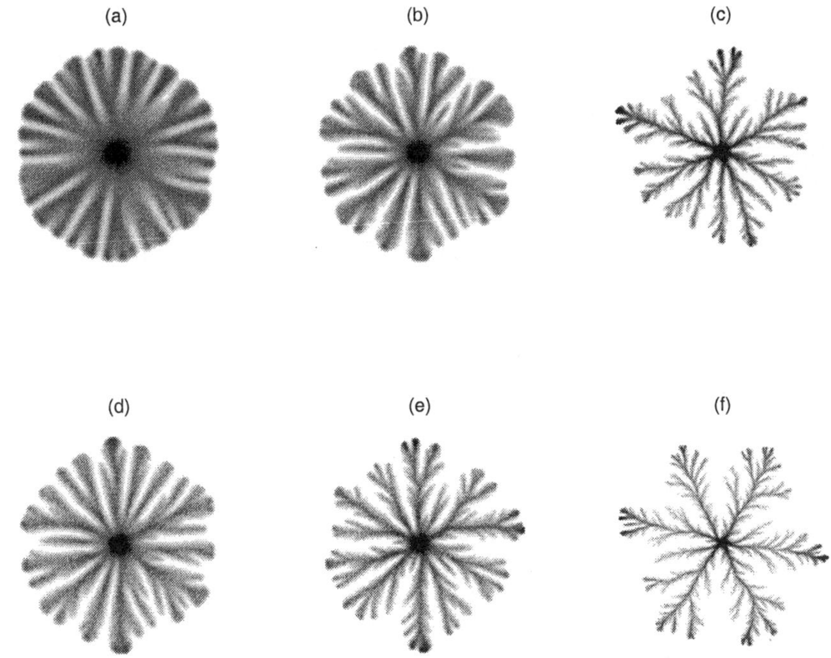

FIG. 14. *Effect of varying λ, the fluid absorption rate, on colony pattern. The fluid production rate Γ is 1 in the upper row and 0.3 in the lower row. In both rows λ increases from left to right: $\lambda = 0.03$ (left), $\lambda = 0.1$ (center), $\lambda = 1$ (right) The patterns become more ramified as λ increases. Decreasing Γ also produces a more ramified pattern. The other parameters are: $D_b = D_l = 1, n_0 = 1.5$.*

$$\frac{\partial n}{\partial t} = \nabla^2 n - bn \tag{3.13}$$

$$\frac{\partial s}{\partial t} = \mu b \tag{3.14}$$

For $k > 0$ the 1D model gives rise to a front "wall", with compact support (i.e. $b = 0$ outside a finite domain). For $k > 1$ this wall has an infinite slope. The model exhibits branching patterns for suitable parameter values and initial conditions, as depicted in Fig. 15. Increasing initial levels of nutrient leads to denser colonies, similar to the observed patterns. Analysis of interface stability in this model can be found in [54].

4. The Communicating Spinors model for the chiral growth of $\mathcal{C}$ morphotype. The Communicating Spinors model was developed to explain the chirality of the $\mathcal{C}$ morphotype colonies. Our purpose is to show that the flagella handedness, while acting as a singular perturbation, leads to the observed chirality. It does so in the same manner in which crystalline anisotropy leads to the observed symmetry of snowflakes [6].

FIG. 15. *Growth patterns of the Non-Linear Diffusion model, for different values of initial nutrient level* n_0. *Parameters are:* $D_0 = 0.1, k = 1, \mu = 0.15$. *The apparent 6-fold symmetry is due to the underlying tridiagonal lattice.*

It is known [34, 95, 87] that flagella have specific handedness. Ben-Jacob *et al.* [14] proposed that the latter is the origin of the observed chirality. In a fluid (which is the state in most experimental setups), as the flagella unfold, the cell tumbles and ends up at a new random angle relative to the original one. The situation changes for quasi 2D motion – motion in a "lubrication" layer thinner than the cellular length. We assume that in this case, of rotation in a plane, the tumbling has a well-defined handedness of rotation. Such handedness requires, along with the chirality of the flagella, the cells' ability to distinguish up from down. Growth in an upside- down petri- dish shows the same chirality. Therefore, we think that the determination of up versus down is done either via the vertical gradient of the nutrient concentration, the vertical gradient of signaling materials inside the substrate, or the friction of the cells with the surface of the agar. The latter is the most probable alternative; soft enough agar enables the bacteria to swim *below* the surface of the agar which leads to many changes in the patterns, including reversing the bias of the branches.

To cause the chirality observed on semi-solid agar, the rotation of tumbling must be, on average, less than 90° and relative to a specific direction. Co-alignment (orientational interaction) limits the rotation. We further assume that the rotation is relative to the local mean orientation of the surrounding cells.

To test the above idea, we included the additional assumed features in the Communicating Walkers model [20], changing it to a 'Communicating Spinors' model (as the particles in the new model have an orientation and move in a quasi-1D random walk). The Communicating Walkers model [20] was inspired by the diffusion-transition scheme used to study solidification from supersaturated solutions [90, 91, 89]. The former is a hybridization of the "continuous" and "atomistic" approaches used in the study of non-living systems. Ben-Jacob *et al.* have presented in the past a version of the Communicating Spinors model for chiral growth [14]. The model we present here is closely related to a previous model of chiral growth, but it differs in two features. The first is the orientation field (see below), which was discontinuous piecewise constant and in this model it is continuous piecewise linear. The second difference is the definition of a single run of a spinor (the stretch between two tumbling events), which was defined as one run per one time unit (i.e. each step is a run) and now is defined as variable number of steps in the same direction.

The representation of bacteria as spinors allows for a close relation to the bacterial properties. The bacterial cells are represented by spinors allowing a more detailed description. At the end of growth in a typical experiment there are 10^8–10^9 bacterial cells in the petri-dish. Thus it is impractical to incorporate into the model each and every cell. Instead, each of the spinors represents about 10–1000 cells, so that we work with 10^4–10^6 spinors in one numerical "experiment".

Each spinor has a position $\vec{r}_i$, direction θ_i (an angle) and a metabolic state ('internal energy') E_i. The spinors perform an off-lattice constrained random walk on a plane within an envelope representing the boundary of the wetting fluid. This envelope is defined on the same tridiagonal lattice where the diffusion equations are solved. To incorporate the swimming of the bacteria into the model, at each time step each of the active spinors (motile and metabolizing, as described below) recalculate its direction θ'_i and moves a step of size d in this direction.

The direction in which each spinor moves is determined in two steps; first the spinor decides whether it should continue the current run, that is to continue in the same direction $\theta'_i = \theta_i$. In the basic version of the model (see Sec. 5.2 for extension of the model) the decision is random with a specific probability p to continue the run. The resulting runs have a geometric distribution of lengths, with mean run length of d/p. Once a spinor decides to change direction, the new direction θ'_i is derived from the spinor's previous direction by

$$\theta'_i = P(\theta_i, \Phi(\vec{r}_i)) + Ch + \xi + \omega \tag{4.1}$$

Ch and ξ represent the new features of rotation due to tumbling. Ch is a fixed part of the rotation and ξ is a stochastic part, chosen uniformly from an interval $[-\eta, \eta]$ (η constant). ω is an orientation term that takes, with equal probabilities, one of the values 0 (forward direction) or π (backward direction). This orientation term gives the spinors their name, as it makes their re-orientation invariant to forward or backward direction. $\Phi(\vec{r}_i)$ is the local mean orientation in the neighborhood of $\vec{r}_i$. P is a projection function that represents the orientational interaction which acts on each spinor to orient θ_i along the direction $\Phi(\vec{r}_i)$. P is defined by

$$P(\alpha, \beta) = \alpha + (\beta - \alpha). \tag{4.2}$$

Once oriented, the spinor advances a step d in the direction θ'_i, and the new location $\vec{r'}_i$ is given by:

$$\vec{r'}_i = \vec{r}_i + d\,(\cos\theta'_i, \sin\theta'_i) \tag{4.3}$$

The movement is confined within an envelope which is defined on the tridiagonal lattice. The step is not performed if $\vec{r'}_i$ is outside the envelope. Whenever this is the case, a counter on the appropriate segment of the envelope is increased by one. When a segment counter reaches N_c, the envelope advances one lattice step and a new lattice cell is occupied. Note that the spinor's direction is not reset upon hitting the envelope, thus it might "bang its head" against the envelope time and time again. The requirement of N_c hits represent the colony propagation through wetting of unoccupied areas by the bacteria. Note that N_c is related to the agar dryness, as more wetting fluid must be produced (more "collisions" are needed) to push the envelope on a harder substrate.

Next we specify the mean orientation field Φ. To do so, we assume that each lattice cell (hexagonal unit area) is assigned one value of $\Phi(\vec{r})$, representing the average orientation of the spinors in the local neighborhood of the center of the cell. The value of Φ is set when a new lattice cell is first occupied by the advancement of the envelope, and then remains constant. We set the value of $\Phi(\vec{r})$ to be equal to the average over the orientations of the N_c attempted steps that led to the occupation of the new lattice cell. The value of Φ in any given point inside the colony is found by linear interpolation between the three neighboring centers of cells. Clearly, the model described above is a simplified picture of the bacterial movement. For example, a more realistic model would include an equation describing the time evolution of Φ. However, the simplified model is sufficient to demonstrate the formation of chiral patterns. A more elaborate model will be presented elsewhere [26].

Motivated by the presence of a maximal growth rate of the bacteria even for optimal conditions, each spinor in the model consumes food at a constant rate Ω_c if sufficient food is available. We represent the metabolic state of the i-th spinor by an "internal energy" E_i. The rate of change of the internal energy is given by

$$\frac{dE_i}{dt} = \kappa C_{consumed} - \frac{E_m}{\tau_R}\,, \tag{4.4}$$

where κ is a conversion factor from food to internal energy ($\kappa \cong 5\cdot 10^3 cal/g$) and E_m represents the total energy loss for all processes over the reproduction time τ_R, excluding energy loss for cell division. $C_{consumed}$ is given by $\min(\Omega_C, \Omega'_C)$, where Ω'_C is the maximal rate of food consumption as limited by the locally available food (Sec. 2). When sufficient food is available, E_i increases until it reaches a threshold energy. Upon reaching this threshold, the spinor divides into two. When a spinor is starved for a long interval of time, E_i drops to zero and the spinor "freezes". This "freezing" represents entering a pre-spore state (starting the process of sporulation, see Section 5.2).

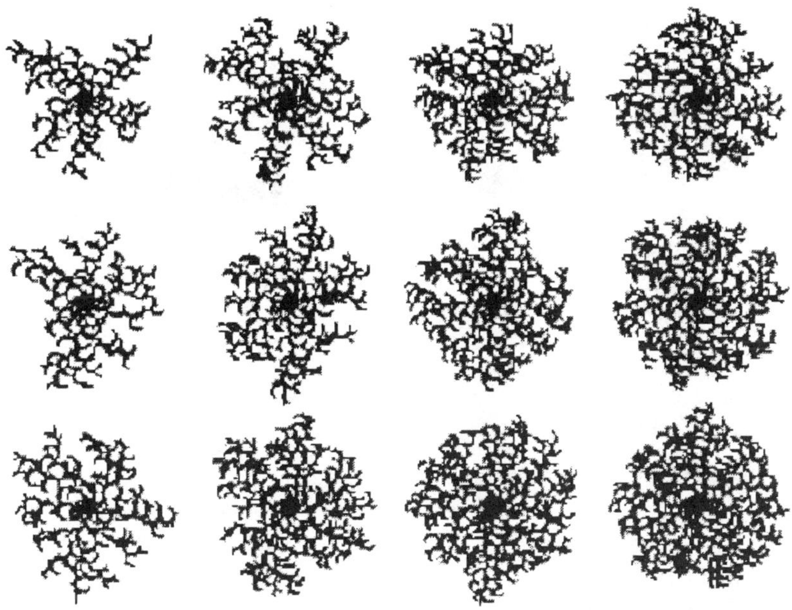

FIG. 16. *A morphology diagram of the Communicating Spinors model for various values of N_c and initial n concentration n_0. $Ch = 6°$, $\eta = 3°$, $d = 0.2$, $p = 0.5$.*

We represent the diffusion of nutrients by solving the diffusion equation for a single agent whose concentration is denoted by $n(\vec{x}, t)$:

$$\frac{\partial n}{\partial t} = D_n \nabla^2 C - bC_{consumed} , \tag{4.5}$$

where the last term includes the consumption of food by the spinors (b is their density). The equation is solved on the same tridiagonal lattice on which the envelope is defined. The length constant of the lattice a_0 must be larger than the size of the spinors' step d. The simulations are started with an inoculum of spinors at the center and a uniform distribution of the nutrient. Both Φ and the spinors at the inoculum are given uniformly distributed random directions.

Results of the numerical simulations of the model are shown in Fig. 16. These results do capture some important features of the observed patterns: the microscopic twist Ch leads to a chiral morphology on the macroscopic level. The growth is via stable tips, all of which twist with the same handedness and emit side-branches. The dynamics of the side-branch emission in the time evolution of the model is similar to the observed dynamics.

For large noise strength η the chiral nature of the pattern gives way to a branching pattern (Fig. 17). This provides a plausible explanation for the

a b

FIG. 17. *When the noise η is increased to $\eta = 180°$ the tumbling of the spinors becomes unrestricted. Their movement becomes like that of the $\mathcal{T}$ bacteria and accordingly the simulated colonial pattern is like that of $\mathcal{T}$ morphotype. On the left $\eta = 3°$, on the right $\eta = 180°$.*

branching patterns produced by $\mathcal{C}$ morphotype grown on high peptone levels, as the cells are shorter when grown on a rich substrate. The orientation interaction is weaker for shorter cells, hence the noise is stronger.

5. The effect of chemotaxis. So far, we have shown that the models can reproduce many aspects of the microscopic dynamics and the patterns in some range of nutrient level and agar concentration, but at least for the $\mathcal{T}$ -like growth, other models can do the same [39, and reference there in]. We will now extend the Non-Linear Diffusion model and the Communicating Spinors model to test for their ability to describe other aspects of the bacterial colonies involving chemotaxis and chemotactic signaling (which are believed to be used by the bacteria [9, 25, 8, 7]).

5.1. Chemotaxis in the Non-Linear Diffusion model. As we mentioned in Section 2.3.2, in a continuous model we incorporate the effect of chemotaxis by introducing a chemotactic flux $\vec{J}_{chem}$:

$$\vec{J}_{chem} \equiv \zeta(b)\chi(R)\nabla R \tag{5.1}$$

$\chi(R)\nabla R$ is the gradient sensed by the bacteria (with $\chi(R)$ having the units of 1 over chemical concentration). $\chi(R)$ is usually taken to be either constant or the "receptor law". $\zeta(b)$ is the bacterial response to the sensed gradient (having the same units as a diffusion coefficient times the units of the bacterial density b). In the Non-Linear Diffusion model the bacterial diffusion is $D_b = D_0 b^k$, and the bacterial response to chemotaxis is $\zeta(b) = \zeta_0 b\left(D_0 b^k\right) = \zeta_0 D_0 b^{k+1}$. ζ_0 is a constant, positive for attractive chemotaxis and negative for repulsive chemotaxis.

We claim that the fine radial branching patterns at low nutrient levels result from repulsive chemotactic signaling. The equation describ-

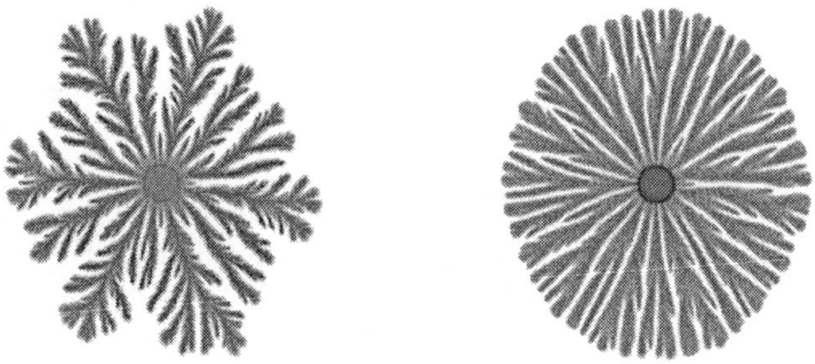

FIG. 18. *Growth patterns of the Non-Linear Diffusion model with food chemotaxis (left, see Section 2) and repulsive chemotactic signaling (right) included.* $\chi_{0f} = 3, \chi_{0R} = 1, D_R = 1, \Gamma_R = 0.25, \Omega_R = 0, \Lambda_R = 0.001$. *Other parameters are the same as in Figure 15. The apparent 6-fold symmetry is due to the underlying tridiagonal lattice.*

ing the dynamics of the chemorepellent contains terms for diffusion, production by pre-spores, decomposition by active bacteria and spontaneous decomposition:

$$\frac{\partial R}{\partial t} = D_R \nabla^2 R + s\Gamma_R - \Omega_R bR - \lambda_R R \tag{5.2}$$

where D_R is the diffusion coefficient for the chemorepellent, Γ_R is the emission rate of repellent by pre-spores, Ω_R is the decomposition rate of the repellent by active bacteria, and λ_R is the rate of self decomposition of the repellent.

Fig. 18 demonstrates the effect of repulsive chemotactic signaling. In the presence of repulsive chemotaxis the pattern becomes much denser with a smooth circular envelope, while the branches are thinner and radially oriented.

5.2. Chemotaxis in the Communicating Spinors model. The colonial patterns of $\mathcal{C}$ morphotype (e.g. Fig. 16) are rarely as ordered as the simulated patterns of the Communicating Spinors model. For example, the branches of the observed colonies usually have varying curvature. In the simulations of $\mathcal{C}$ morphotype shown in Fig. 16 all the branches have a uniform curvature. One of the reasons for this difference is the simplifications taken during the model's development. A more elaborate model that we will present [26] will be a better description of the colony. However, some of the observed features can be explained in the context of the Communicating Spinors model. In some of the observed patterns (Fig. 11(b)), the curvature of the branches has a distinct relation to the branch's radial orientation (the orientation relative to the radial direction): the curvature is smaller when the branch is in the radial orientation and larger when the branch is orthogonal to that orientation. This brings to mind the radial organization of branches in the $\mathcal{T}$ morphotype, and indeed we were able to explain the chiral pattern with the aid of the same concept – repulsive chemotaxis.

Chemotaxis was introduced in previous versions of the Communicating Walkers model by varying, according to the chemical gradient, the probability of moving in different directions [20, 7]. Modulating the directional probability is not the way bacteria implement chemotaxis – they modulate the length of runs. However, the growth of $\mathcal{T}$ morphotype is insensitive to the details of the movement. Modulating the directional probability is as good an implementation of chemotaxis as many other implementations (it was chosen for computational convenience). The pattern of the $\mathcal{C}$ morphotype is based on amplification of microscopic effects (singular perturbation) such as the left bias in the bacterial tumbling. Small differences in the microscopic dynamics of chemotaxis might affect the global pattern. Indeed we found that modulating the directional probability yield unrealistic results in the simulations of $\mathcal{C}$ morphotype. We had to resort to the bacterial implementation of chemotaxis – modulating the length of runs according to the chemical gradient.

When modulating the length of runs of walkers or spinors one must be careful not to change the particles' speed. Such a change is not observed in experiments [23, 84] and it has far reaching effects on the dynamics. Changing the particles' speed is like changing the diffusion coefficient of the bacterial density field, a change that can have undesirable effects on the pattern.

Modulating the length of spinors' runs without changing their speed can be done by modulating the number of steps that compose a single run (that was our motivation for dividing the runs into steps). Since the mean number of steps in a run is determined by the reorientation probability p, chemotaxis should modulate this probability. For chemotaxis, the probability of changing direction by the i-spinor in one time step is (for a repellent R):

$$p^* = p + \chi(R)\partial_{\theta_i} R \tag{5.3}$$

where R is measured at the spinor position $\vec{r}_i$, $\chi(R)$ is the same as in the continuous model (either constant or the "receptor law") and ∂_{θ_i} is the directional derivative in the spinor's direction θ_i. p^* is truncated to within the range $[0, 1]$ as it is a probability. The length of the resulting runs will depend on the runs' direction, where a spinor moving up the gradient of the repellent will have shorter *mean* run length than the same spinor moving down the gradient.

The production and dynamics of the repulsive chemotactic signaling in the Communicating Spinors model is the same as in the Non-Linear Diffusion model, see Eq. (5.2) (with s representing the density of spinors that "froze"). The patterns resulting from including repulsive chemotaxis in the model have indeed branches with variable curvature, as can be seen in Fig. 19. The curvature is smaller for branches in the radial direction.

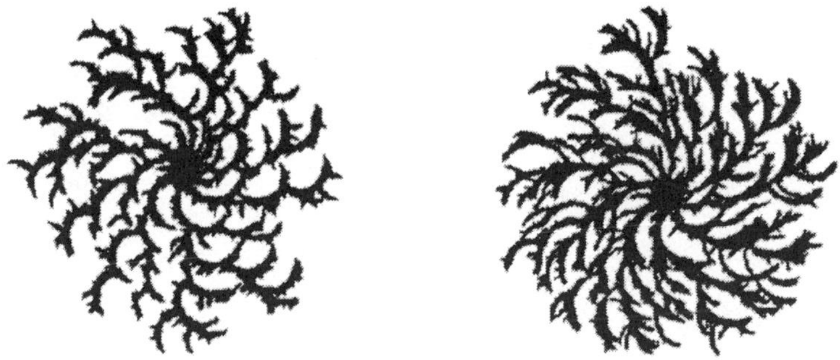

a b

FIG. 19. *The effect of repulsive chemotactic signaling on the Communicating Spinors model. a) Without chemotaxis. b) With repulsive chemotaxis. The Spinors are repelled from the inner parts of the colony. The resulting curvature of the branches is reduced when they are in the radial direction. In spite of the reversed handedness, the pattern resembles Fig. 11(b).*

Food chemotaxis also varies the branches' curvature, but in a less ordered manner, not similar to the observed bacterial patterns.

Under different growth conditions the $\mathcal{C}$ morphotype can produce very different patterns. As mention above, if the agar is soft enough the bacteria can move inside it. In such cases, the bias in the bacterial movement might change or even reverse, and it is manifested in the curvature of the branches. Widely changing curvature of the branches can be seen in Fig. 11(a). The agar hardness was tuned such that in the beginning of the growth the bacteria could swim inside the agar, but they are forced to swim *on* the agar by the end of growth due to the marginal water evaporation during growth. In Fig. 20 we demonstrate the models' ability to explain such patterns by changing the spinors' bias Ch during the simulation. Ch is set to be a continuous random function of the colonial size, which is constrained only at the beginning and end of growth to have certain values. The function for Ch is the same in all the images of Fig. 20, but various types of chemotaxis are used. As can be seen, repulsive chemotactic signaling is needed to explain the observed bacterial patterns.

6. Weak chirality in $\mathcal{T}$ morphotype. Colonies of $\mathcal{T}$ morphotype grown on hard substrate (above 2.0% agar concentration) exhibit branching patterns with a global twist with the same handedness, as shown in Figs. 6 and 21. Similar observations during growth of other bacterial strains have been reported by Matsuyama *et al.* [65, 64]. We refer to such growth patterns as having weak chirality, as opposed to the strong chirality exhibited by the $\mathcal{C}$ morphotype.

In [14], Ben-Jacob *et al.* proposed that in the case of $\mathcal{T}$ morphotype, it is the high viscosity of the "lubrication" fluid during growth on a hard

a b c

FIG. 20. *The snake-like branches observed in Fig. 11(a) can be reproduced by the Communicating Spinors model. Ch is a continuous function of the colony's radius (the same function in a, b, and c). Maximal value of Ch is* 8°, *minimal value is* -2°. *(a) With repulsive chemotactic signaling. (b) Without chemotaxis. (c) With food chemotaxis. The best resemblance to the observed colony is obtained with repulsive chemotactic signaling.*

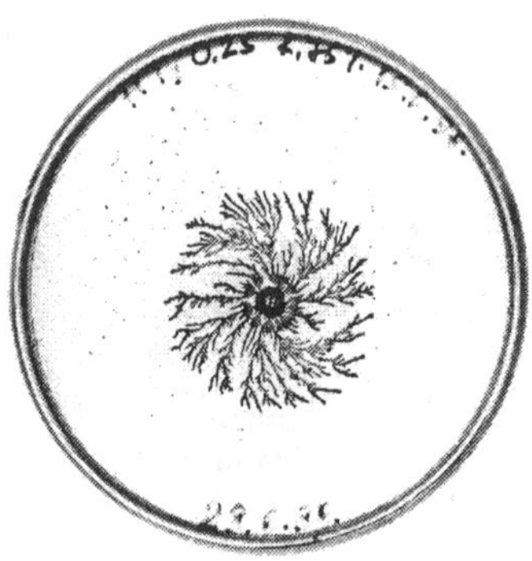

FIG. 21. *Weak chirality (global twist of the branches) exhibited by the* $\mathcal{T}$ *morphotype for a peptone level of 0.25g/l peptone level and agar concentration of 1.75%.*

surface that replaces the cell-cell co-alignment of the $\mathcal{C}$ morphotype that limits the rotation of tumbling. They further assumed that the rotation should be relative to a specified direction. They used the gradient of a chemotaxis signaling field (specifically, the long-range repellent chemotaxis) as a specific direction, rather than the local mean orientation field which is used in the case of $\mathcal{C}$ morphotype. It was shown in [14] that inclusion of the above features in the Communicating Walkers model indeed leads to a weak chirality which is highly reminiscent of the observed one. The idea above also provides a plausible explanation of the observations of weak chirality by Matsuyama *et al.* [65] in strains defective in production of "lubrication" fluid.

6.1. Weak chirality – the Non-Linear Diffusion model. In the reaction-diffusion model, weak chirality can be obtained by modifying the

FIG. 22. *Growth patterns of the Non-Linear Diffusion model with a "squinting" repulsive chemotactic signaling, leading to weak chirality. Parameters are as in the previous picture, $\theta = 43^o$.*

chemotactic mechanism and causing it to twist: We alter the expression for the chemotactic flux $\vec{J}_{chem}$ (Eq. 2.4) so that it is no longer oriented with the chemical gradient (∇R). Instead it is oriented with a rotated vector $\hat{\mathbf{R}}(\theta)\nabla R$, where $\hat{\mathbf{R}}(\theta)$ is the two-dimensional rotation operator and θ is the rotation angle. The chemotactic flux is thus written:

$$\vec{J}_{chem} = \zeta(b)\chi(R)(\hat{\mathbf{R}}(\theta)\nabla R) \tag{6.1}$$

The effect of rotating the repulsive chemotaxis, as depicted in Fig. 22, is to make the pattern chiral, with the degree of chirality determined by the rotation angle θ.

One must note that adding a similar rotation to the food chemotaxis does not have the same effect, because the nutrient gradients do not posses the long-range, radial nature of the chemorepellent gradients.

6.2. Weak chirality – the Communicating Spinors model. As was demonstrated in Section 4, the Communicating Spinors model is robust enough to reproduce patterns of $\mathcal{T}$ morphotype, as well as patterns of $\mathcal{C}$ morphotype. Here we use it to model patterns of weak chirality. Two simulated $\mathcal{T}$ -like colonies are shown in Fig. 23. Fig. 23(a) shows a colony with radial branches while Fig. 23(b) shows a colony with weak chirality and thinner branches. In the two simulations the spinors have exactly the same response to chemotaxis and the same bias $Ch = 9°$. The two runs differ in the freedom of rotation η; in Fig. 23(a) the spinors can rotate freely ($\eta = 180°$) while in Fig. 23(b) spinor rotation is somewhat limited ($\eta = 35°$, while for the colony of strong chirality $\eta = 5°$).

It seems that both models – the Non-Linear Diffusion model and the Communicating Spinors model – can capture the essential features of the observed weak chirality. Yet a closer examination reveals that the de-

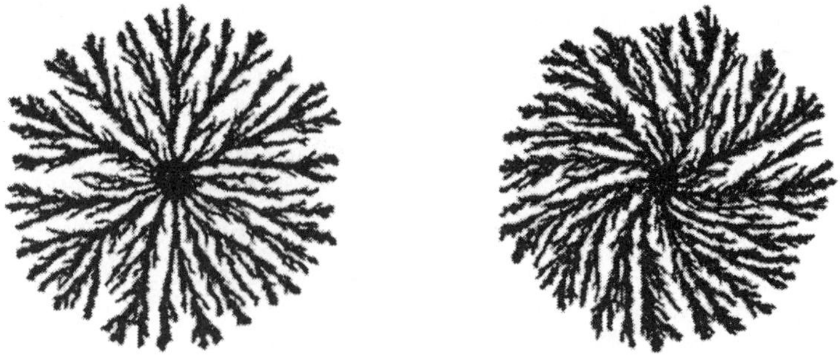

a b

FIG. 23. *Weak chirality of the $\mathcal{T}$ morphotype is modeled by the Communicating Spinors model. Both simulations are with repulsive chemotactic signaling and with bias in the walkers rotation. a) With free rotation ($\eta = 180°$) the pattern is branched, without apparent chirality expressed. b) With constrained rotation ($\eta = 35°$) weak chirality is expressed.*

scription of the two models is incompatible. In the Non-Linear Diffusion model the bias from the direction of the gradient is through the chemotaxis process. The spinors, like the bacteria, cannot modulate their runs as a function of the difference between their direction and that of the gradient; they do not know what is the direction of the gradient, only the directional derivative along their path. As was demonstrated in figure 23, one of the key features for the weak chirality in the spinors model is the correlation in orientation of the spinors (through Φ). In fact in this model the twist of the branches stems from the deviation of the run directions from the orientation of neighboring spinors. The twist of the branches is related only indirectly, through the neighbors' orientation, to the chemorepellent's gradient. A continuous model of such processes should include information about the mean orientation of the bacterial cells. It should include chemotaxis without rotation, anisotropic diffusion (smaller diffusion coefficient in orientations orthogonal to the mean orientation of neighbors) and a rotation on the diffusion operator. Such a model will be presented elsewhere [26].

The discrete spinors model allows for a detailed representation of the bacterial properties. The macroscopic dynamics and resulting patterns, however, are similar in both models – apparently the growth does not amplify the difference in the microscopic dynamics. Thus the 'unrealistic' microscopic description of the Non-Linear Diffusion model does not rule it out as an approximation to the growth dynamics of bacterial colonies with weak chirality.

6.3. Chirality measure. All through this manuscript we referred to chirality as a two-valued property – either the pattern is chiral or it is not. There are various attempts in the literature to quantify chirality with a continuous measure. See, for example, the method of Avnir *et al.* [102, 47], who applied their method also to large disordered objects. While this method is general and can quantify with a single number the measure of asymmetry of any given object, it is not quite satisfactory for our purpose. We would like to know the time evolution of the chirality of a colony, and not just "mean" chirality given by a single number. We sacrifice the generality of the measure to that end.

Since the growth velocity of the colonies (both experimental and simulated) is constant, we measure the chirality as a function of radius instead of a function of time. Thus we can work on the chirality of an image, not of a process. The image can be a scanned picture of the real colony or the result of a computer simulation. We look for a mapping of the image to a new one, which in some sense does not distinguish left from right (the ambiguity stems from the fact that a large random object will not have, in general, reflection symmetry, thus there is no trivial definition for chirality of such objects). The mapping is defined by:

$$(r, \theta) \rightarrow (r, \theta + \Delta\theta(r)) \tag{6.2}$$

where each point in the image is described by the polar coordinates (r, θ), measured from the center of the colony. Thus, each point is rotated by an increment $\Delta\theta$ which depends on the radius r (i.e. the distance from the center).

Working on many experimental patterns, as well as simulated patterns, we have learned that in most cases a linear dependence of $\Delta\theta$ on r is sufficient to give quite satisfactory results, that is, to transform a chiral pattern into a "normal" branching pattern. The rotating angle is thus written:

$$\Delta\theta(r) = \left(\frac{r}{r_{max}}\right)\theta_{max} \tag{6.3}$$

where r_{max} is the radius of the colony, and θ_{max} is the rotation angle at that radius.

The fact that this linear angular mapping suffices to "de-chiral" the simulated patterns may not be of much importance (in the case of the continuous model, at least, this is almost a direct result of the way in which we introduce the weak chirality). The same transformation works for images of real colonies of $\mathcal{T}$ morphotype, but *does not* work for chiral colonies of other bacteria (see Sec. 7). This strengthens our belief in the models.

7. Conclusions. We first briefly reviewed experimental observations of colonial patterns formed by bacteria of the species *Paenibacillus dendritiformis.* We described both the tip-splitting growth of the $\mathcal{T}$ morphotype and the chiral, twisted-branches growth of $\mathcal{C}$ morphotype. Both colonial patterns and optical microscope observations of the bacteria dynamics were presented.

In this manuscript we presented observations of various forms of chiral patterns in bacterial colonies. Our goal was to explain the various aspects of chirality. We used two types of models: continuous reaction-diffusion models which deal with bacterial density, and a hybrid semi-discrete model which deals with properties of the individual bacterium. From a comparison of model simulations and experimental observations we conclude that chemotactic signaling plays an important part in the development of colonies of the two types. We also estimate how sensitive the growth is to the details of the microscopic dynamics, demonstrating that the more 'complex' the pattern is, the more sensitive the growth is to the small details.

We would like to note that the *P. dendritiformis* is not the only bacteria whose colonies exhibit chirality. Ben-Jacob *et al.* discussed in [10] the formation of colonies of *Paenibacillus vortex*, where each branch is produced by a leading droplet and emits side branches, each with its own leading droplet. Each leading droplet consists of hundreds to millions of bacterial cells that circle a common center (a vortex) at a cellular speed of about 10μm/s (*P. vortex* is not a close relative of *P. dendritiformis* and its movement on the agar is swarming, not swimming). In Fig. 24 we show a colonial pattern of these bacteria. The chirality we termed 'weak chirality' is evident in this figure. In this case the chirality is not related to the handedness of the flagella, but to the rotation of the vortices. When "pushed" by repulsive chemotaxis, Magnus force acts on the vortices and drives them side-ways from the radial direction of the chemorepellent's gradient. This difference in mechanisms is expressed in the global pattern: The colonial patterns of *P. vortex* cannot be "de-chiraled" by the transformation (linear angular mapping) that "de-chiral" the $\mathcal{T}$ morphotype. The fact that the models for weak chirality match in this respect the the weak chirality of $\mathcal{T}$ morphotype and not the 'weak chirality' of *P. vortex* is another support for their success in describing the bacterial colonies.

We hope we have convinced the reader that chirality in patterns of bacterial colonies gives important clues to the underlying dynamics. The processes leading to such patterns are more complex than those leading to non-twisted branching patterns. The chiral patterns are more sensitive to the underlying dynamics and as such they require more accurate models. This reflects on the success of the models we presented as being a good description of the colonies.

Acknowledgments. Identifications of the Paenibacillus dendritiformis and genetic studies are carried in collaboration with the group of D. Gutnick. Presented studies are supported in part by a grant from the Israeli Academy of Sciences grant no. 593/95, by the Israeli-US Binational

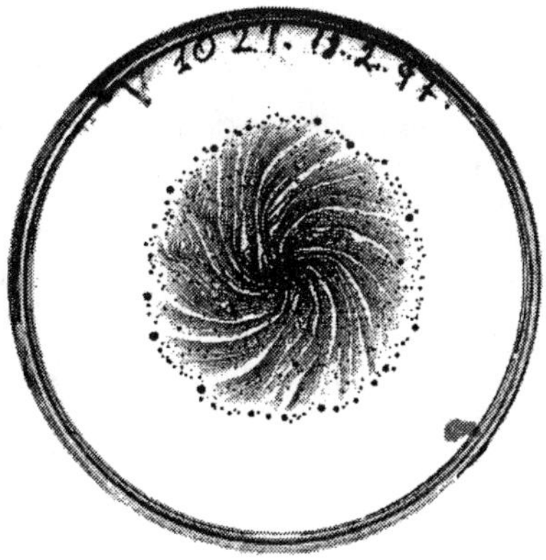

FIG. 24. *A colony of Paenibacillus dendritiformis on 10 g/l peptone level and 2% agar concentration. The dots at the tips of the branches are bacterial vortices – each is composed of up to millions of bacterial cells rotating around a common center. The twist of the branches results from a Magnus force induced by repulsive chemotactic signaling.*

Science Foundation BSF grant no. 00410-95 and by a grnat from IMK Almene Fond. Two of us, E. Ben-Jacob and I. Golding, thank the IMA for hospitality during part of this project. One of us, I. Cohen, thanks The Colton Scholarships for their support.

REFERENCES

[1] J. Adler. Chemoreceptors in bacteria. *Science*, **166**:1588–1597, 1969.

[2] L. Alberti and R.M. Harshey. Differentiation of *Serratia marcescens* 274 into swimmer and swarmer cells. *J. Bact.*, **172**:4322–4328, 1990.

[3] G. Albrecht-Buehler. In defense of "nonmolecular" cell biology. *Int. Rev. Cytol.*, **120**:191–241, 1990.

[4] V.A. Avetisov, V.I. Goldanskii, and V.V. Kuzmin. Handedness, origin of life and evolution. *Phys. Today*, **44**(7):33–41, 1991.

[5] M.Y. Azbel. Survival-extinction transition in bacteria growth. *Europhys. Lett.*, **22**(4):311–316, 1993.

[6] E. Ben-Jacob. From snowflake formation to the growth of bacterial colonies, Part I: Diffusive patterning in non-living systems. *Contemp. Phys.*, **34**:247–273, 1993.

[7] E. Ben-Jacob. From snowflake formation to the growth of bacterial colonies, Part II: Cooperative formation of complex colonial patterns. *Contemp. Phys.*, **38**:205–241, 1997.

[8] E. Ben-Jacob and I. Cohen. Cooperative formation of bacterial patterns. In J.A. Shapiro and M. Dworkin, editors, *Bacteria as Multicellular Organisms.* Oxford University Press, New-York, 1997.

[9] E. Ben-Jacob, I. Cohen, and A. Czirók. Smart bacterial colonies. In *Physics of Biological Systems: From Molecules to Species*, Lecture Notes in Physics, pp. 307–324. Springer-Verlag, Berlin, 1997.

[10] E. Ben-Jacob, I. Cohen, A. Czirók, T. Vicsek, and D.L. Gutnick. Chemomodulation of cellular movement and collective formation of vortices by swarming bacteria and colonial development. *Physica A*, **238**:181–197, 1997.

[11] E. Ben-Jacob, I. Cohen, and D. Gutnick. Cooperative organization of bacterial colonies: From genotype to morphotype. *Annu. Rev. Microbiol.*, **52**:779–806, 1998.

[12] E. Ben-Jacob, I. Cohen, and H. Levine. Cooperative self-organization of microorganisms. *Adv. Phys.*, 1999 (in press).

[13] E. Ben-Jacob, I. Cohen, O. Shochet, I. Aronson, H. Levine, and L. Tsimering. Complex bacterial patterns. *Nature*, **373**:566–567, 1995.

[14] E. Ben-Jacob, I. Cohen, O. Shochet, A. Czirók, and T. Vicsek. Cooperative formation of chiral patterns during growth of bacterial colonies. *Phys. Rev. Lett.*, **75**(15):2899–2902, 1995.

[15] E. Ben-Jacob and P. Garik. The formation of patterns in non-equilibrium growth. *Nature*, **343**:523–530, 1990.

[16] E. Ben-Jacob, H. Shmueli, O. Shochet, and A. Tenenbaum. Adaptive self-organization during growth of bacterial colonies. *Physica A*, **187**:378–424, 1992.

[17] E. Ben-Jacob, O. Shochet, and A. Tenenbaum. Bakterien schließen sich zu bizarren formationen zusammen. In A. Deutsch, editor, *Muster des Ledendigen: Faszination inher Entstehung und Simulation.* Verlag Vieweg, 1994.

[18] E. Ben-Jacob, O. Shochet, A. Tenenbaum, and O. Avidan. Evolution of complexity during growth of bacterial colonies. In P.E. Cladis and P. Palffy-Muhoray, editors, *Spatio-Temporal Patterns in Nonequilibrium Complex Systems*, Santa-Fe Institute studies in the sciences of complexity, pp. 619–634. Addison-Weseley Publishing Company, 1995.

[19] E. Ben-Jacob, O. Shochet, A. Tenenbaum, I. Cohen, A. Czirók, and T. Vicsek. Communication, regulation and control during complex patterning of bacterial colonies. *Fractals*, **2**(1):15–44, 1994.

[20] E. Ben-Jacob, O. Shochet, A. Tenenbaum, I. Cohen, A. Czirók, and T. Vicsek. Generic modeling of cooperative growth patterns in bacterial colonies. *Nature*, **368**:46–49, 1994.

[21] E. Ben-Jacob, A. Tenenbaum, O. Shochet, and O. Avidan. Holotransformations of bacterial colonies and genome cybernetics. *Physica A*, **202**:1–47, 1994.

[22] H. C. Berg. *Random Walks in Biology.* Princeton University Press, Princeton, N.J., 1993. Expanded ed.

[23] H.C. Berg and E.M. Purcell. Physics of chemoreception. *Biophysical Journal*, **20**:193–219, 1977.

[24] I. Cohen. Mathematical modeling and analysis of pattern formation and colonial organization in bacterial colonies, 1997. M.Sc. thesis, Tel-Aviv University, ISRAEL.

[25] I. Cohen, A. Czirók, and E. Ben-Jacob. Chemotactic-based adaptive self organization during colonial development. *Physica A*, **233**:678–698, 1996.

[26] I. Cohen, I. Golding, and E. Ben-Jacob. Models of chiral bacterial growth, 2000 (in preparation).

[27] I. Cohen, I. Golding, Y. Kozlovsky, and E. Ben-Jacob. Continuous and discrete models of cooperation in complex bacterial colonies. *Fractals*, **7**:235–247, 1999.

[28] A. Czirók, E. Ben-Jacob, I. Cohen, and T. Vicsek. Formation of complex bacterial colonies via self-generated vortices. *Phys. Rev. E*, **54**:1791–1801, 1996.

[29] E. Deak, I. Szabo, A. Kalmaczhelyi, Z. Gal, G. Barabas, and A. Penyige. Membrane-bound and extracellular beta-lactamase production with developmental regulation in *Streptomyces griseus* NRRL B-2682. *Microbiol.*, **144**:2169–2177, 1998.

[30] D.J. DeRosier. The turn of the screw: The bacterial flagellar motor. *Cell*, **93**:17–20, 1998.

[31] J.D. Desai and I.M. Banat. Microbial production of surfactants and their commercial potential. *Microbiol. Mol. Biol. Rev.*, **61**:47–64, 1997.

[32] M. Dworkin. *Developmental biology of the bacteria.* Benjamin/Cummings Publishing Company, Reading, 1985.

[33] M. Dworkin. Recent advances in the social and developmental biology of the myxobacteria. *Microbiol. Rev.*, **60**:70–102, 1996.
[34] M. Eisenbach. Functions of the flagellar modes of rotation in bacterial motility and chemotaxis. *Molec. Microbiol.*, **4**(2):161–167, 1990.
[35] M. Eisenbach. Control of bacterial chemotaxis. *Mole. Microbiol.*, **20**:903–910, 1996.
[36] J. Feder. *Fractals.* Plenum, New York, 1988.
[37] H. Fujikawa and M. Matsushita. Fractal growth of *Bacillus subtilis* on agar plates. *J. Phys. Soc. Jap.*, **58**:3875–3878, 1989.
[38] H. Fujikawa and M. Matsushita. Bacterial fractal growth in the concentration field of nutrient. *J. Phys. Soc. Jap.*, **60**:88–94, 1991.
[39] I. Golding, Y. Kozlovsky, I. Cohen, and E. Ben-Jacob. Studies of bacterial branching growth using reaction-diffusion models of colonial development. *Physica A*, **260**(3, 4):510–554, 1998.
[40] H. Haken. *Information and self-organization.* Springer-Verlag, Berlin, 1988.
[41] R.M. Harshey and T. Matsuyama. Dimorphic transition in *Escherichia coli* and *Salmonella typhimurium* – surface-induced differentiation into hyperflagellate swarmer cells. *Proc. Natl. Acad. Sci. USA*, **91**:8631–8635, 1994.
[42] R.A. Hegstrom and D.K. Kondepudi. The handedness of the universe. *Sci. Am.*, **262**:108–115, 1990.
[43] J. Henrichsen. Bacterial surface translocation: A survey and a classification. *Bac. Rev.*, **36**:478–503, 1972.
[44] T.H. Henrici. *The Biology of Bacteria: The Bacillaceae.* D. C. Heath & company, 3rd edition, 1948.
[45] E. Hoiczyk and W. Baumeister. The junctional pore complex, a prokaryotic secretion organelle, is the molecular motor underlying gliding motility in cyanobacteria. *Curr. Biol.*, **8**:1161–1168, 1998.
[46] J. Horgan. From complexity to perplexity. *Sci. Am.*, pp. 74–79, June 1995.
[47] O. Katzenelso, H.Z. Hel-Or, and D. Avnir. Chirality of large random supramolecular structures. *Chem.-Euro. J.*, **2**:174–181, 1996.
[48] K. Kawasaki, A. Mochizuki, M. Matsushita, T. Umeda, and N. Shigesada. Modeling spatio-temporal patterns created by bacillus-subtilis. *J. Theor. Biol.*, **188**:177–185, 1997.
[49] D.A. Kessler, J. Koplik, and H. Levine. Pattern selection in fingered growth phenomena. *Adv. Phys.*, **37**:255, 1988.
[50] D.A. Kessler and H. Levine. Pattern formation in *dictyostelium* via the dynamics of cooperative biological entities. *Phys. Rev. E*, **48**:4801–4804, 1993.
[51] D.A. Kessler, H. Levine, and L. Tsimring. Computational modeling of mound development in *dictyostelium. Physica D*, **106**(3, 4):375–388, 1997.
[52] J.O. Kessler. Co-operative and concentrative phenomena of swimming microorganisms. *Cont. Phys.*, **26**:147–166, 1985.
[53] J.O. Kessler and M.F. Wojciechowski. Collective behavior and dynamics of swimming bacteria. In J.A. Shapiro and M. Dworkin, editors, *Bacteria as Mullticellular Organisms*, pp. 417–450. Oxford University Press Inc., New York, 1997.
[54] S. Kitsunezaki. Interface dynamics for bacterial colony formation. *J. Phys. Soc. Jpn*, **66**(5):1544–1550, 1997.
[55] Y. Kozlovsky, I. Cohen, I. Golding, and E. Ben-Jacob. Lubricating bacteria model for branching growth of bacterial colonies. *Phys. Rev. E*, **59**:7025–7035, 1999.
[56] J.M. Lackiie, editor. *Biology of the chemotactic response.* Cambridge Univ. Press, 1986.
[57] J.S. Langer. Dendrites, viscous fingering, and the theory of pattern formation. *Science*, **243**:1150–1154, 1989.
[58] S.A. Mackay. Computer simulation of aggregation in dictyostelium discoideum. *J. Cell. Sci.*, **33**:1–16, 1978.
[59] B.B. Mandelbrot. *The Fractal Geometry of Nature.* Freeman, San Francisco, 1977.

[60] B.B. Mandelbrot. *Fractals: Form, Chance and Dimension.* Freeman, San Francisco, 1977.
[61] M. Matsushita and H. Fujikawa. Diffusion-limited growth in bacterial colony formation. *Physica A*, **168**:498–506, 1990.
[62] M. Matsushita, J. Wakita, H. Itoh, I. Rafols, T. Matsuyama, H. Sakaguchi, and M. Mimura. Interface growth and pattern formation in bacterial colonies. *Physica A*, **249**:517–524, 1998.
[63] M. Matsushita, J.-I. Wakita, and T. Matsuyama. Growth and morphological changes of bacteria colonies. In P.E. Cladis and P. Palffy-Muhoray, editors, *Spatio-Temporal Patterns in Nonequilibrium Complex Systems*, Santa-Fe Institute studies in the sciences of complexity, pp. 609–618. Addison-Weseley Publishing Company, 1995.
[64] T. Matsuyama, R.M. Harshey, and M. Matsushita. Self-similar colony morphogenesis by bacteria as the experimental model of fractal growth by a cell population. *Fractals*, **1**(3):302–311, 1993.
[65] T. Matsuyama and M. Matsushita. Fractal morphogenesis by a bacterial cell population. *Crit. Rev. Microbiol.*, **19**:117–135, 1993.
[66] T. Matsuyama and Y. Nakagawa. Bacterial wetting agents working in colonization of bacteria on surface environment. *Colloids Surf. B: Biointerfaces*, **7**:207–214, Nov. 1, 1996.
[67] N.H. Mendelson. Helical *Bacillus subtilis* macrofibers: Morphogenesis of a bacterial multicellular macroorganism. *Proc. Natl. Acad. Sci. USA*, **75**(5):2478–2482, 1978.
[68] N.H. Mendelson. Bacterial macrofibres: The morphogenesis of complex multicellular bacterial forms. *Sci. Progress*, **74**:425–441, 1990.
[69] N.H. Mendelson, A. Bourque, K. Wilkening, K.R. Anderson, and J.C. Watkins. Organized cell swimming motions in *Bacillus subtilis* colonies: Patterns of short-lived whirls and jets. *J. Bact.*, **181**:600–609, 1999.
[70] N.H. Mendelson and S.L. Keener. Clockwise and counterclockwise pinwheel colony morphologies of *Bacillus subtilis* are correlated with the helix hand of the strain. *J. Bacteriol.*, **151**(1):455–457, 1982.
[71] N.H. Mendelson and J.J. Thwaites. Cell wall mechanical properties as measured with bacterial thread made from *Bacillus subtilis. J. Bacteriol.*, **171**(2):1055–1062, 1989.
[72] M. Mimura, H. Sakaguchi, and M. Matsushita. A reaction-diffusion approach to bacterial colony formation. *Preprint*, 1997.
[73] J.D. Murray. *Mathematical Biology.* Springer-Verlag, Berlin, 1989.
[74] L.K. Nakamura. *Bacillus thiaminolyticus* sp. nov., nom. rev. *Int. J. Syst. Bacteriol.*, **40**:242–246, 1990.
[75] G. Nicolis and I. Prigogine. *Exploring Complexity.* W.H. Freeman and company, New-York, 1989.
[76] H. Parnas and L. Segel. A computer simulation of pulsatile aggregation in *Dictyostelium discoideum. J. Theor. Biol.*, **71**:185–207, 1978.
[77] P. Pelce and A. Pocheau. geometrical approach to the morphogenesis of unicellular algae. *J. Theor. Biol.*, **156**:197–214, 1992.
[78] F. Peypoux, J.M. Bonmatin, and J. Wallach. Recent trends in the biochemistry of surfactant. *Appl. Microbiol. Biotech.*, **51**:553–563, 1999.
[79] M. Doudoroff R.Y. Stainer and E.A. Adelberg. *The Microbial World.* Prentice-Hall and Inc., N. J., 1957.
[80] I. Rafols. *Formation of Concentric Rings in Bacterial Colonies.* M.Sc. thesis, Chuo University, Japan, 1998.
[81] O. Rauprich, M. Matsushita, C.J. Weijer, F. Siegert, S.E. Esipov, and J.A. Shapiro. Periodic phenomena in *proteus mirabilis* swarm colony development. *J. Bact.*, **178**:6525–6538, 1998.
[82] E. Rosenberg, editor. *Myxobacteria: Development and Cell Interactions.* Springer series in molecular biology. Springer-Verlag, 1984.

[83] L.M. Sander. Fractal growth processes. *Nature*, **322**:789–793, 1986.

[84] J.E. Segall, S.M. Block, and H.C. Berg. Temporal comparisons in bacterial chemotaxis. *Proc. Natl. Acad. Sci. USA*, **83**:8987–8991, 1986.

[85] J.A. Shapiro. Bacteria as multicellular organisms. *Sci. Am.*, **258**(6):62–69, 1988.

[86] J.A. Shapiro and D. Trubatch. Sequential events in bacterial colony morphogenesis. *Physica D*, **49**:214–223, 1991.

[87] C.H. Shaw. Swimming against the tide: Chemotaxis in *Agrobacterium*. *BioEssays*, **13**(1):25–29, 1991.

[88] N.J. Shih and R.G. Labbe. Characterization and distribution of amylases during vegetative cell growth and sporulation of *Clostridium perfringens*. *Can. J. Microbiol.*, **42**:628–33, 1996.

[89] O. Shochet. *Study of late-stage growth and morphology selection during diffusive patterning*. PhD thesis, Tel-Aviv University, 1995.

[90] O. Shochet, K. Kassner, E. Ben-Jacob, S.G. Lipson, and H. Müller-Krumbhaar. Morphology transition during non-equilibrium growth: I. Study of equilibrium shapes and properties. *Physica A*, **181**:136–155, 1992.

[91] O. Shochet, K. Kassner, E. Ben-Jacob, S.G. Lipson, and H. Müller-Krumbhaar. Morphology transition during non-equilibrium growth: II. Morphology diagram and characterization of the transition. *Physica A*, **187**:87–111, 1992.

[92] P.A. Spiro, J.S. Parkinson, and H.G. Othmer. A model of excitation and adaptation in bacterial chemotaxis. *Proc. Natl. Acad. Sci. USA*, **94**:7263–7268, 1997.

[93] A.M. Spormann. Gliding motility in bacteria: Insights from studies of *Myxococcus xanthus*. *Microbiol. Molec. Biol. Rev.*, **63**:621–, 1999.

[94] S.J. Stahl, K.R. Stewart, and F.D. Williams. Extracellular slime associated with *Proteus mirabilis* during swarming. *J. Bacteriol.*, **154**(2):930–937, 1983.

[95] J.B. Stock, A.M. Stock, and M. Mottonen. Signal transduction in bacteria. *Nature*, **344**:395–400, 1990.

[96] T. Matsuyama, K. Kaneda, Y. Nakagawa, K. Isa, H. Hara-Hotta, and I. Yano. A novel extracellular cyclic lipopeptide which promotes flagellum-dependent and -independent spreading growth of *Serratia marcescens*. *J. Bacteriol.*, **174**:1769–1776, 1992.

[97] M. Tcherpakov, E. Ben-Jacob, and D.L. Gutnick. *Paenibacillus dendritiformis* sp. nov., proposal for a new pattern-forming species and its localization within a phylogenetic cluster. *Int. J. Syst. Bacteriol.*, **49**:239–246, 1999.

[98] D. van Sinderen, R. Kiewiet, and G. Venema. Differential expression of two closely related deoxyribonuclease genes, nucA and nucB, in *Bacillus subtilis*. *Mol. Microbiol.*, **15**:213–223, 1995.

[99] T. Vicsek. *Fractal Growth Phenomena*. World Scientific, New York, 1989.

[100] N. Wiener. *Cybernetics: Control and communication in the animal and machine*. Wiley, New-York, 1948.

[101] G.M. Young, M.J. Smith, S.A. Minnich, and V.L. Miller. The *Yersinia enterocolitica* motility master regulatory operon, flhDC, is required for lagellin production, swimming motility, and swarming motility. *J. Bact.*, **181**:2823–2833, 1999.

[102] H. Zabrodsky and D. Avnir. Continuous symmetry measures chirality. *J. Am. Chem. Soc.*, **117**:462–473, 1995.

MODELING SELF-PROPELLED DEFORMABLE CELL MOTION IN THE *DICTYOSTELIUM* MOUND; A STATUS REPORT

WOUTER-JAN RAPPEL*, HERBERT LEVINE*, ALASTAIR NICOL*, AND WILLIAM F. LOOMIS†

Abstract. A population of *Dictyostelium* discoideum amoebae changes its behavior dramatically upon starvation. Cell division ceases, and a developmental program which enables the cells to aggregate and exhibit multicellular behavior follows. Developing a computational model which can address collective cell motion and self-organization in the multicellular state is obviously of interest and will undoubtedly have applications to biological systems other than *Dictyostelium*. In this report, we describe our recent attempts to generate such a model and to thereafter apply it to some recent experimental findings.

1. Introduction. The developmental process wherein *Dictyostelium* amoebae change from being solitary to cooperatively forming a multicellular "organism" [1] has long been a favorite system for mathematical modeling [2]. The reason for this is quite clear. In the early stages of this process, cells communicate with each other via a fairly well-understood chemical wave mechanism [3]. This cAMP chemical wave is similar to other instances of excitable waves such as the BZ reaction [4] and electrical impulses in heart tissue [5], and hence can be studied with familiar techniques. The fact that the cell-cell interaction is long-ranged allows one to disentangle the complicated cell behavior (in detecting, transmitting and responding to the cAMP signal) from the wavefield evolution dynamics and from the cell density dynamics. So, while there are still controversies and questions, we have a good idea as to the types of models which can help us understand the aggregation phase.

As the system progresses to the next developmental stage, that of mound development, cell sorting and tip formation, the previously independent cells begin life as components of a rudimentary multicellular organism. This leads to a profound proliferation of complexity. Now, one has to take into account cell-cell adhesion, interactions between the cells and the extracellular matrix (a set of secreted proteins which can form a type of gel external to the cells) including the slime sheath, possible signaling due to various chemicals, etc. Progress here should directly translate to progress in understanding similar developmental processes in more complex eukaryotes such as ourselves. In what follows, we provide a brief review of the phenomenology of this state. Subsequently, we will describe our efforts to construct a new class of models suitable for simulating multicellular behavior.

*Dept. of Physics, University of California, San Diego, La Jolla, CA 92093-0319; This work was supported in part by NSF DBI-95-12809.

†Dept. of Biology, University of California, San Diego, La Jolla, CA 92093.

As cells enter the mound, they differentiate into two major cell types, pre-spore and pre-stalk. Most evidence to date [6] suggests that this differentiation happens in a spatially random manner, leading to a mixed population of 80% pre-spore and 20% pre-stalk. Exactly which signals are used to coordinate differentiation and to ensure proper proportioning is at present unknown. For a recent modeling effort we refer to [7].

Over a period of roughly 2–4 hours, the cells sort such as to place most of the pre-stalk cells at a point of the periphery of the mound. These pre-stalk cells then form a protruding tip, distending the covering sheath as they do so. Mutants that cannot form a specialized subclass of pre-stalk cells do not show tip formation [8]. Other mutants show sorting defects and lead to papillated mounds [9]. Almost nothing is known for certain about how these events come about, to some extent because of the technically difficulties involved in tracking cell motion in three dimensions in a volume the size of the mound for the entire period of time in question with enough temporal resolution [10]. Later, we will discuss a new protocol for sorting and tip formation in two dimensional aggregates which gets around this experimental bottleneck.

Over the years, there have been many observations of coherent "swirling" motion in the mound stage [11]. One interpretation has been that cells continue to move chemotactically and the swirling is caused by cells responding to circularly propagating waves of cAMP. In support of this hypothesis, there have been observations [12] of darkfield waves in mounds; these darkfield waves have been conclusively linked to cAMP signals for aggregation stage amoebae and one might hope that a similar correspondence holds in mounds. Once one accepts this, one is also led to hypothesize that cAMP wave guidance plays a critical role in sorting and in tip formation. On the other hand, mutant cells created by Wang and Kuspa [13] have no detectable levels of external cAMP yet still sort and form functional tips.

It is clear that adhesion plays a role in mound dynamics [14]. During this developmental stage, cells express adhesion proteins which are distinct from those expressed in early aggregation. Mutants such as LagC [15] which are not "sticky" have mounds which fall apart (presumably due to the random motion of the cells) after formation. There is some evidence in favor of differential adhesion [16], namely that there is a difference in interaction between cells depending on their type. It has been known for a long time that differential adhesion can by itself lead to cell sorting [17]. Examples of this can be seen in the recent work of Glazier and co-workers [18]. However, differential adhesion on its own is unlikely to be responsible for the observed swirling which cannot be explained as a set of random shape changes exploring the adhesiveness "landscape". Also, it is very hard to understand why tips would form under purely adhesive forces. One oft-mentioned possibility is that the dynamics incorporates both differential adhesion and some type of chemical signaling. This forms the basis for the simulation studies of Jiang and collaborators [19] and Savil and Hogeweg [20]. We will briefly review the findings of those studies below. Another possibility, more consistent with data from a recent set of "two-dimensional" experiments [21, 22], invokes self-organization of self-propelled entities together with differential adhesion as the primary effectors of multicellular movement in the mound. New simulation studies of a model based on these assumptions are presented in a later section of this paper.

2. The models. The models that we have investigated are based on an encoding of the cell configuration as a type of "spin" system [23]. That is, each cell is treated as a finite number of sites on a n-dimensional square lattice (n can be either two or three). To distinguish the cells, each site of the lattice is given a spin σ. The value of σ specifies to which cell the site belongs and can take on the value between 1 and N, the total number of cells.

Cell motion is introduced in the model by allowing the cells to fluctuate. This type of fluctuation-dominated kinetics was pioneered by Glazier and Graner [23] for the study of sorting in mixtures of cells derived from early chicken embryos. They defined an effective free energy which contained two types of terms:

$$\mathcal{H} = \mathcal{H}_{vol} + \mathcal{H}_{coh} \tag{1}$$

The first term constrains the cells to a more-or-less constant area:

$$\mathcal{H}_{vol} = \lambda \sum_{\sigma} (A(\sigma) - A_{target})^2 \tag{2}$$

where λ is a parameter that determines the strength of the constraint, $A(\sigma)$ is the area of cell σ (measured in number of lattice sites) and A_{target} is the target area. The medium is not subject to an area constraint.

The second term describes the cohesive energy between cells and between cells and medium:

$$\mathcal{H}_{coh} = \sum_{\sigma,\sigma'} J_{\tau(\sigma)\tau(\sigma')}(1 - \delta_{\sigma,\sigma'}) \tag{3}$$

where $J_{\tau(\sigma)\tau(\sigma')}$ represents an energy cost [24] to have a boundary between cell σ of type $\tau(\sigma)$ and cell σ' of type $\tau(\sigma')$. Apart from the medium (m) we will consider here two other cell types corresponding to pre-stalk and pre-spore cells. We will designate them as dark ($\tau(\sigma) = d$) and light ($\tau(\sigma) = l$) respectively. Since there might be different bond energies associated with boundaries between these three different cell types, we need to define J_{ll}, J_{dd}, J_{ld}, J_{lm} and J_{dm} for each simulation. Note that a larger $J_{\tau(\sigma)\tau(\sigma')}$ corresponds to an energetically less favorable bond and means weaker adhesivity between cell σ and σ'.

The cell configuration is dynamically updated using a standard Monte Carlo algorithm. First, a site of the lattice is chosen at random. Next we provisionally change the spin of this site to the spin of a randomly selected nearest or second-nearest neighboring site. Then, we calculate the difference in energy $\Delta\mathcal{H} = \mathcal{H}_{new} - \mathcal{H}_{old}$ between the old configuration and the new configuration. If $\Delta\mathcal{H} < 0$ the change would lead to a smaller global energy and we accept the change unconditionally. If $\Delta\mathcal{H} > 0$ we accept it with a probability $exp(-\Delta\mathcal{H}/T)$ where T is an effective temperature which determines the amplitude of the allowed fluctuations. We define one Monte Carlo step (MCS) to be as many trial steps as there are lattice points.

Without any further complication, this energy can be used to study sorting via differential adhesion. To get tip formation, however, one must supplement this model with additional mechanisms. As already mentioned, one scenario involves a cooperation between adhesion and chemotaxis. Models of this sort for *Dictyostelium* were studied by Savil and Hogeweg [20] and subsequently by Jiang and coworkers [19]. They both assumed that the stochastic cell was coupled to an underlying deterministic chemical dynamics. These studies produced various plausible scenarios for sorting and mound morphogenesis. But there are problems. One relates to the Wang-Kuspa mutant mentioned above. Another is that the swirling motion necessitates a chemical wave pinwheel and this is not clearly consistent with the simple notion of tip formation via inward chemotaxis, as seen in these simulations.

3. Self-propelled entities. We initiated a new set of experiments to look at multicellular motions in a two-dimensional geometry. The details of the protocol are given elsewhere [22], but roughly speaking, the developmental process takes place on a glass cover-slip with a thin agar sheet overlaid on the cells. This leads to at most a few cell monolayers for the vertical extent of the aggregate which allows for detailed cell tracking during both the swirling motion phase and the subsequent sorting and "tip" formation processes.

Our basic observations are as follows. There is a long period of time in which the cell aggregates exhibit coherent circular motion. These self-organized "vortices" can be either disk-like or toroidal (see Fig. 3). These vortices form also in a non-signaling strain which indicates that the rotational motion is *not* driven by cAMP. At subsequent times, the motion becomes disturbed and sorting ensues; exactly how long it takes for this to happen is strain dependent. What appears to happen is that some of the pre-stalk cells stop moving and form the nucleus of the eventual pre-stalk dominant tip region. Depending on exactly where inside the aggregate this occurs, this can lead to a global re-organization of the swirling motion. Our results are in given in detail in ref. [22].

We now discuss efforts devoted to modeling these observations, namely the existence of coherent vortices [21] and the detailed nature of the sorting process. One thing that is very clear upon observing the cells is that do not just passively deform to maximize binding energy. This is hardly surprising.

Each cell contains an active cytoskeleton which can generate forces by cycles of front protrusions and back retractions [25]. At any given moment, then, the cell is trying to move in a particular direction determined internally. Our approach models this propulsive tendency by introducing for each cell a linear potential $U(\sigma)$ that has its origin at the center of mass (CM) of the cell and that is linearly decreasing in the direction of the propulsion force $\hat{\mathbf{F}}$:

$$U(\sigma) = -\sum \hat{\mathbf{F}} \cdot \mathbf{r} \tag{4}$$

Here, the sum is over all sites of the cell and $\mathbf{r}$ is the vector pointing from the CM to the site. The fact that U is decreasing in the direction of $\hat{\mathbf{F}}$ ensures a propulsive force in the direction of $\hat{\mathbf{F}}$. The propulsion term in the Hamiltonian can then simply be written as

$$\mathcal{H}_{prop} = C \sum_{\sigma} U(\sigma) \tag{5}$$

where C is a positive force constant which determines the strength of the propulsive force.

Finally, we need to specify how the self-propulsive force is updated. Our basic assumption is that each cell adjusts its own propulsive force to match the cohesive force due to their neighbors' movement. This is roughly similar to ideas that have been put forth to study flocking behavior [26]. We have investigated several mechanisms whereby this could be incorporated in our model. They all lead to identical qualitative behavior. The mechanism we used in this paper is as follows: we first calculate the number of neighbor sites $M_{\sigma,\sigma'}$ between σ and all other cells. We then calculate the new direction of $\mathbf{F}$ as the vector sum of the velocities $\mathbf{v}_\sigma$ of the neighboring cells weighted by the number of neighboring sites:

$$\hat{\mathbf{F}} = \frac{\sum_{\sigma'} M_{\sigma,\sigma'} \mathbf{v}_{\sigma'}}{\left|\sum_{\sigma'} M_{\sigma,\sigma'} \mathbf{v}_{\sigma'}\right|} \tag{6}$$

In the next section, we describe some results obtained with this new model. We will see that there is indeed a self-organized rotating disc structure, although we have not as yet succeeded in finding its toroidal cousin. Also, sorting will occur if adhesivities vary, but the details of how this occurs do not appear to match those of the experimental data.

4. Results. As a first test of our model we have simulated the movement of a single self-propelled particle, as shown in Fig. 1. The self-propelled cell, displayed in black, is surrounded by non-propelled cells. The force is directed towards the upper right-hand corner of the picture. The snapshots are taken every 800 MCS, and clearly illustrates the deformability of the cell and reproduces qualitatively the observed movement of a *Dictyostelium* cell. This can be seen more dramatically in the mpeg movie crawl.mpg of the simulation where the frames are taken every 100 MCS. This, and additional mpeg movies of the simulations and experiments reported here, can be viewed or downloaded from `http://herbie.ucsd.edu/levine/dicty.html`.

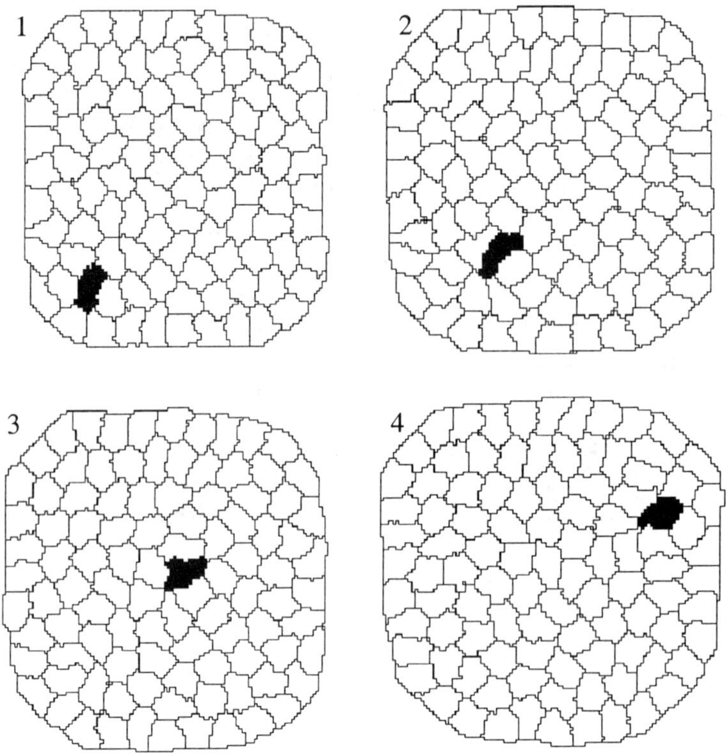

FIG. 1. *Movement of a single self-propelled cell (black) surrounded by non-propelled cells. Snapshots are taken every 800 MCS.*

To study the generation of coherent vortex states we started with either 100 or 400 square cells, each containing 100 sites and of type l. The cells were stacked in a square, surrounded by medium and made "sticky" by choosing $J_{lm} > J_{ll}$. The initial force direction of each cell was chosen at random. After a transient of roughly 100–1000 MCS for $N = 100$ and 1000–10000 MCS for $N = 400$ the cells form a roughly circular patch and are rotating around the center of the patch. A typical final state for $N = 100$ is shown in Fig. 2. It shows the boundaries of the cells and the force direction as a line which starts at the CM (shown as a dot) of the cell and which points in the direction of the force. Depending on the initial conditions, the cells will rotate either clockwise or anti-clockwise. The corresponding mpeg movies (rotation_n100.mpg and rotation_n400.mpg) show clearly the transient and the resulting vortex state. For $N = 100$ (rotation_n100.mpg) the frames are shown every 50 MCS while for $N = 400$ (rotation_n400.mpg) the frames are shown every 100 MCS.

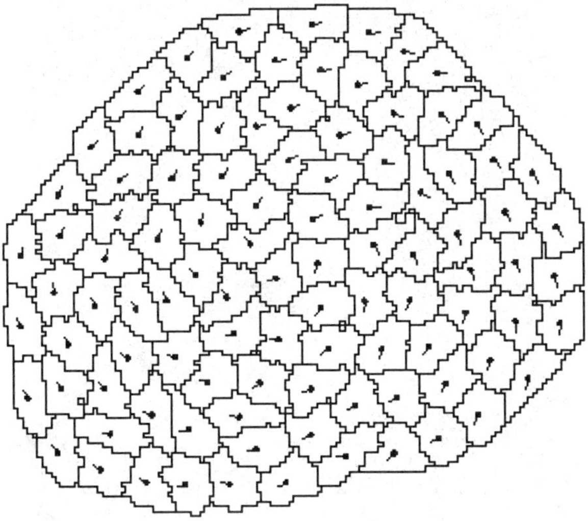

FIG. 2. *Snapshot of a typical final state of the model. The solid lines within each cell start at the CM of the cell, shown as a solid dot, and point in the direction of the force. The parameter values for this simulation are:* $N = 100$, $J_{ll} = 5$, $J_{lm} = 15$ *and* $C = 1.0$. *Throughout this paper the lattice contains 200x200 sites (for* $N = 100$*) and 300x300 sites (for* $N = 400$*), the force direction is updated every 2.5 MCS,* $A_{target} = 100$, $\lambda = 10$ *and* $T = 5$.

It is worth pointing out that we have obtained vortices for a wide range of parameters. The choice of J_{lm} and J_{ll} is critical however, since these parameters control the cohesion forces between the cells. For a vortex state to develop we found that it was necessary to choose J_{lm} sufficiently larger than J_{ll}. This corresponds to "sticky" cells where cells prefer to be bordered by other cells.

Examples of experimental pictures of a vortex state is shown in Fig. 3. As in the simulations the cells aggregate in a rotating "pancake" structure. The rotation can be either clockwise or counter-clockwise depending on the initial conditions. In the majority of the experiments the final structure was a compact disc of the type shown in Fig. 3a. In what follows, we will compare the numerical and experimental results for these structures and argue that our model does a good job in accounting for the data. However, we occasionally also observed a toroidal like structure where the center of the pancake was free of cells. The model, however, has not been able to produce the toroidal structure; we hope to address this in future work.

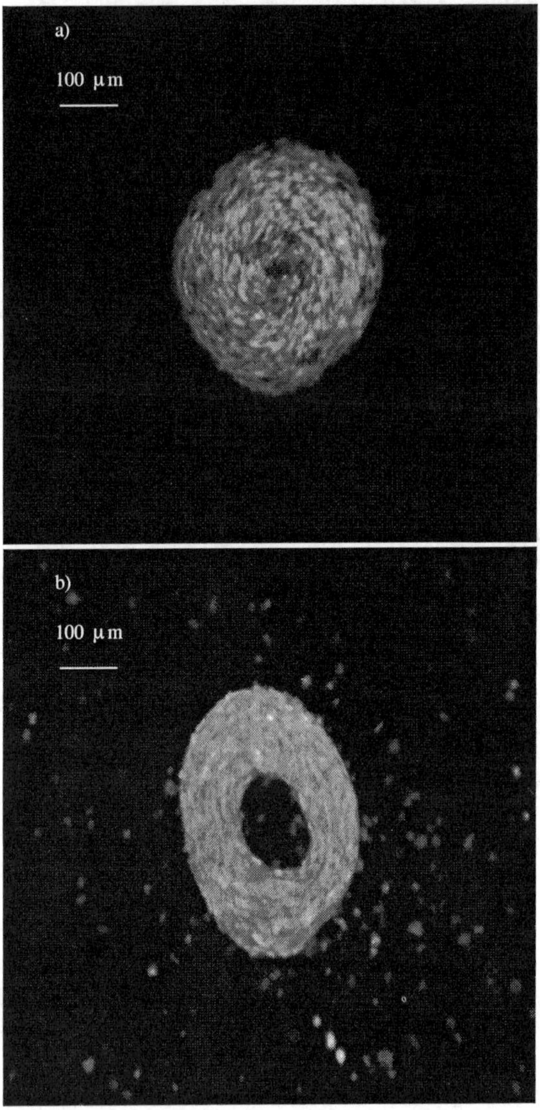

FIG. 3. *Pictures of the vortex state in the experiments. In (a) the final state is a compact structure while in (b) the final state is a toroidal structure.*

The angular velocity of the cells in the simulations can be compared to the ones obtained in experiments. In the experiments we were able to follow individual cells using a strain in which the gene for green fluorescent protein [27] has been fused to the CAR1 (cyclic AMP receptor) gene [28]; the expression of this gene leads to a membrane-localized fusion protein which causes the cell to be fluorescently outlined. We tracked cells in six separate sequences of 15 min. each and measured the angular velocity every 8 s. During each sequence the radius of the cells changed little. Next, we grouped the data in radius intervals of 4 μm and calculated the average velocity and the standard deviation for each interval. The data is shown in Fig. 4 where the vertical bars represent one standard deviation.

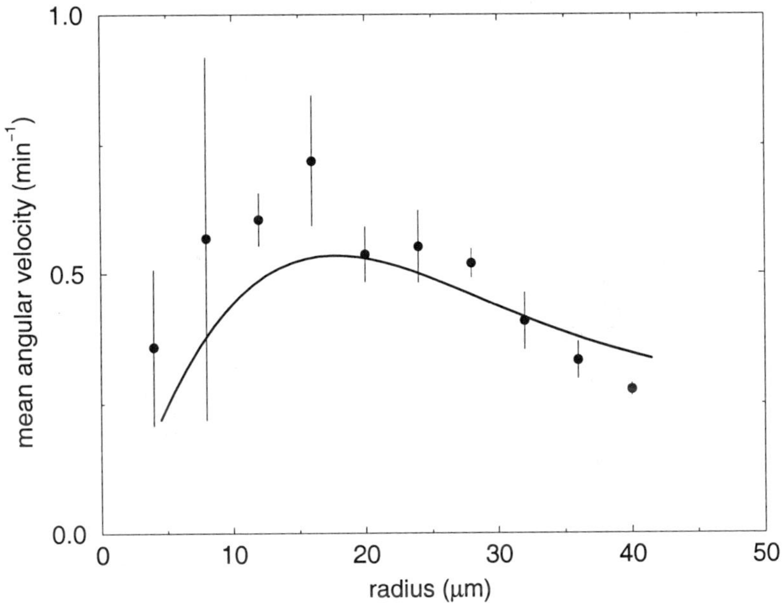

FIG. 4. *The angular velocity as a function of the radius measured in the experiments (solid circles) and calculated using the model (solid line). The overall time scale in the simulations was adjusted to provide the best fit of the model to the data.*

Comparison of experimental angular velocity data to the model requires the identification of a time-scale in the simulations. This was obtained by adjusting the overall time scale corresponding to a MCS was to provide the best fit of the model (solid line) to the data. As a consistency check, we note that this gives an isolated cell velocity of 8 μm/min, which is very close to the experimentally reported value of 10 μm/min [29].

We now turn to cell sorting. To examine the effect of sorting on cell motion we used as an initial condition the vortex state described above. We then changed 20% of the cells at random from l to d, corresponding to prestalk cells, and continued the simulation. To facilitate sorting we chose the energy cost between a dark and light boundary to be larger than between a dark-dark and light-light boundary (i.e. $J_{ld} > J_{dd}$ and $J_{ld} > J_{ll}$). This means that cells prefer to be surrounded by cells of the same type. For a more detailed discussion of the possible choices of the energy costs and the resulting patterns we refer to [23]

Fig. 5 shows a sequence of snapshots for the case where the propulsive force is turned off. As a consequence of the different adhesion properties between light and dark cells, dark cells try to minimize their boundary with light cells and rapidly form small clusters. Further sorting occurs on a much longer time scale as clusters of dark cells diffuse through the light cells and merge with other clusters of dark cells. The two mpeg movies for $N = 100$ (sort_c=0_n100_short.mpg and sort_c=0_n100_long.mpg) show the sorting with a time difference between frames of 5000 MCS and 50000 MCS respectively. The sorting in the $N = 400$ takes considerably longer and is not displayed here.

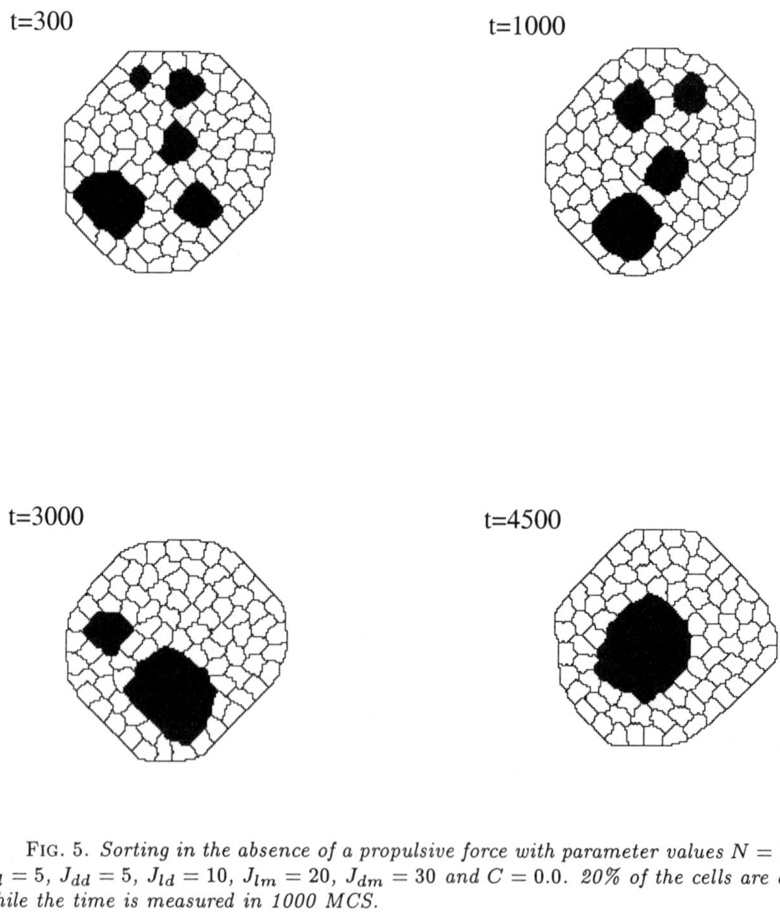

FIG. 5. *Sorting in the absence of a propulsive force with parameter values* $N = 100$, $J_{ll} = 5$, $J_{dd} = 5$, $J_{ld} = 10$, $J_{lm} = 20$, $J_{dm} = 30$ *and* $C = 0.0$. *20% of the cells are dark while the time is measured in 1000 MCS.*

Fig. 6 shows a similar sequence of snapshots for $N = 100$ for identical parameters values but now in the presence of a propulsive force. The sorting takes considerable less amount of time which can also clearly be observed on the animations. sort_c=1_n100.mpg shows the sequence for $N = 100$ with 50 MCS between each frame. sort_c=1_n400.mpg shows the equivalent animation for $N = 400$ and 500 MCS between each frame. However, the mechanism for sorting appears to be different than the one observed in experiment. In the simulations, clusters of dark cells form rapidly. The angular speed of these clusters depends on their size: bigger clusters have a larger angular velocity than smaller ones. As a consequence, large clusters catch up and merge with smaller ones. In the experiments on the other hand, a region consisting predominantly of pre-stalk cells appear to suddenly stop rotating and blocking the movement of cells behind it. After sorting out, the pre-stalk cluster starts to slowly migrate towards the outer edge of the disc.

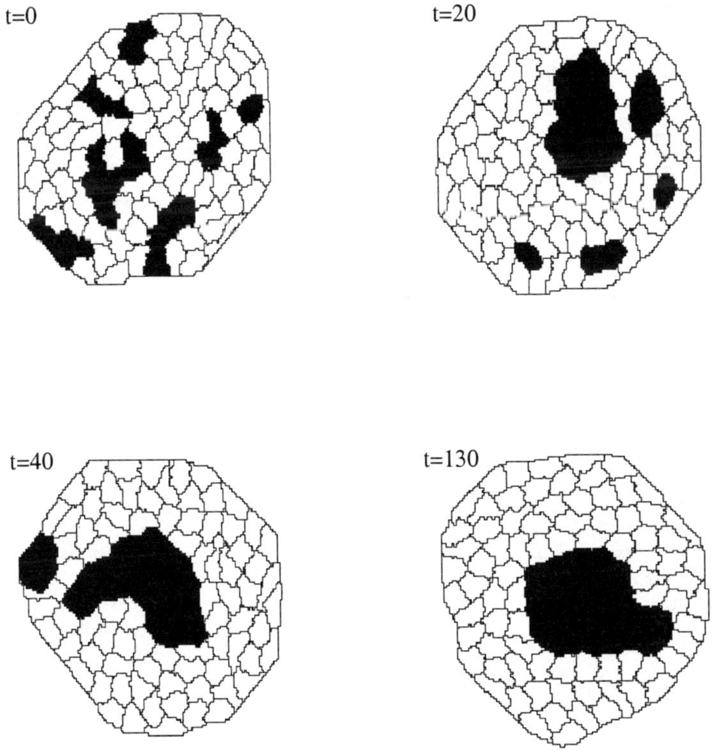

FIG. 6. *Sorting in the vortex state. Parameters and initial conditions are as in Fig. 5 but with $C = 1.0$, corresponding to the presence of self-propulsion. Time is again measured in 1000 MCS.*

If one assumes that our computational approach correctly predicts the macroscopic structure from a given set of microscopic interactions, it is clear that differential adhesion in its simplest form is inconsistent with the experimental findings. One must therefore search for other possible mechanisms, all the while continuing the process of checking the overall validity of our methodology. Recently [22], we speculated on a possible explanation which involves a change in the spatial expression of adhesion molecules on the cell surface of pre-stalk cells. Upon entering the aggregate cells appear adhesive at their ends and appear to move in circulating files. These files could be broken up if the lateral adhesion of pre-stalk cells, but not pre-spore cells, increases due to the expression of new adhesion molecules (initially) evenly distributed over the cell surface, This could result in a domain dominated by pre-stalk cells as observed in the experiments. Modeling efforts that are capable of addressing spatially specific adhesion are currently under way.

5. Outlook. The developmental dynamics of *Dictyostelium* provides a rich phenomenology upon which one can test notions of how to apply ideas from the field of non-equilibrium pattern-formation physics to processes involving biological complexity. The leading edge of this effort involves, in our opinion, understanding what is the appropriate class of models for studying the initial stages of multicellular behavior. In this status report, we have attempted to explain our own approach, including both its successes and its failures. We are hopeful that our efforts and those of others will eventually converge on a reliable computational strategy for going from cell-cell interactions to collective multicellular behavior.

We would like to thank A. Kuspa and P. Devreotes for providing some of the *Dictyostelium* strains used in the experiments. Also, one of us (HL) acknowledges useful conversations with E. Ben-Jacob, J. Glazier and Y. Jiang.

REFERENCES

[1] For a general introduction, see W. Loomis, *The Development of Dictyostelium discoideum* (Academic, New York, 1982).
[2] For reviews, see H.G. Othmer and P. Schaap, *Comments on Theor. Biol.* **5**, 175 (1998) and H. Levine, *Physica A* **249**, 53 (1998).
[3] P.N. Devreotes, *Science* **245**, 1054 (1989); *Neuron* **12**, 235 (1994).
[4] G.S. Skinner H.L. and Swinney, *Physica D* **48**, 1 (1991); A. Winfree, *J. Phys. Chem.* **93**, 740 (1989).
[5] A.T. Winfree, *When Time Breaks Down*, Princeton, NJ, 1987.
[6] W. Loomis, *Microbiol. Rev.* **60**, 135 (1996).
[7] P. Schaap, Y. Tang and H.G. Othmer, *Differentiation* **60**, 1 (1996).
[8] G. Shaulsky, A. Kuspa and W.F. Loomis, *Genes and Development* **9**, 1111 (1995).
[9] J.T. Stege, G. Shaulsky and W.F. Loomis, *Debv. Biol.* **185**, 34 (1997).
[10] For some efforts at 3-d imaging, see K.W. Doolittle. I. Reddy, and J.G. McNally, *Dev. Biol.* **167**, 118 (1995); K.A. Kellerman and J.G. McNally, *Dev. Biol* **208**, 416 (1999).

[11] J. Rietdorf, F. Siegert, and C.J. Weijer *Dev. Biol.* **177**, 427 (1996) and references therein.

[12] F. Siegert and C.J. Weijer, *Curr. Biol* **5**, 937 (1995).

[13] B. Wang and A. Kuspa, *Science* **277**, 251 (1997); their construct bypasses the need for internal cAMP and thereby allows for development without the presence of adenyl cyclase (ACA), the enzyme which manufactures cAMP.

[14] S. Bozzaro and E. Ponte, *Experientia* **51**, 1175 (1995).

[15] J. L. Dynes. A.M. Clark, G. Shaulsky, A. Kuspa, W.F. Loomis and R.A. Firtel, *Genes and Dev.* **8**, 948 (1994).

[16] I. Takeuchi, T. Kakutani, and M. Tasaka, *Devel. Genetics* **9**, 607 (1988); C. Siu, B.D. Roches, and T.Y. Lam, *Proc. Natl. Acad. Sci.* **80**, 6596 (1983).

[17] M.S. Steinberg, *Science* **141**, 401 (1963).

[18] J.C.M. Mombach, J.A. Glazier, R.C. Raphael and M. Zajac, *Phys. Rev. Lett.*, **75** 2244 (1995).

[19] Y. Jiang, J. Glazier and H. Levine, *Biophys. J*, **75**, 2615 (1998).

[20] N. Savil and P. Hogeweg, *J. Theor. Biol.* **184**, 229 (1997).

[21] W.-J. Rappel, A. Sarkissian, A. Nicol, H. Levine and W.F. Loomis, *Phys. Rev. Lett.* **83** 1247 (1999).

[22] A. Nicol, W.-J. Rappel, H. Levine and W.F. Loomis, *J. cell sci.* **112**, 3923–3929 (1999).

[23] J.A. Glazier and F. Graner, *Phys. Rev. E* **47** 2128 (1993).

[24] In recent (unpublished) work, Glazier and co-workers have reversed the sign convention of the cohesion term, thereby making cell-cell adhesive energies more negative (rather than less positive) than cell-medium terms. This choice must then be supplemented by an inextensibility constraint on the cell membrane. This new choice is in some sense more physical, and we are in the process of implementing it in our modeling efforts. While we expect some quantitative changes, we do not expect any qualitative conclusions to depend on this level of detail.

[25] T.P. Stossel, *Am. Scient.* **78**, 408 (1990).

[26] T. Vicsek, A. Czirók, E. Ben-Jacob, I. Cohen, O. Shochet and A. Tenenbaum, *Phys. Rev. Lett.* **75**, 1226 (1995).

[27] W. Ward in "Photochemical and Photobiological Reviews", K. Smith, ed. Plenum, NY (1979).

[28] Z. Xiao, N. Zhang, D.B. Murphy, P.N. Devreotes, J. Cell Biol. **139**, 365-74 (1997).

[29] R. Escalante, D. Wessels, D.R. Soll and W.F Loomis *Mol. Biol. Cell* **8**, 1763 (1997).

A MINIMAL MODEL OF LOCOMOTION APPLIED TO THE STEADY GLIDING MOVEMENT OF FISH KERATOCYTE CELLS

A. MOGILNER*, E. MARLAND†, AND D. BOTTINO‡

Abstract. In this paper we present the quantitative analysis of the basic mechanisms underlying the phenomenon of animal cell motility. We describe plausible mechanisms of actin-based protrusive force generation at the cell's leading edge. We also demonstrate that the dynamics of self-alignment and contraction of the actin-myosin network can explain forward translocation of the cell body. Regulation of graded adhesion between the substrata and the ventral surface of the cell is then discussed. Finally, we derive a one-dimensional mathematical model of cell locomotion applied to fish keratocyte cells.

1. Introduction. Many animal cells possess the fundamental ability to crawl; this form of locomotion is essential for morphogenesis and wound healing. Mechanochemical models of cell tissue pattern formation (pioneered by J.D. Murray, G.F. Oster and their colleagues) are based on the ability of early embrionic cell populations to move in a coordinated fashion (see Ch. 17 of [44] for review). Before we can study in detail interacting populations, we must first understand individual cell migration. Despite recent advances in cell biology, biochemistry and biophysics relating to cell motility [3], we still do not have a clear picture of how animal cells move over surfaces. In this paper we discuss various approaches to the quantitative description of motility and introduce a model of the steady motion of a cell.

Two decades ago, Abercrombie described animal cell motion as a five-step process [1, 36, 55, 52, 14]: (1) protrusion of the cell's leading edge, (2) adhesion of this protrusion to the substratum, (3) forward translocation of the cell body, (4) detachment of the cell's rear edge, (5) retraction of the cell's rear edge. Since that initial observation, fish and amphibian keratocyte cells have become one of the most popular model locomotion systems because of the propensity of these cells to move spontaneously when placed on a surface and the relative simplicity of their motile appendages.

The fish keratocyte is one of the most rapidly moving eucaryotic cell types. When a single keratocyte is placed on a surface, it can move a few tenths of a micron per second. All five steps of cell motion occur simultaneously so that the cell appears to glide steadily while maintaining a fanlike shape (Figure 1). The front, fanlike part of the migrating cell - the "lamellipod" - is a broad, flat cytoskeletal protrusion devoid of organelles. The lamellipod is a few tenths of a micron thick and 10-15 microns long

---

*Dept. of Mathematics and Inst. of Theoretical Dynamics, University of California, Davis, CA 95616.

†Inst. of Theoretical Dynamics, University of California, Davis, CA 95616.

‡College of Natural Resources, University of California, Berkeley, CA 94720.

and wide. Behind the lamellipod is a roundish cell body, a few microns in size, containing the nucleus and organelles.

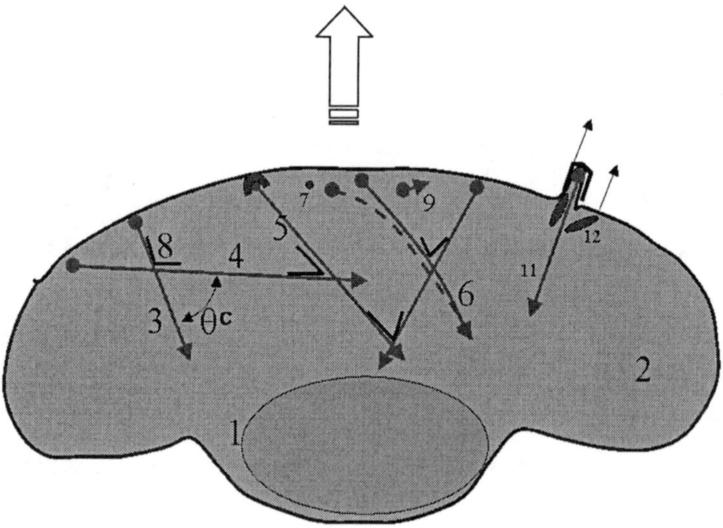

FIG. 1. *The fan-like shape of the migrating cell: view from above. The cell body (1) is preceded by the lamellipod (2). Long actin filaments (3-6) are solidified into the lamellipodial network by crosslinking proteins (8). Fibers normal to the cell boundary (3) cannot bend effectively and do not generate the force of protrusion, nor do filaments incident on the membrane at the critical angle θ_c (4). Some polymers (5) are bound to membrane associated proteins and generate force against protrusion. Slightly bent filaments incident on the membrane at moderate angles (6) generate protrusive force. Such filaments grow when actin monomers (7) intercalate into the gap that appears due to thermal writhing (dashed). A hypothetical mechanism of myosin powered protrusion: filopodial actin (11) serves as a track for myosin I motors (12), some of which push forward the cell membrane, while others push forward a "protein cap" creating space for actin growth. Short filaments (9) do not participate in protrusive force generation.*

It is now widely accepted that the lamellipod is the basic engine for gliding and crawling locomotion [36]. The cell body and posterior seem to be mechanically passive structures pulled forward entirely by the lamellipod's action. Of the three major polymer components of the eucaryotic cell cytoskeleton, intermediate filaments are not known to take part in locomotion [3]. Although microtubules play an important role in the locomotion of many animal cells, keratocytes can move without microtubules [36]. Therefore, the lamellipod of keratocyte can be thought of as a flat network of filamentous actin [3, 14, 35, 55].

Many of the details are known, but the interactions of the individual mechanisms resulting in cell motion are not well understood. A huge variety of molecular mechanisms are involved in locomotion, which leads to *multiplicity and redundancy in force generation machineries and signaling pathways* in the same cell. Thus we should investigate simple model motility systems before we can understand the full complex phenomenon. One such system is the propulsion of the parasitic bacterium *Listeria*, which exploits an actin based protrusion mechanism [2]. Another is the amoeboid migration of the actin-free *Ascaris* sperm [21]. The fish keratocyte is perhaps the next simplest system because it can be dissected into just three, rather than Abercrombie's five, substeps of motility. First, growth of the actin network leads to the extension of the leading edge of the cell. Secondly, graded substratum-coupled anchoring is developed, so that at the front the lamellipodium adheres to the surface much more firmly than at the cell's rear. Finally, the lamellipodial cytoskeleton is contracted, causing forward translocation of the cell body (sometimes called "traction" [36]).

The idea that these three basic mechanisms - protrusion, forward translocation and adhesion - acting together are necessary and sufficient to produce highly effective cell movement has been expressed repeatedly in recent biological literature. The corresponding kinematic principles underlying the movement of these cells has already been described within the "graded radial extension model" [29]. This model concluded that lamellipodial extension and retraction in keratocytes must be highly coordinated to maintain the cell's simple shape. Protrusion and retraction take place locally at the cell's edge in the direction normal to the cell's boundary. The local rates of protrusion and retraction are graded, varying in such a way that the simple semicircular shape of the keratocyte is maintained. In addition to kinematic analysis, imaging of traction forces generated by the lamellipodial cytoskeletal contraction has been done [45]. Force distributions reveal very little contraction at the front part of the lamellipodium and significant strains and stresses *normal to the direction of motion* at the sides of the cell. Interference microscopy studies of the attachments between the ventral surface of the cell and the substratum found the highest density of adhesion sites to be at the frontal periphery of the lamellipodium [30].

The mechanical dynamics is just one aspect of cell migration. Equally important is the spatiotemporal regulation of the mechanical properties of the lamellipodial cytoskeleton by signaling pathways involving a large variety of actin associated proteins [3, 55, 19, 27, 35, 51].

The experimental work and theoretical modeling attempt to dissect the complex process of motility into simpler mechanical phenomena (protrusion, traction and adhesion) and structural elements of the lamellipodium (cytoskeletal, adhesion and membrane systems). In this paper we build a minimal model of keratocyte cell locomotion. The model is based on known biochemical, structural and mechanical data. We analyze plausible molecular mechanisms of force generation, cytoskeletal regulation and self-organization. Compared to existing models, our approach is similar to the model [11], which was the earliest comprehensive attempt to incorporate the theoretical description of all three major elements of the locomotory process into a model of the whole cell.

In the next section, models for protrusion of the cell's leading edge are introduced. Section 3 is devoted to the analysis of the one-dimensional mathematical model of the dynamic contraction of the depolymerizing actomyosin network. In Section 4 the dynamic contraction model is coupled with models of protrusion and adhesion to produce a minimal self-consistent description of steady cell propulsion. Finally, future research directions and the interplay between theory and experiment are discussed in Section 5.

2. Protrusion of the cell's leading edge. The dynamics of a moving cell's leading edge is probably the best understood example of the molecular basis of motility [6, 35]. In this section we will concentrate on the analysis of protrusive force generation, postponing the problem of regulation until Section 4.

Actin, the main constituent of lamellipodia, exists either in a monomeric (G-actin) or polymeric (F-actin) form. Each actin monomer can bind ATP, which is hydrolyzed to ADP after incorporation of the monomer into the polymer. These polymers are highly dynamic polarized structures. The plus (barbed) end grows more quickly than the minus (pointed) end. *In vitro*, a stable equilibrium concentration ratio between G- and F-actin occurs for which the net rate of polymerization at the plus ends is equal to the net rate of depolymerization at the minus ends. At this equilibrium, actin fibers are said to undergo "treadmilling" because they translocate in the plus direction without actual physical motion. *In vivo*, monomeric actin is present at concentrations well above what is required for rapid polymerization, and a host of actin sequestering, capping, severing, nucleating, and depolymerizing proteins control F-actin assembly.

2.1. Structure of the lamellipodial actin cytoskeleton. The structure of the lamellipodial actin cytoskeleton has several distinguishing characteristics. Close to the ventral surface, 2-D sheets made of actin fibers several microns long span the length of the lamellipodium (Figure 1). Most of the filaments are oriented with their barbed ends toward the leading edge, indicating that filament growth occurs predominantly at the inner membrane surface at the leading edge of the cell. Earlier electron micrographs of the cytoskeleton at the leading edge suggested that long actin fibers are arranged into a square lattice [54]. However, more recent observations show that filaments are distributed evenly over a wide range of angles, approximately between $-60°$ and $+60°$ relative to the direction of cell motion [56].

The 2-D actin sheets are interspersed by long, tight filament bundles which form filopodial protrusions and shorter 'microspike' and 'rib' bundles. Filopodial unipolar bundles, oriented mostly perpendicular to the leading edge, consist of $\sim$ 20–30 filaments oriented with their plus ends toward the front [53]. There is also a 3-D isotropic actin meshwork of shorter transverse fibers spanning the dorsal-ventral surfaces [31]. The actin network mesh size is $\sim$ 10–50 nm.

The architecture of the lamellipodial cytoskeleton and the similarity in magnitude of cell extension rates and actin polymerization rates at physiological concentrations clearly indicate that actin polymerization at the leading edge is likely to account for lamellipodial protrusion. The central question for protrusion, however, is: *what is the nature of the force that generates protrusion*? Is polymerization alone able to account for the protrusive force, or does some other molecular mechanism push the cell membrane away from polymer tips, allowing the polymerizing actin to fill the newly created gap?

2.2. Possible mechanisms of protrusive force generation. Besides polymerization acting alone [6, 48], the following mechanisms for protrusive force generation have been proposed:
1) hydrostatic pressure generated by cortical contraction *away from the leading edge* [9],
2) osmotic pressure generated by actin gel swelling *at the leading edge* [47],
3) myosin powered protrusion [53, 52, 8, 36].
None of these mechanisms can be excluded from a mechanical point of view. The maximal force of protrusion can be measured only very approximately because when a crawling cell is stalled by an obstacle, it tends to develop a protrusion elsewhere and turn around the obstacle. Experimental estimates for the protrusive force F_p range from 10^4 to 10^5pN [46]. Cortical contraction in dividing sea urchin eggs was theoretically estimated to generate pressure of up to $1Atm = 0.1\text{pN/nm}^2$ [15]. If 10% to 100% of this pressure is used to apply hydrostatic pressure to a $0.2\mu\text{m} \times 5\mu\text{m}$ strip of the leading edge, the experimentally measured protrusive force would be achieved. The number of myosin molecules in the lamellipodium can be realistically estimated as $\sim 10^4$ [56]. A single myosin head develops a power stroke of 1–2 pN [42], so myosin powered protrusion is also possible.

Locally generated osmotic/hydrostatic pressure is the most difficult to analyze because several different molecular events can be responsible for this phenomenon. Gel swelling pressure has never been accurately computed, but its order of magnitude can be estimated as follows. The elastic stress of polyelectrolyte actin gel is balanced by its osmotic pressure, which arises from the entropic motion and counterion pressure of actin strands. When the gel is solated, its elastic modulus decreases and the osmotic pressure causes the gel to expand. A long actin filament can be significantly bent by a $\sim$ 10pN force and completely broken by a $\sim$ 100pN force [24]. Realistically, therefore, the elastic force cannot be more than 10pN per fiber. Estimating the number of fibers in contact with the leading edge to be $0.2\mu\text{m} \times 5\mu\text{m}/400\text{nm}^2 = 2500$ (the contact area for one fiber is the mesh size $\sim$ 20nm squared), the upper bound for the total gel swelling force would once again be $\sim 10^4$pN. Osmotic pressure is partially caused by cytoplasmic ions other than actin counterions. Measurements of the cytoplasmic osmolarity of nerve growth cones [4] revealed the hydrostatic pressure at the leading edge to be $0.1Atm = 0.01\text{pN}/\text{nm}^2$. This pressure would result in a lower protrusive force than what is observed, but the discrepancy is only one order of magnitude.

Although all these mechanisms are mechanically plausible and might occur in various cell types, there is compelling evidence for ruling them out in the case of fish keratocyte protrusion. Keratocyte cells continue to move when the membrane is perforated with antibiotics, relieving the internal hydrostatic pressure. Lamellipodia subjected to a hypertonic environment continue to expand at an unchanged rate [4]. Thus, both non-local hydrostatic pressure and non-counterion osmotic pressure are unlikely to be the significant driving mechanisms. The classical gel swelling pressure mechanism requires the existence of a rubber-like entropic meshwork of long, flexible fibers [16] that expands dramatically upon partial solation [47]. However, the front of the lamellipodial network consists of semi-stiff, heavily cross-linked polymers, only the complete solation of which would cause swelling.

According to the hypothesis for protrusion by a myosin powered mechanism [53, 52, 8, 36], bundled actin polymerizes into a space created by myosin motors which push 'protein caps' at the tips of the filopodia forward (Figure 1). The lamellipodial sheet is then pulled forward by myosin on the filopodial 'tracks' immobilized by adhesions to the substrate. In keratocytes, however, filopodial bundles do not always exist and generally do not protrude beyond the leading edge. Also, the predicted slippage of the lamellipodium relative to the filopodia is not observed.

Evidence for the sufficiency of actin polymerization alone comes from the much studied bacteria *Listeria* [2]. This pathogenic organism propels itself inside a host cell by assembling a comet-like tail from the host cell's actin. Because of the absence of traction and adhesion, *Listeria* propulsion is a convenient protrusion model. Since this movement takes place far from the cell membrane and can be reproduced in cytoplasmic extract, hydrostatic pressure cannot be utilized for *Listeria* propulsion [36]. Finally, this system does not stain for known myosins, and known myosin inhibitors do not affect bacterial propulsion [36].

2.3. Protrusive force generated by polymerization alone. Direct experimental proof that *actin polymerization alone can produce the necessary force* was recently obtained by [38] from observations of protrusive growth in giant liposomes. The free energy of polymerization can be used to generate protrusive force by the "thermal brownian ratchet" mechanism [48]. In [39, 40] the brownian ratchet idea was generalized to the following "elastic polymerization ratchet" mechanism. When a long actin fiber crosslinked into the lamellipodial network and incident on the cytoplasmic face of the membrane bends, it exerts an elastic force of protrusion (Figure 1). When this filament undergoes thermal bending undulations (actin bends much more easily than it compresses), monomers intercalate into the emerging gap and assemble onto the tip of the filament. This results in the effective advancement of the leading edge by $\delta\cos(\theta)$. Here $\delta = 2.7$nm is one-half the size of an actin monomer (filaments consist of two strands), and θ is the incidence angle relative to the direction of cell motion. The corresponding work W of protrusion is $f\delta\cos(\theta)$, where f is the resistive force. The effective polymerization rate can be derived easily near thermodynamic equilibrium, when the filament growth is almost stalled. The condition of near-thermodynamic equilibrium is satisfied if relatively few filaments grow against significant resistance, which is often the case. The rate $k_{\text{on}}M$ of actin assembly, where k_{on} is the polymerization constant and M is the local effective concentration of polymerizable monomeric actin, is modified by the Boltzmann factor $exp(-W/k_BT)$ [16]. If the depolymerization constant k_{off} is assumed to be force independent (there are no indications otherwise), then the net rate of actin assembly is $(k_{\text{on}}M exp(-W/k_BT) - k_{\text{off}})$, and the corresponding effective rate of growth of the filament is:

$$(2.1) \qquad V_p = \delta\cos\theta(k_{\text{on}}M \; exp(-f\delta\cos\theta/k_BT) - k_{\text{off}}).$$

The maximal force that can be developed by a single actin filament at stall ($V_p \simeq 0$) is $f_s \simeq (k_BT/\delta)\ln(k_{\text{on}}M/k_{\text{off}})$. An important feature of this expression is that this force is not sensitive to the concentration of polymerizable monomeric actin (which is difficult to estimate) since M appears only in the logarithmic term. Under physiological conditions, $k_{\text{on}}M/k_{\text{off}} \sim 100$, which modifies the force $k_BT/\delta \simeq 1.5$pN by the logarithmic factor $\simeq 4$. Thus, *one polymerizing actin filament can develop a protrusive force up to* $6pN$, of the same order of magnitude as the myosin motor. Using the same estimate we used previously of 2500 filament tips along the leading edge, the elastic polymerization ratchet mechanism can account for the observed protrusive force of $\sim 10^4$pN.

These estimates demonstrate the plausibility of the elastic polymerization ratchet mechanism, but to estimate the force of polymerization and rate of protrusion more accurately, we have to consider the whole ensemble of the lamellipodial actin and the nature of the resistive force. Lamellipodial actin fibers can be approximated as *independent 'assembly' motors*. They can be assumed to generate the force independently because thermal modes of fluctuation of the tips of the fibers have amplitudes and frequencies large enough to be more effective than collective modes of vibration of the entire lamellipodial network.

The best-supported assumption about the establishment of the lamellipodial network is that Arp2/3 protein complexes are activated by membrane associated protein complexes at the leading edge and attach to the sides of the pre-existent crosslinked long actin filaments [43]. The important question of how a high concentration of activating proteins at the front is achieved is discussed in Section 4. Activated Arp2/3 complexes nucleate actin filaments and cap their minus ends. The plus ends then grow, and the lamellpodial network assumes its characteristic branching structure [56]. Fibers not parallel to the substrate would grow at most a few hundred nanometers in length, at which point their tips would be stalled by either the dorsal or ventral membrane planes.

The filament orientations are strongly correlated locally [35], but initially the network is globally isotropic. However, in the flat sheet of long fibers, those fibers almost parallel to the leading edge soon lag behind and do not contribute to the force of polymerization. Filaments that are growing exactly in the direction of cell motion are effectively rigid [39]; as a result they cannot bend sufficiently and are therefore unable to generate force. Thus, only filaments oriented between angles θ_0 and θ_c can generate force. The angle $\theta_0 \simeq (10\text{–}15)^\circ$ is determined by the elastic properties of actin [39, 40]. The 'cutoff' angle θ_c is determined by the kinematics of protrusion: if $V_0 = k_{\text{on}}M$ is the free polymerization velocity (actin disassembly can be neglected because normally $k_{\text{on}}M >> k_{\text{off}}$), then fibers oriented at the cutoff angle grow freely with the rate V_0, but extend in the direction of cell motion with the rate of protrusion V_p (Figure 1). Thus $V_p = V_0 \cos\theta_c$, and

$$\theta_c = \arccos(V_p/V_0). \tag{2.2}$$

One of the conclusions from our theory is that at higher resistance, when the rate of protrusion decreases, the cutoff angle θ_c increases.

All filaments incident on the membrane extend forward with the same rate V_p. From (2.1) and (2.2), a single such filament generates the force $f \simeq (k_B T/\delta\cos\theta)\ln(\cos\theta/\cos\theta_c)$. The total force of protrusion generated by the long lamellipodial fibers is:

$$F_p = \frac{k_B T}{\delta}\int_{\theta_0}^{\theta_c} d\theta \frac{n(V_p,\theta)}{\cos\theta}\ \ln\frac{\cos\theta}{\cos\theta_c}. \tag{2.3}$$

Here $n(V_p,\theta)$ is the velocity-dependent angular filament density per unit length that depends also on the processes of regulation; it will be derived in Section 4. According to our theory, stiff filopodial bundles are ineffective as force generators, and their function is not mechanical.

The resistance to protrusion is mainly due to two sources: membrane surface tension and association between the actin cortex and the membrane. The viscous drag force of protrusion is negligible [39, 40]. There seems to be confusion in the literature because of various uses of the term "surface tension." Here we use this term for a bending modulus determined by the splay of the outer membrane leaflet and compression of the inner leaflet [12].

The corresponding force density, $\simeq 0.035$pN/nm [13] causes resistance of ~ 0.035pN/nm $\times 5\mu$m $\sim (1.5\text{–}2) \times 10^3$pN.

The second source of resistance to protrusion comes from the viscous dissipation that occurs when the membrane flows over the cytoskeleton (Figure 1). Actin filaments which are part of the crosslinked cortex bind to integral membrane proteins, for example, ponticulin [3]. When uncapped filament tips push the membrane forward, these bonds break, dissipating mechanical energy. Such resistance force was not measured for keratocytes, but for other cells the corresponding values are of the order of 10pN/$(0.2\mu\text{m})^2$ [26, 10]. For the leading edge of the keratocyte lamellipodium, this would account for $\sim 250\text{pN}/\mu\text{m}^2 \times 0.2\mu\text{m} \times 5\mu\text{m} = 250\text{pN}$, slightly augmenting the membrane bending resistance. The velocity dependence of the resistance force is very weak [17].

The estimated resistance force, $\sim 10^3$pN, is an order of magnitude less than the observed maximal force of protrusion, $\sim 10^4$pN. This means that only a small percentage of filaments incident on the membrane generate force, while others have their tips bound to membrane proteins and resist the protrusion rather than generate it. Thus, we predict that the rate of protrusion, which increases with the number of free filament tips, is inversely proportional to the force of separation between the membrane and actin cortex, which is in turn proportional to the number of bound tips. This prediction is in agreement with recent observations [49]. The force F_p of protrusion is balanced by the resistive force F_r, which we will approximate as a velocity and density independent constant: $F_r = 10^3$pN. In Section 4 we will obtain the equation for the function $n(V_p, \theta)$ and substitute it into formulae (2.2) and (2.3) to find an expression for the velocity-dependent force of protrusion, $F_p(V_p)$. Finally, we will solve the equation $F_p(V_p) = F_r$ to determine the rate of protrusion.

3. Forward translocation of the cell. The next step after the extension of the cell's leading edge and its firm adhesion to the substratum is the forward translocation of the cell body, or, assuming that the cell body is a passive cargo, lamellipodial traction. We will assume that adherence at the very front of the lamellipodia is infinitely strong, so that forces of traction do not cause any rearward slippage of the leading edge that would produce retrograde flow. This is a good approximation for keratocyte cells [58, 29, 30]. Traction is an essential part of motility and the least well understood [34, 36, 7]. Here we will review plausible traction mechanisms and then present a minimal model of dynamic contraction of the actomyosin network.

3.1. Review of models for traction. An obvious question about traction is: can protrusion alone generate a cell's forward motion? This would be the case if the cell membrane were a non-stretchable bag enveloping the treadmilling polarized actin network. The protrusion force of $\sim 10^3$pN would be more than sufficient given the weak adhesion of the cell rear, because viscous resistance to cell motion is on the order of $\sim$ 10pN. This simplest mechanism does not work, however. The cell membrane is capable of changing its surface area thousands of times through regulated endo- and exocytosis [10, 49]. In general, the membrane does not play active mechanical role in traction [25]. The fact that short forward motions of the cell body without simultaneous protrusion are observed in keratocytes indicates independence of protrusion and traction. Lin and Forscher [33] demonstrated this independence directly in nerve growth cone lamellipodia.

Note that the main "goal" of lamellipodial traction may not be locomotion; it may be the deformation and re-modelling of the substratum, as is the case in wound healing. Indeed, the lamellipodium of the keratocyte is able to generate intense traction stress of $\sim 10^{-3}$pN/nm^2 [45]. An area of $\sim 10\mu$m $\times$ (1–2) μm at the rear of the lamellipod is involved in generating a total traction force of $\sim 10^4$pN. Interestingly, the traction force is of the same order of magnitude as the maximal force of protrusion.

There exists a mechanism of lamellipodial traction not requiring any new elements except those already involved in protrusion: actin filaments and crosslinking proteins. This mechanism is based on the phenomenon of entropic contraction of the depolymerizing polymer network [40]. Long actin filaments of the lamellipodia are straight [54] because in the process of growth they get crosslinked into the network before bending significantly; the persistence length of actin polymer, $\geq 1\mu$m [23, 18], is an order of magnitude greater than the likely distance $\sim$ 10–100 nm between crosslinkers. As spontaneous or induced depolymerization (see below) takes place across the lamellipodia, distances between neighboring crosslinks increase, energy stored in the process of crosslinking is released, and, due to thermal writhing equilibrium, the distance between the remaining crosslinks decreases. This leads to the development of an entropic, contractile stress and subsequent contraction of the lamellipodial network. In [40] the estimate for the maximal entropic contractile force for one filament of $k_B T\lambda/l_c^2 \sim$ (1–10) pN was obtained, where $\lambda \sim$ (1–10) μm is the persistence length of actin polymer and $l_c \sim$ (50–100) nm is the plausible inter-crosslink distance at the leading edge. Therefore an estimated 2500 long lamellipodial filaments can account for the observed traction force.

There are, however, strong indications that lamellipodial traction is myosin-powered in keratocytes, fibroblasts, and a number of other cells. This was demonstrated directly by inhibition of myosin and the subsequent debilitation of contraction in *Aplysia* neuronal growth cones [33, 32], the lamellipodia of which are very similar to those of keratocytes.

The myosin powered mechanism is a plausible candidate for the observed traction force; $\sim 10^4$ myosin motors making $\sim$ 1pN powerstrokes each would account for a total force of $\sim 10^4$pN. A related system in which thousands of myosin motors are consolidated and employed with extreme efficiency is striated muscle [3]. The main feature of muscle cells is the presence of permanent highly ordered arrays of actin filaments arranged so that myosin powered contraction is maximally effective. Non-muscle cells lack such permanent structures, and the main goal of this modeling is to examine plausible alternative self-organizing lamellipodial traction structures.

One possibility is the "transport mechanism," in which the cell body is carried forward by myosin motors gliding on pre-existing actin lamellipodial "tracks." There do not appear to be any such tracks close to the cell body in the keratocyte, however [56]. A contraction mechanism based on the myosin-powered antiparallel sliding of actin filaments is another possibility. Since there is no shear between different populations of microfilaments in the keratocyte lamellipodia [58], this mechanism is not likely to be based on the superposition of small local contractions taking place throughout the lamellipodia. In the next subsection we present a model of dynamic actomyosin contraction of the small rear part of the lamellipod which generates forward translocation of the entire cell. The model is based on the experiments of [56] and the calculations of [41].

3.2. 1-D model of dynamic contraction of actomyosin network. At the front of the lamellipod, actin fibers are rigidly crosslinked by actin binding proteins; the probable distance between the crosslinkers along a fiber is much less than F-actin persistence length. Adherence of the cytoskeleton to the substratum is strong. Thus the actin network near the front can be considered to be rigid and stationary relative to the surface; low contractility and high adhesivity at the front favor extension.

Myosin II molecules dispersed throughout the cytoplasm polymerize into clusters bound to actin. The experiments of [56] showed that small myosin clusters at the front are stationary relative to the network and surface. The likely reason for this is that just a few myosin heads cannot develop torque powerful enough to re-direct crosslinked actin fibers, while motors cannot walk efficiently along many filaments which are not oriented in parallel with the myosin cluster axis [50, 57]. Small myosin clusters nucleated close to the cell's leading edge continue to grow, and farther away from the front their size is great enough to produce a torque bending the minus ends of actin polymers into small bipolar bundles favorable for actomyosin gliding (Figure 2). Those actin minus ends are abundant throughout the network due to the breaking of filaments at random points (assisted by ADF/cofilin complexes [35]) taking place uniformly across the lamellipod [58]. At the rear of the lamellipod, actin disassembly weakens the network so that evolving actomyosin bundles pull the remaining filaments into one muscle-like bundle oriented parallel to the leading edge.

Myosin power strokes contract the bundle. The stress from the contraction pulls long lamellipodial fibers into the bundle. This thrust moves the actomyosin bundle and cell body connected to it in the direction of protrusion. Far from the leading edge, where the adhesion is weak, there is almost no resistance to this forward thrust. The bundle contracts and can

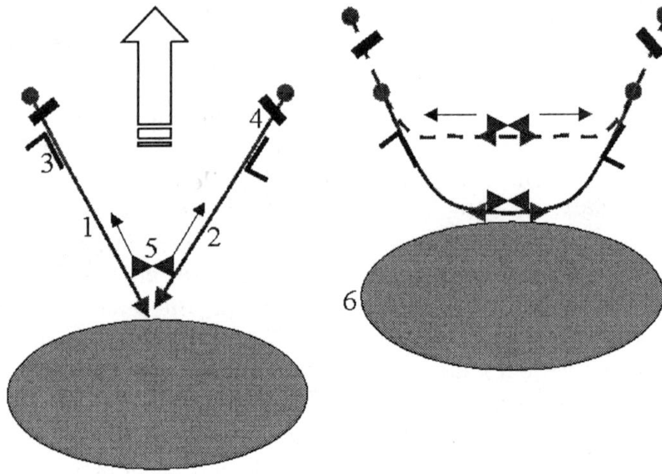

FIG. 2. *The dynamic actomyosin contraction mechanism. Initially (left) long actin filaments bound to crosslinking (3) and adhesion (4) proteins are straight. Myosin II clusters (5) bind to the filaments and try to move toward their plus ends, thereby creating a bending torque that pulls the minus ends into an anti-parallel bundle (right). Subsequent myosin gliding toward plus ends contracts the bundled parts of the fibers (dashed), thus pulling the actin bundle and with it the cell body (6) forward. Meanwhile, the front parts of the fibers continue to grow. Note that the positions of the plus ends along the leading edge change with growth (lateral flow).*

move forward faster than the lamellipod can extend. Closer to the front, however, firm adhesion slows this motion. This leads to a stable steady motion of the bundle at the rate of protrusion.

This model (Figure 2) explains the long standing problem of how myosin II can function in the isotropic actin network of the lamellipodium. Its function is two-fold: (i) ordering the filamentous network into the bipolar array at the rear of the lamellipod, and (ii) developing muscle-like contraction of the actin bundle, which is translated into the forward translocation force.

Quantitatively, we describe the dynamics of the myosin clusters by a linear density $m(x,t)$ of polymerized myosin. Here t is time; the x-axis is directed backward and has its origin at the leading edge. The dynamics of myosin clusters is governed by the equation:

$$\frac{\partial m}{\partial t} = k_a \bar{m} - k_d m - V_p \frac{\partial m}{\partial x} + \frac{\partial}{\partial x}(vm), \tag{3.1}$$

$$\bar{m}L + \int_0^L m\, dx = \bar{M}. \tag{3.2}$$

The first term in the right hand side of (3.1) is responsible for the assembly of myosin into clusters at the rate k_a. The density $\bar{m}$ of unpolymerized myosin molecules dispersed evenly throughout the lamellipod of width L can be found from the conservation of the total amount of lamellipodial myosin, $\bar{M}$. The second term in (3.1) describes the depolymerization of myosin with the rate k_d. The third term accounts for the kinematic drift of the clusters relative to the leading edge, while the last term describes forward translocation with rate v.

The length densities of network and bundled actin filaments are given by $L_n(x,t)$ and $L_b(x,t)$, respectively. The following equations govern their dynamics:

$$\frac{\partial L_n}{\partial t} = -\gamma L_n - V_p \frac{\partial L_n}{\partial x} + \alpha v \frac{\partial L_n}{\partial x}, \tag{3.3}$$

$$\frac{\partial L_b}{\partial t} = -\gamma L_b - V_p \frac{\partial L_b}{\partial x} - \alpha v \frac{\partial L_n}{\partial x} + \frac{\partial}{\partial x}(v L_b). \tag{3.4}$$

The first terms in the right hand side of both equations account for actin depolymerization with the corresponding rate γ. The second terms are responsible for the kinematic drift relative to the leading edge. The third terms describe the "sink" for network actin due to the bending of the actin minus ends into the bipolar bundle and the corresponding "source" of bundled actin. Finally, the last term in (3.4) accounts for the forward translocation of the bundled actin with rate v synchronously with the motion of the myosin clusters. The parameter α is the relative area covered by the myosin clusters. The form of the source and sink terms is based on the fact that the density of the termination points (either minus ends or bending points) of the network fibers is $-\partial L_n/\partial x$. When the relative part α of such points moves forward with the rate of myosin drift v, it increases the length of the bundled actin at the rate $-\alpha v \partial L_n/\partial x$. The parameter α is defined as follows:

$$\alpha = \begin{cases} m/m_{max} & , \quad m < m_{max} \\ 1 & , \quad m \geq m_{max}, \end{cases} \tag{3.5}$$

where m_{max} is the polymerized myosin concentration at which the clusters cover the whole area.

We now derive a key element in this model, the constitutive relation for the rate $v(x,t)$ of forward actomyosin gliding. The force-velocity relation of the myosin motor can be approximated by the linear function [5]: $f \simeq f_m(1 - v/V_m)$, where f_m is the force developed by stalled myosin, and V_m is the speed of free myosin gliding. A mass-action approximation for the reaction of attachment/detachment between myosin molecules and bundled fibers gives the density of pulling motors $\sim mL_b$, and the density of contractile force $\sim mL_b f_m(1 - v/V_m)$. The relative motion of bundled and network filaments causes breaking of the crosslinks, the density of which is proportional to the concentration of the points of overlapping between the two subpopulations of actin. The corresponding density of the resistance force is $\sim (1-\alpha)L_n L_b$. An additional source of resistance is the breaking of adhesion bonds, the concentration of which is proportional to the density of integrin molecules $i_v(x,t)$ (introduced in the next section). The mass-action approximation for the reaction of attachment/detachment between adhesion complexes and bundled fibers gives the density of the corresponding force: $\sim (1-\alpha)i_v L_b$. Balancing myosin-generated force with both resistance forces and solving the corresponding algebraic equation, we obtain the constitutive relation for the rate of forward actomyosin gliding:

$$v = V_m\Big(1 - \frac{(1-\alpha)}{m}(b_1 L_n + b_2 i_v)\Big). \tag{3.6}$$

Here $b_{1,2}$ are phenomenological coefficients depending on various protein concentrations, reaction rates and mechanical properties of intercytoskeleton and cytoskeleton-adhesion links.

The boundary conditions at the leading edge are such that the concentrations of myosin clusters and actin bundles vanish and the length density of network actin is fixed:

$$m(0) = 0, \qquad L_b(0) = 0, \qquad L_n(0) = \bar{L}. \tag{3.7}$$

In the next section we analyze the models for protrusion and adhesion, and we calculate the dynamic parameters V_p, $\bar{L}$ and integrin density i_v, thereby closing the mechanical model (3.1–7). Finally, we report the results of numerical studies based on this model.

4. Minimal 1-D model of cell locomotion. Here we introduce the missing links between protrusion and traction and derive a self-consistent model of locomotion. These links are (i) nucleation of actin filaments and (ii) adhesion between the cytoskeleton and substratum. In this paper, we do not analyze regulation of motion by outside signals and the corresponding signaling pathways. Instead, we assume that the cell is no longer symmetric and that the polarized lamellipodial actin network already exists.

4.1. Graded adhesion. To migrate, cells must use cytoskeletal forces to exert propulsive traction on the substratum; thus the formation of adhesive contacts between the ventral cell surface and the substratum is essential for locomotion. Mitchison and Kirschner [37] introduced the idea that the adhesive system may work as a "molecular clutch" transforming lamellipodial contraction into forward thrust. In order for the cell to move forward, propulsive traction at the front must be greater than the friction at the rear. One possible way to develop such graded traction is to have asymmetric contraction, which would result in directed motion even with homogeneous adhesion. Another way is to build the asymmetry into the adhesion. Keratocytes, as well as other animal cells, use both mechanisms.

Integrins are key transmembrane proteins which mediate the anchorage of the microfilament system to the substratum [28, 20]. They are crucial components in "close contacts" and "focal contacts" - protein complexes responsible for links between actin and the extracellular environment. Besides integrins, talin, vinculin and α-actinin molecules participate in the assembly of close contacts. Paxillin and FAK are present in the more mature focal contacts. Lee and Jacobson [30] observed a highly asymmetric distribution of the adhesions over the cell's ventral surface. There is a very high concentration within a band a few microns wide at the front of the lamellipod; the concentration decreases significantly 5–10 μm behind the leading edge.

Lauffenburger and Lindeman [28] list the following means to achieve the observed asymmetry in adhesion: (i) polarized receptor trafficking, (ii) localized proteolysis of attachments, and (iii) localized covalent modification of adhesion receptors. Our model includes only the first mechanism. The initial step in the formation of adhesions is the aggregation of integrins at the leading edge and their binding to the substratum. Then, talin molecules are incorporated into the integrin clusters. Talin is an actin binding protein, so integrin-talin-actin complexes constitute the simplest close contacts. The relevant association/dissociation constants are on the order of seconds, so the adhesions are established within the front rim $\sim$ 500nm wide. Further away, close contacts are reinforced by vinculin and α-actinin molecules. As this reinforcement takes place, dissociation constants for adhesion complexes probably increase to minutes.

This scenario depends on a high concentration of integrins at the front, which requires their recycling from close contact dissociation sites. Recycling of integrins is possible by a forward transport mechanism involving either myosin, vesicle transport pathways to the front, or both [52]. Experiments demonstrate co-localization of a myosin I isoform to forward moving particles in lamellipodia (reviewed in [7]). Forward transport of integrins on the dorsal cell surface with rates 0.1–0.5 $\mu m/s$ [26] has been observed. Integrins probably dissociate often from myosin carriers and then undergo lateral diffusion on the dorsal surface with rates (2–4) $\times 10^{-2} \mu\text{m}^2/s$ [22].

We now derive a simple mathematical description of the graded adhesion based on this scenario (Figure 3). Let us introduce linear densities of integrin molecules $i_v(x,t)$ and $i_d(x,t)$ on the ventral and dorsal surfaces, respectively. We will assume that adhesion complex assembly at the leading edge coincides with binding to the substratum, and that free integrin molecules do not exist on the ventral surface. Thus integrins on the ventral surface are static relative to the substratum and drifting away from the leading edge at the rate V_p. Taking into account the process of integrin detachment from the adhesion complexes with rate κ, we can write a conservation law for the amount of integrin on the ventral surface:

$$\frac{\partial i_v}{\partial t} = -V_p \frac{\partial i_v}{\partial x} - \kappa i_v. \tag{4.1}$$

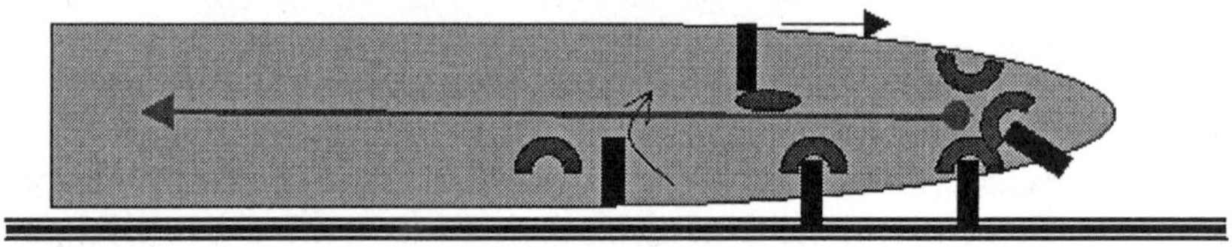

FIG. 3. *The graded adhesion mechanism. Long actin filaments are nucleated, crosslinked and polymerize in the proximity of membrane associated proteins (of those, only talin is shown). Talin (bent rods), localized at the leading edge due to the membrane curvature, binds to integrins (straight rods). Integrin-talin complexes mediate cytoskeleton binding to the surface. When the adhesion complexes disassemble, myosin I (oval) gliding to actin plus ends creates integrin traffic in the forward direction, increasing their concentration at the leading edge. As a result, adhesions are abundant only at the front.*

Little is known about how integrin molecules are recycled, so for simplicity we will assume that, after adhesion disassembly, integrins are transfered immediately from the ventral to the dorsal surface. On the dorsal surface, they undergo lateral diffusion at the rate D and drift forward at the rate V_t relative to the lamellipodial actin network. Taking into account that the network's slippage in the keratocyte cell is kept to a minimum, the drift rate of dorsal integrins relative to the leading edge is approximately equal to $V_p - V_t$. The resulting conservation law for the amount of integrin on the dorsal surface is:

$$\frac{\partial i_d}{\partial t} = D\frac{\partial^2 i_d}{\partial x^2} + (V_t - V_p)\frac{\partial i_d}{\partial x} + \kappa i_v. \tag{4.2}$$

Values of the model parameters (Table 1) can be obtained from the literature. We assume that at the leading edge most integrin molecules are assembled into the adhesion complexes. This boundary condition, together with the conservation law for the total amount of integrins I, allows us to find stationary solutions of the Equations (4.1–2). At the given values of the model parameters, the stationary distribution of the integrins on the ventral surface can be approximated by:

$$i_v(x) \simeq \frac{I\kappa}{V_t}\Big(\frac{V_t}{V_p} - 1\Big)exp\Big(-\frac{\kappa x}{V_p}\Big). \tag{4.3}$$

This expression will be used to close the system of equations of locomotion.

4.2. Nucleation of actin at the leading edge. One of the issues of leading edge extension is which mode of the actin network growth is responsible for the extension. One possibility is that elongation of preexisting long actin filaments causes protrusion [54]. Another possibility is

TABLE 1
Model parameters.

Symbol	Value	Meaning	Reference
$k_B T$	$4.1\ pN \cdot nm$	thermal energy	[47]
δ	$2.7nm$	half size of actin monomer	[3]
F_r	$2 \times 10^3 pN$	force of resistance to protrusion	estimated using data from [13]
V_0	$0.2\mu m/sec$	net free actin polymerization rate at the plus end	order of magnitude from [58]
V_m	$0.5\mu m/sec$	free sliding rate for myosin molecule	order of magnitude from [50, 42]
V_t	$0.2\mu m/sec$	rate of integrin traffic on dorsal surface	order of magnitude from [26]
M	10^4	number of myosin motors molecules	estimated using data from [56]
γ	$0.02 sec^{-1}$	effective rate of actin disassembly	estimated using data from [56]
κ	$0.02 sec^{-1}$	rate of disassembly of close contacts	estimated using data from [30]
η	$0.04 sec^{-1}$	effective rate of actin nucleation per integrin molecule	estimated roughly using data from [30, 20]
k_a	$0.01 sec^{-1}$	rate of assembly of myosin clusters	estimated using data from [56]
k_d	$0.01 sec^{-1}$	rate of disassembly of myosin clusters	estimated using data from [56]
b_1	$1\mu m^{-1}$	crosslinking friction coefficient	estimated roughly using data from [3, 24, 28, 56]
b_2	1	adhesion friction coefficient	estimated roughly using data from [3, 24, 28, 56]
I	10^3	number of integrins involved in adhesion	estimated roughly using data from [28, 30]
l	$5\mu m$	length of the central part of the leading edge	[29, 58]
D	$(2\text{–}4) \times 10^{-2} \mu m^2/s$	lateral diffusion coefficient of integrins	[22]

that nucleation and release of short filaments expands the leading edge [58]. In fact, elements of both processes take place.

Let us consider actin filaments crosslinked into the network and incident on the membrane at angles $\pm\theta$. We assume that these angles are more acute than the cutoff angle, so that the filament tips are always near the leading edge. As the leading edge extends with the rate V_p, the tips move sideways at the rate $\pm V_p \tan(\theta)$. The linear density of polymer tips $n(y,t)$, where y is the coordinate along the leading edge, satisfies the equation:

$$\frac{\partial n(y,t)}{\partial t} = \pm V_p \tan(\theta) \frac{\partial n(y,t)}{\partial y} + N(V_p, \theta), \tag{4.4}$$

where $N(V_p, \theta)$ is the nucleation rate. On the strip of the leading edge of length $l = 5\mu$m with boundary conditions of no influx of the filaments at the sides of the strip, solutions of this equation give the constant stationary linear density of the polymer tips:

$$n(V_p, \theta) = lN(V_p, \theta)\ \cot(\theta)/V_p. \tag{4.5}$$

If there is no nucleation of new filaments, $N = 0$, lateral flow of actin tips causes tip density to vanish, and protrusion ceases. To estimate the rate of nucleation, we have to consider multiple processes of regulation at the leading edge. We will restrict the analysis to two plausible pathways.

According to recent results of [43], nucleation of the filaments takes place on Arp2/3 protein complexes attached to the sides of the existing fibers. This is preceded by the activation of Arp2/3 at the leading edge, most probably as a result of interaction with Ena/VASP/zyxin protein complexes associated with integrin molecules [35, 2]. In addition to being actin crosslinkers, talin molecules are also actin nucleating proteins [20], so the formation of integrin-talin adhesion complexes may coincide with nucleation. Thus the following positive feedback loop can explain the localization of polymerization at the leading edge. First, an external signal induces integrin aggregation. Next, integrin-associated nucleating and crosslinking protein complexes initiate the growth of the polarized semi-stiff actin network. As the degree of actin polarization grows, the effective rate of myosin I forward drift increases. This leads to a higher integrin concentration at the leading edge, and with it faster protrusion and greater actin polarization.

If binding to integrins is the rate limiting factor of the described nucleation pathways, then the resulting rate of nucleation, which can be assumed to be angle independent, will be proportional to the concentration of the integrin molecules at the front:

$$N(V_p, \theta) = \eta i_v(0) \simeq (\eta I \kappa / V_t)((V_t/V_p) - 1)). \tag{4.6}$$

Here η is a phenomenological coefficient dependent on the concentrations and kinetics of nucleation related proteins. We first substitute the previous expression into (4.5) and the result of this into (2.3). After some asymptotic expansions of the integral in (2.3) and numerical integration, we obtain the following algebraic equation for the rate of protrusion in dimensionless form:

$$\frac{V_t}{V_p}\Big(\frac{V_t}{V_p} - 1\Big) \ln\Big(\frac{V_0}{V_p}\Big) \simeq s, \qquad s = \Big(\frac{F_r \delta}{k_B T}\Big)\Big(\frac{V_t}{l\kappa}\Big)\Big(\frac{V_t}{\eta l I}\Big). \tag{4.7}$$

The previous equation can be solved numerically to obtain the effective rate of the leading edge extension V_p. Then the linear density $n(0)$ of actin network at the leading edge can be found using Equations (2.2–3). The corresponding estimate for the length density is:

$$\bar{L} \simeq \frac{l^2 \eta I \kappa}{V_t V_p}\left(\frac{V_t}{V_p} - 1\right). \tag{4.8}$$

4.3. Results. Here we report some preliminary results on the work in progress on the self-consistent 1-D model of a steadily moving cell. The core of this model is the system of non-linear hyperbolic reaction-advection Equations (3.1–7) for the dynamic distributions of actin and myosin over the width of the central part of the lamellipod. This model of the traction mechanism is completed by Equations (2.2–3, 4.4–7) describing the protrusion mechanism, and Equations (4.1–3) describing the adhesion mechanism. The characteristic time scales for all three principal mechanisms are of the same order of magnitude, so that all of the equations have to be solved simultaneously. This will be done in a future paper.

In a simplified model, a constant protrusion rate $V_p \simeq 0.1\mu\text{m}/sec$ and a length density of network actin at the leading edge $\bar{L} \simeq 1600$ were estimated analytically from (4.6–8). These parameters and the stationary distribution of integrins on the ventral surface (4.3) were used to numerically solve the traction model (3.1–7). Values of model parameters used in these calculations can be found in Table 1. The details of the numerical procedure will be described in [41].

Asymptotically stable distributions of the network and bundled actin and myosin across the width of the lamellipod are shown in Figure 4. The main qualitative result of our model is that it predicts a finite width of the lamellipod of the correct order of magnitude. Quantitative results, such as the distance from the leading edge to the front edge of the dense actin bundle ($\simeq 12\mu$m), width of the actomyosin bundle ($\simeq 3\mu$m) and ratio of network and bundle actin ($\simeq 3.5$) are in agreement with experimental data of [56].

5. Summary and discussion. Our model describes quantitatively the fast, steady migration of the lamellipod according to the following scenario. Polymerization of the actin network at the leading edge both creates the force of protrusion and is responsible for extension. Polarity of the actin network and myosin I based transport at the leading edge are responsible for high concentrations of membrane associated proteins enhancing adhesion and actin growth at the leading edge. Myosin II molecules polymerize into clusters that become very dense toward the rear of the lamellipodium, where the actin network is largely depolymerized. Dense myosin clusters collapse the actin network into the actomyosin bundle, and the resulting muscle-like contraction leading to the forward translocation of the cell body completes the cycle of migration.

The 1-D model presented in this paper is a caricature that captures the essential features of the locomotory process. To incorporate important

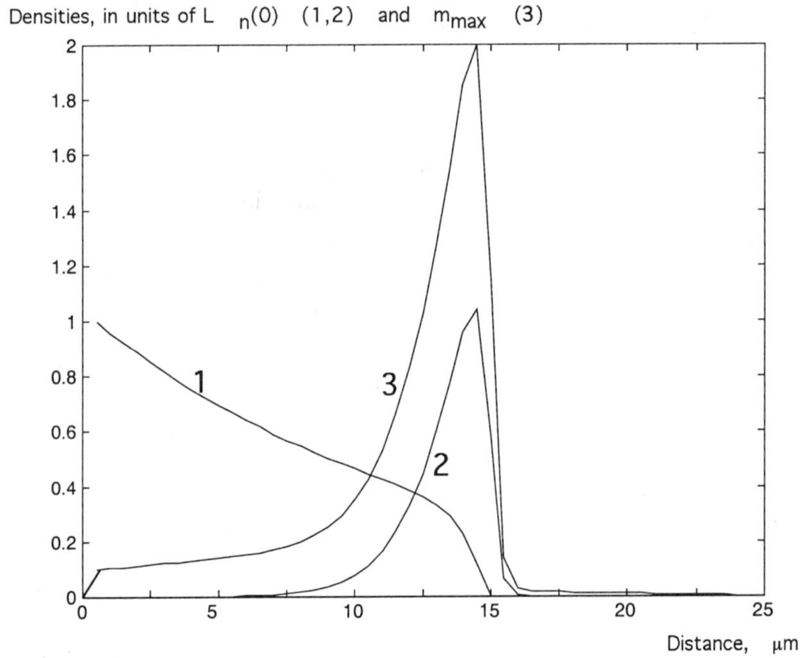

FIG. 4. *Results of the numerical solution of the 1-D mathematical model of the keratocyte's lamellipod. Asymptotically stable stationary distributions of actin and myosin as functions of distance from the leading edge are shown. Length densities of the network actin (1) and bundled actin (2) are plotted in units of the leading edge actin density. Myosin concentration (3) is scaled so that the maximal possible concentration* $m_{max} = 2$.

mechanical, kinetic and structural data, we shall generalize the model to 2-D. One possibility is to consider continuous dynamics of spatio-angular-temporal densities of actin network and actomyosin bundles and spatio-temporal densities of adhesions and small myosin clusters. The resulting system of non-linear partial integrodifferential equations will have to be solved on a domain with a moving boundary. The corresponding boundary conditions can be derived from the microscopic dynamics of protrusive and adhesive machinery at the leading edge. Needless to say, the analysis of such a multidimensional continuous problem (two spatial plus one angular degrees of freedom) promises to be demanding. Two other approaches can provide alternative quantitative descriptions of the whole cell.

In our 1-D traction model we assume that the lamellipodial actin network is static relative to the substratum. Bundled actin moves in the laboratory coordinate system, however. If we neglect the relatively small density of bundled actin in the lamellipodial interior, then all of the dynamics are confined to a narrow ($\sim 1\mu$m wide) band at the cell boundary. Then we can approximate the angular density of actin with the first two momenta of the angular distribution, thereby restricting ourselves to *scalar* boundary densities of network and bundled actin and adhesion-nucleating proteins. The resulting effectively 1-D model of the moving cell boundary would be very valuable for simulating cellular responses to external signals.

Another approach (Bottino and Mogilner, in progress) is based on Monte Carlo simulations of ensembles of actin and myosin filaments. The computational model consists of actin strands that are strings of nodes. At each time step the bending moment at each interior node is computed and the nodes are moved according to Stokes' approximation. The bending is myosin-driven and takes place when myosin molecules, modeled as short rods, attach to two non-parallel actin strands. Polymerization is modeled by the growth of the link joining the first two nodes of each strand and the insertion of new nodes as needed. Depolymerization is modeled by periodic deletion of a random number of the minus-end nodes of a strand. Myosin binding and gliding kinetics are based on the rules described in Section 3.2. Actin strands are generated at random orientations at the cell's boundary, which is defined as the convex hull of the set of nodes. Figure 5 illustrates the results of a simulation of one growing filament bent by myosin mediated interactions with several fixed, crosslinked strands. The dynamic filament is polymerizing at the cell boundary, which in this simulation is considered to be rigid. As the filament undergoes 'rearward flow', it intersects with the fixed strands and is bent by myosin molecules 'walking toward' the plus ends of the fixed fibers. The dynamic filament locally becomes perpendicular to the network. As more filaments are freed from the crosslinks, they will be pulled in parallel with the first one, and the actomyosin bundle perpendicular to the average network polarization should develop.

We considered very few regulation mechanisms of the cell's motile machinery. The main methodological challenge modelers are facing is how one should model multiple redundant signaling pathways coupled to lamellipodial mechanics.

Acknowledgements. Alex Mogilner and Eric Marland were supported by National Science Foundation Grant DMS 9707750 and NSF RTG Grant DBI-9602226. Dean Bottino was supported by National Science Foundation Grant DMS 9805494 and National Institute of Health Grant GM 29123. G. Oster was a constant source of inspiration and generator of ideas for us. We would also like to thank G. Borisy, A. Verkhovsky, K. Jacobson, P. Janmey, M. Sheetz, J. Italiano, M. Dembo, G. Isenberg, J. Condeelis, J. Theriot, P. Forscher, V. Small, J. Tang, J. Lee and R.D. Mullins for valuable discussions.

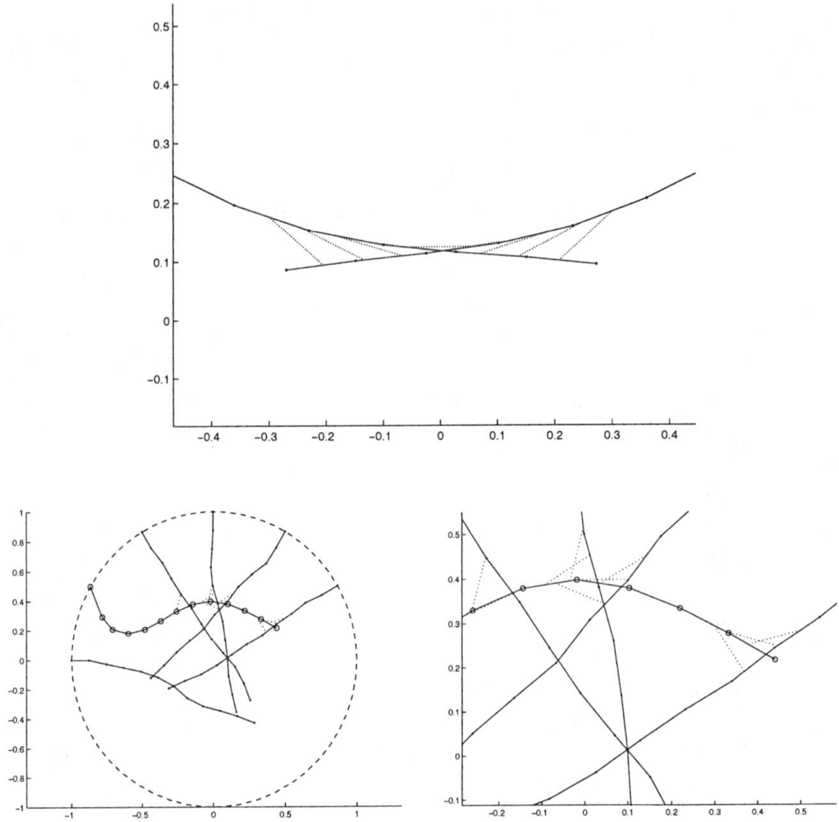

FIG. 5. *Monte Carlo simulations of actin - myosin interactions. Top: two actin filaments (solid curves) which were initially straight are being bent by myosin molecules (dotted lines) into an antiparallel bundle. Bottom left: a growing filament (curve with circular nodes) with its plus end clamped in the circumferential boundary is bent by myosin-mediated interactions with the stiff crosslinked actin network (other solid curves). Bottom right: enlargement of the interacting part of the growing fiber.*

REFERENCES

[1] M. ABERCROMBIE. The crawling movement of metazoan cells. *Proc. R. Soc. Lond. (Biol.)*, **207**:129–147, 1980.

[2] M.C. BECKERLE. Spatial control of actin filament assembly: lessons from *Listeria*. *Cell*, **95**:741–748, 1998.

[3] D. BRAY. *Cell Movements.* Garland, New York, 1992.

[4] D. BRAY, N. MONEY, F. HAROLD, AND J. BAMBURG. Responses of growth cones to changes in osmolality of the surrounding medium. *J. Cell Sci.*, **98**:507–515, 1991.

[5] S. CHAEN, K. OIWA, T. SHIMMEN, H. IWAMOTO, AND H. SUGI. Simultaneous recordings of force and sliding movement between a myosin-coated glass microneedle and actin cables in vitro. *Proc. Natl. Acad. Sci. U.S.A.*, **86**:1510–1514, 1989.

[6] J. CONDEELIS. Life at the leading edge: the formation of cell protrusions. *Annu. Rev. Cell Biol.*, **9**:411–444, 1993.

[7] L.P. CRAMER. Molecular mechanism of actin-dependent retrograde flow in lamellipodia of motile cells. *Frontiers in Bioscience*, **2**:260–270, 1997.

[8] L.P. CRAMER AND T.J. MITCHISON. Myosin is involved in postmitotic cell spreading. *J. Cell Biol.*, **131**:179–189, 1995.

[9] C. CUNNINGHAM. Actin polymerization and intracellular solvent flow in cell surface blebbing. *J. Cell Biol.*, **129**:1589–1599, 1995.

[10] J. DAI AND M.P. SHEETZ. Mechanical properties of neuronal growth cone membranes studied by tether formation with laser optical tweezers. *Biophys. J.*, **68**:988–96, 1995.

[11] P.A. DIMILLA, K. BARBEE, AND D.A. LAUFFENBURGER. A mathematical model for the effects of adhesion and mechanics on cell migration speed. *Biophys. J.*, **60**:15–37, 1991.

[12] E. EVANS AND R. SKALAK. *Mechanics and Thermodynamics of Biomembranes.* CRC Press, Boca Raton, 1980.

[13] E. EVANS AND A. YEUNG. Apparent viscosity and cortical tension of blood granulocytes determined by micropipet aspiration. *Biophys. J.*, **56**:161–160, 1989.

[14] A. GREBECKI. Membrane and cytoskeleton flow in motile cells with emphasis on the contribution of free-living amoebae. *Int. Rev. Cytol.*, **148**:37–80, 1994.

[15] H. HE AND M. DEMBO. On the mechanics of the first cleavage division of the sea urchin egg. *Exp. Cell Research*, **233**:252–273, 1997.

[16] T.L. HILL. *An Introduction to Statistical Thermodynamics.* Addison-Wesley Publ. Co, London, 1960.

[17] R. HOCHMUTH, J. SHAO, J. DAI, AND M. SHEETZ. Deformation and flow of membrane into tethers extracted from neuronal growth cones. *Biophys. J.*, **70**:358–369, 1996.

[18] H. ISAMBERT, P. VENIER, A. MAGGS, A. FATTOUM, R. KASSAB, D. PANTALONI, AND M. CARLIER. Flexibility of actin filaments derived from thermal fluctuations: effects of bound nucleotide, phalloidin, and muscle regulatory proteins. *J. Biol. Chem.*, **270**:11437–11444, 1995.

[19] G. ISENBERG. New concepts for signal perception and transduction by the actin cytoskeleton at cell boundaries. *Semin. Cell Dev. Biol.*, **7**:707–715, 1996.

[20] G. ISENBERG AND W.H. GOLDMANN. Peptide-specific antibodies localize the major lipid binding sites of talin dimers to oppositely arranged *n*-terminal 47 kda subdomains. *FEBS Letters*, **426**:165–170, 1998.

[21] J.E. ITALIANO, T.M. ROBERTS, M. STEWART, AND C.A. FONTANA. Reconstitution in vitro of the motile apparatus from the amoeboid sperm of *Ascaris* shows that filament assembly and bundling move membranes. *Cell*, **84**:105–114, 1996.

[22] K. JACOBSON, A ISHIHARA, AND R. INMAN. Lateral diffusion of proteins in membranes. *Ann. Rev. Physiol.*, **49**:163–175, 1987.

[23] J. KAS, H. STREY, M. BARMANN, AND E. SACKMANN. Direct measurement of the wave-vector-dependent bending stiffness of freely flickering actin filaments. *Europhys. Lett.*, **21**:865–870, 1995.

[24] A. KISHINO AND T. YANAGIDA. Force measurements by micromanipulation of a single actin filament by glass needles. *Nature*, **334**:74–76, 1988.

[25] D.F. KUCIK, E.L. ELSON, AND M.P. SHEETZ. Cell migration does not produce membrane flow. *J. Cell Biol.*, **111**:1617–1622, 1990.

[26] D.F. KUCIK, S.C. KUO, E.L. ELSON, AND M.P. SHEETZ. Preferential attachment of membrane glycoproteins to the cytoskeleton at the leading edge of lamella. *J. Cell Biol.*, **115**:1029–1036, 1991.

[27] D. LAUFFENBURGER AND A. HORWITZ. Cell migration: a physically integrated molecular process. *Cell*, **84**:380–399, 1996.

[28] D.A. LAUFFENBURGER AND J.J. LINDEMAN. *Receptors: Models for Binding, Trafficking and Signalling.* Oxford University Press, Oxford, 1993.

[29] J. Lee, A. Ishihara, J.A. Theriot, and K. Jacobson. Principles of locomotion for simple-shaped cells. *Nature*, **362**:467–471, 1993.

[30] J. Lee and K. Jacobson. The composition and dynamics of cell-substratum adhesions in locomoting fish keratocytes. *J. Cell Sci.*, **110**:2833–2844, 1997.

[31] A.K. Lewis and P.C. Bridgman. Nerve growth cone lamellipodia contain two populations of actin filaments that differ in organization and polarity. *J. Cell Biol.*, **119**:1220–1243, 1992.

[32] C.H. Lin, E.M. Esprefacio, M.S. Mooseker, and P. Forscher. Myosin drives retrograde f-actin flow in neuronal growth cones. *Neuron*, **16**:769–782, 1996.

[33] C.H. Lin and P. Forscher. Growth cone advance is inversely proportional to retrograde F-actin flow. *Neuron*, **14**:763–71, 1995.

[34] C.H. Lin, C.A. Thompson, and P. Forscher. Cytoskeletal reorganization underlying growth cone motility. *Cur. Opin. Neurobiol.*, **4**:640–647, 1994.

[35] L.M. Machesky. Cell motility: complex dynamics at the leading edge. *Current Biology*, **7**:R164–R167, 1997.

[36] T.J. Mitchison and L.P. Cramer. Actin-based cell motility and cell locomotion. *Cell*, **84**:371–379, 1996.

[37] T.J. Mitchison and M. Kirschner. Cytoskeletal dynamics and nerve growth. *Neuron*, **1**:761–772, 1988.

[38] H. Miyata, S. Nishiyama, K.-I. Akashi, and K. Kinosita. Protrusive growth from giant liposomes driven by actin polymerization. *Proc. Natl. Acad. Sci. U.S.A.*, **96**:2048–2053, 1999.

[39] A. Mogilner and G. Oster. Cell motility driven by actin polymerization. *Biophys. J.*, **71**:3030–3045, 1996.

[40] A. Mogilner and G. Oster. The physics of lamellipodial protrusion. *Eur. Biophys. J.*, **25**:47–53, 1996.

[41] A. Mogilner, T. Svitkina, A. Verkhovsky, and G. Borisy. Mathematical model for myosin driven forward translocation in keratocytes. *In Preparation*, 1999.

[42] J.E. Molloy, J.E. Burns, J. Kindrick-Jones, R.T. Tregear, and D.C.S. White. Movement and force produced by a single myosin head. *Nature*, **378**:209–212, 1995.

[43] R.D. Mullins, J.A. Heuser, and T.A. Pollard. The interaction of Arp2/3 complex with actin: nucleation, high affinity pointed end capping, and formation of branching networks of filaments. *Proc. Natl. Acad. Sci. U.S.A.*, **95**:6181–6186, 1998.

[44] J.D. Murray. *Mathematical Biology*. Springer-Verlag, New York, 1989.

[45] T. Oliver, M. Dembo, and K. Jacobson. Traction forces in locomoting cells. *Cell Motil. Cytoskel.*, **31**:225–240, 1995.

[46] T. Oliver, J. Lee, and K. Jacobson. Forces exerted by locomoting cells. *Semin. Cell Biol.*, **5**:139–147, 1995.

[47] G. Oster and A. Perelson. Cell protrusions. In S. Levin, editor, *Frontiers in Mathematical Biology*, volume 100 of *Lectures Notes in Biomathematics*, pp. 53–78. Springer, Berlin, 1994.

[48] C. Peskin, G. Odell, and G. Oster. Cellular motions and thermal fluctuations. *Biophys. J.*, **65**:316–324, 1993.

[49] D. Raucher and M.P. Sheetz. Membrane expansion increases endocytosis rate during mitosis. *J. Cell Biol.*, **144**:497–506, 1999.

[50] K. Saito, T. Aoki, and T. Yanagida. Movement of single myosin filaments and myosin step size on an actin filament suspended in solution by a laser trap. *Biophys. J.*, **66**:769–777, 1994.

[51] A. Schmidt and M.N. Hall. Signalling to the actin cytoskeleton. *Annu. Rev. Cell Dev. Biol.*, **14**:305–338, 1998.

[52] M. Sheetz. Cell migration by graded attachment to subsrates and contraction. *Sem. Cell Biol.*, **5**:149–155, 1994.

[53] M.P. Sheetz, D.B. Wayne, and A.L. Pearlman. Extension of filopodia by motor-dependent actin assembly. *Cell Motil. Cytoskeleton*, **22**:160–169, 1992.
[54] J.V. Small. Getting the actin filaments straight: nucleation-release or treadmilling? *Trends Cell Biol.*, **5**:52–55, 1995.
[55] T. Stossel. On the crawling of animal cells. *Science*, **260**:1086–1094, 1993.
[56] T.M. Svitkina, A.B. Verkhovsky, K.M. McQuade, and G.G. Borisy. Analysis of the actin-myosin II system in fish epidermal keratocytes: Mechanism of cell body translocation. *J. Cell Biol.*, **139**:397–415, 1997.
[57] H. Tanaka, A. Ishijima, M. Honda, K. Saito, and T. Yanagida. Orientation dependence of displacements by a single one-headed myosin relative to the actin filament. *Biophys. J.*, **75**:1886–1894, 1998.
[58] J. Theriot and T. Mitchison. Actin microfilament dynamics in locomoting cells. *Nature*, **352**:126–131, 1991.

COMPUTER SIMULATIONS OF MECHANOCHEMICAL COUPLING IN A DEFORMING DOMAIN: APPLICATIONS TO CELL MOTION

DEAN C. BOTTINO*

Abstract. The coordinated response of cytoskeletal components resulting in the polarization and chemotaxis of crawling cells is likely to require the asymmetric distribution of signal molecules. The asymmetrically distributed signals are thought to locally regulate protrusive and/or contractile force generation in the cytoskeleton. The author's previous research [4, 5] involves computational simulation of the cytoskeleton as a dynamic network of immersed nodes connected by passive and active force-generating elastic elements. This paper describes recent progress toward incorporating reaction-diffusion-advection (RDA) equations for intracellular signaling into the existing mechanical model. A general method using Voronoi diagrams for solving the RDA equations on an irregularly shaped, nonconvex deforming domain is described. Test runs indicate that the method is a promising tool for this class of problems as well as for modeling signaling and mechanical interactions among many cells. The incorporation of this method into the existing mechanical model, as well as future implementation of the same method for modeling multicellular interactions, is discussed.

AMS(MOS) subject classifications. Voronoi diagrams, reaction-diffusion-advection equations, intracellular signaling.

1. Introduction. Directional cell movement is essential for nerve distribution in embryonic development, the migration of keratinocytes across a wound site, the pursuit and elimination of foreign particles by neutrophils, and the invasion of cancerous cells into healthy tissue [7]. Although understanding of the molecular components that contribute to taxis has improved in recent years, little is known of how the individual molecular mechanisms work together to coordinate the reorganization, polarization, and activation of the cytoskeleton to produce directed locomotion in response to an external signal.

Such a coordinated response is likely to require an asymmetric distribution of intracellular signal molecules; there are several candidates for unevenly distributed signals. For example, oscillations of intracellular calcium ($[Ca^{2+}]_i$) have been correlated temporally to chemotaxis both in human neutrophils [7] and in the well-studied model organism *Dictyostelium discoideum* (Dd). Yumura et al. [27] have observed the formation of a $[Ca^{2+}]_i$ gradient in chemotactically stimulated Dd cells, with the highest $[Ca^{2+}]_i$ occurring in the portion of the cell *opposite* from the point of receptor stimulation by the chemoattractant cAMP. Parent et al. [18] have determined that the activation of certain proteins in the G protein-linked chemotactic signaling pathway of Dd occurs selectively beneath the stimulated portion of the cell membrane. It has been proposed that the redistribution of receptors results in taxis in human keratinocytes [12].

*Department of Mathematics, University of Utah, Salt Lake City, UT 84112-0900; dean_bottino@mail.com.

There are several theories for the coordination by intracellular signals of cytoskeletal dynamics in Dd. One theory suggests that an internal signal initiated by differential binding of chemoattractant to membrane receptors causes a resetting of the cell's "intrinsic oscillator," resulting in a resynchronization of the cell components into a chemotactic response toward the attractant source [24]. The observation that $[Ca^{2+}]_i$ increases in the zone opposite to the stimulated region of the cell has led some to suggest that the observed $[Ca^{2+}]_i$ increase in turn stimulates posterior myosin–II accumulation and subsequent actomyosin contraction, resulting in the forward propulsion of the cell contents [27]. Observations of the three–dimensional pseudopodial dynamics of chemotacting Dd [26] suggest that the formation of lateral pseudopodia off the substrate may be a mechanism by which the cell can integrate temporal information obtained from the growing pseudopod to sense the static chemoattractant gradient surrounding the cell.

A detailed simulation model incorporating spatiotemporal intracellular signal dynamics and their effects on active force generation in the cytoskeleton can be used to better understand the proposed mechanisms of chemotaxis. Such a model can address the following questions regarding pseudopodial or lamellipodial taxis: (1) What minimal set of mechanisms, such as localization of chemoattractant receptors, establishment of an internal signal gradient, coordination of transmembrane adhesions, and spontaneous and signal-controlled cytoskeletal dynamics, can result in a response similar to the chemotactic response observed in living cells? (2) Which of the existing theories of signaling and mechanics are sufficient to generate the observed chemotactic behavior of cells? (3) Do these theories have testable predictions that were not apparent before their incorporation into the simulation model?

Computational models of the mechanical aspects of ameboid cell motion have been developed by Dembo [9, 8, 10] and Bottino [3, 4, 5]. Both formulations rely on an assumed "cybernetic factor" to spatially coordinate cytoskeletal activity. The Bottino model uses the immersed boundary method [19] to represent actin crosslinks among actin clusters ("nodes") as immersed springs with resting lengths that change depending on local signal concentration. This paper describes techniques recently developed to model the reaction and diffusion of intracellular signals in the context of the existing immersed boundary model.

In Section 2, we describe a method whereby the model cell interior can be partitioned into a Voronoi diagram, with each actin node acting as a center for each Voronoi polygon. Section 2.1 details the approximation of reaction-diffusion-advection (RDA) equations in terms of the Voronoi polygons. In Section 2.2 we see that the Voronoi tesselation can be updated at each time step by a local computation so that the computational expense of the algorithm scales linearly with the number of nodes used. Section 2.3 summarizes the computational scheme.

Preliminary results are described in Section 3. The test simulation in 3.1 consists of an irregularly shaped deforming region. Reaction-diffusion equations for IP_3 controlled $[Ca^{2+}]_i$ dynamics are solved on the deforming domain using the Voronoi diagram scheme. In Section 3.2 we demonstrate the versatility of the code by applying it to a caricature simulation of multicellular mechanical and chemical interactions.

These results, while preliminary in nature, are promising, and it is likely that this method will be incorporated into the immersed boundary cell movement model. A simplified, purely Lagrangian intracellular signaling and mechanics model may also result from this research. The intercellular movement and signaling test simulation described in this paper is also likely to be expanded into a more accurate model for coordinated multicellular interactions. These possiblities are discussed in Section 4.

2. Computational model and methods. The two-dimensional mechanical cell simulation developed by the author in [4, 5] consists of a network of actin "nodes" immersed in an aqueous cytosol and surrounded by an impermeable model cell membrane. Figure 1 illustrates the components of the model cell. While the simulation model was able to reproduce directed motion in response to a constant external chemoattractant gradient [5], the active force–generating properties of the elastic interconnections among the actin nodes, as well as the dynamics of substrate adhesion, were based on *ad hoc* rules. These rules were devised to properly coordinate protrusive and contractile events with the formation and relinquishment of adhesions from the model cell to the substrate, which in turn resulted in the movement of the cell up the external chemoattractant gradient.

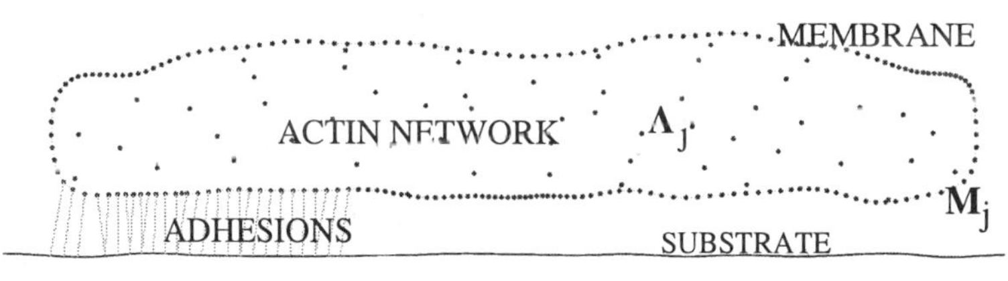

FIG. 1. *Immersed boundary formulation of ameboid cell crawling [3, 5]. The cell is modeled as a set of deformable structures immersed in a fluid medium. The fluid inside the model membrane represents cytosol while the fluid outside represents the extracellular fluid medium. Actin "nodes"* $\{\mathbf{A}_j\}$ *inside the cell are interconnected by elastic filaments (not shown) which give the model cytoskeleton viscoelastic behavior. Active force generation (corresponding to protrusion due to polymerization or contraction due to actomyosin sliding) also takes place among the actin nodes. The actin network is space filling rather than cortically concentrated because the model is two-dimensional. The cell membrane is modeled as an easily extensible but impermeable loop of points* $\{\mathbf{M}_j\}$.

The incorporation of signaling to cytoskeletal events would be a useful extension of the existing immersed boundary simulation model. Such an extension would allow us to use the simulation model as a test bed for existing theories of how coupling between the spatiotemporal reaction-diffusion dynamics of intracellular signals and the mechanical activity of the actin cytoskeleton results in the polarization and directed motion of crawling cells.

To extend the model in such a way we need to solve reaction diffusion advection (RDA) equations of the form

$$\frac{\partial c}{\partial t} + \mathbf{u} \cdot \nabla c = D\Delta c + f(c) \tag{2.1}$$

$$\hat{\mathbf{n}} \cdot \nabla c = 0, \tag{2.2}$$

in the interior of the irregularly shaped, nonconvex, deforming model cell, where $c(x,t)$ is the signal concentration, D is the diffusion coefficient, $\mathbf{u}$ is an incompressible flow field governing the motion of the nodes, and $f(c)$ is the reaction term. Equation (2.2) corresponds to a no-flux (Neumann) boundary condition.

A similar problem has been solved by Dillon and Othmer, also in the context of the immersed boundary method [11]. In their model, they solve the diffusion of growth factors in the interior of an irregularly shaped growing limb bud. They do so by solving the RDA equations on the regular fluid lattice that is used to solve the Navier Stokes equations. The fluid velocity data $\mathbf{u}$ is available at those nodes, so the advection term $\mathbf{u} \cdot \nabla c$ can be computed at each grid node. The diffusion term $D\Delta c$, however, must be handled differently near the boundary. At each time step the code determines what subset of the grid nodes lies inside the boundary. These interior nodes are further divided into sub-cases depending on the number of five-point stencil nodes that lie inside the domain. The resulting scheme is $\mathcal{O}(\Delta x)$ accurate, where $\Delta \mathbf{x}$ is the fluid lattice size.

The approach described above is well-suited to the limb bud model formulation because the growth factor concentration data are required at the fluid grid nodes to determine the magnitude of the fluid source term that represents growth in the model [11]. In the context of the active cell motion model, however, the signal concentration data are needed at the irregularly spaced actin nodes $\{\mathbf{A}_j\}$. Use of Dillon's method would require interpolation of concentration data to the actin nodes, and in the case of stress-induced signal release, interpolation of data from the nodes back to the regular grid. Furthermore, we wish to leave open the possibility of using a meshless method for the mechanics instead of the immersed boundary method. In this case, the velocity is defined only at the irregularly spaced actin nodes and is unavailable for computing the advection terms in the RDA equations.

In the context of the deforming cell model, therefore, it seems more appropriate to solve the RDA equations on the irregular grid defined by the actin nodes $\{\mathbf{A}_j\}$ themselves. The concentration data is available at the nodes, removing the need for interpolation of data between the actin nodes and the regular grid nodes. The boundary of the cell does not need to be tracked since it is fixed in computational space; that is, the actin nodes initially embedded in the cell boundary remain there. The advection term $\mathbf{u} \cdot \nabla c$ disappears from the RDA equations since the actin nodes represent a Lagrangian (moving) frame of reference instead of the Eulerian (fixed) frame. Figure 2 compares the structured and unstructured grid approaches.

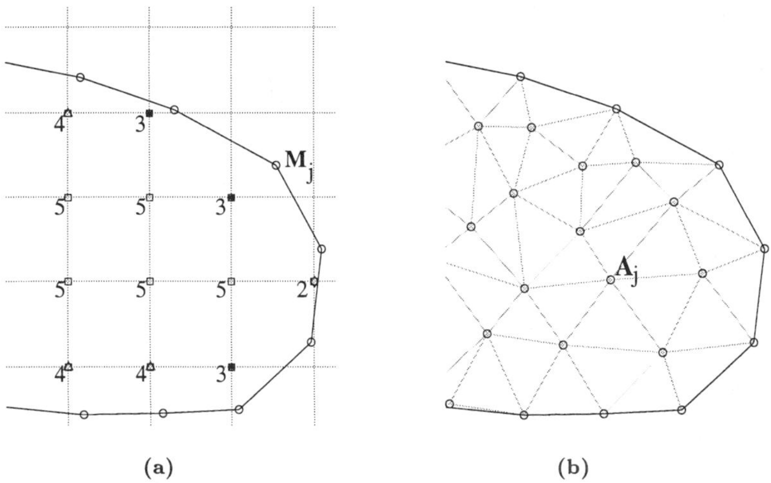

FIG. 2. *Comparison of solving the reaction-diffusion-advection (RDA) equations on structured (a) and unstructured (b) meshes. In the method (a) used by Dillon in [11], at each time step the subset of the fluid lattice nodes that lie inside the domain boundary* $\{\mathbf{M}_j\}$ *is determined. These are split into subcases depending on whether the five-point stencil for the Laplace operator has 2, 3, 4, or 5 interior points. In the unstructured grid method the chemoattractant concentration is solved on the irregular grid defined by the actin nodes* $\{\mathbf{A}_j\}$.

2.1. Voronoi diagrams to the rescue. For each node $\mathbf{A}_j$ we define Ω_j to be the set of all points in the cell interior Ω closer to node $\mathbf{A}_j$ than to any other node. The collection $\{\Omega_j\}$ is called the **Voronoi tesselation** of Ω based on the nodes $\{\mathbf{A}_j\}$. We call two nodes $\mathbf{A}_i$ and $\mathbf{A}_j$ **Delaunay neighbors** if Ω_i and Ω_j share an $(n-1)$ dimensional face. The graph defined by the Delaunay neighbor relationships gives a dual tesselation of the space, called the Delaunay triangulation. The Delaunay edges do not necessarily intersect the Voronoi faces, but they are always perpendicular to the Voronoi faces [16]. Figure 3 illustrates these concepts. Let $\mathcal{N}_j$ denote the index set of the Delaunay neighbors of node j. We approximate Δc on each Voronoi tile[1] by the average of Δc on the tile, and by applying the

[1] We have dropped down to two dimensions although the same argument is valid in any number of dimensions.

divergence theorem we obtain:

$$\Delta c(\mathbf{A}_j) \approx \frac{1}{A(\Omega_j)} \int_{\partial\Omega_j} \hat{\mathbf{n}} \cdot \nabla c \, ds \tag{2.3}$$

$$\approx \frac{1}{A(\Omega_j)} \sum_{i\in\mathcal{N}_j} \frac{c_i - c_j}{||\mathbf{A}_i - \mathbf{A}_j||} ||\mathbf{s}_{ij}|| \tag{2.4}$$

$$\equiv \mathcal{L}c, \tag{2.5}$$

where $c_j = c(\mathbf{A}_j)$, $A(\Omega_j)$ denotes the area of the tile Ω_j, $||\mathbf{s}_{ij}||$ is the length of the face shared by Ω_i and Ω_j, and the summation is over all Delaunay neighbors $\mathbf{A}_i$ of node $\mathbf{A}_j$. The Neumann boundary condition (2.2) can be enforced directly because each summand in (2.4) corresponds to the flux through the face $\mathbf{s}_{ij}$. This method is a special case of the finite volume method, which is described in [13, 14].

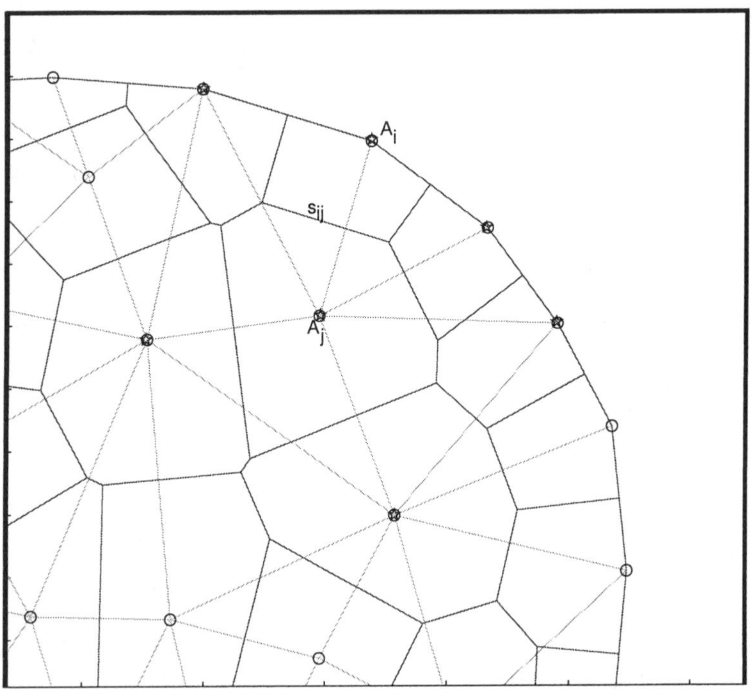

FIG. 3. *Voronoi tesselation (dark lines) and Delaunay triangulation (light lines) of the interior of the model cell.*

In 1986 Börgers and Peskin [2] obtained the same expression as in (2.4) for the approximation of Δc in terms of a Voronoi tesselation. They showed that $\mathcal{L}$ is weakly consistent with Δ of first–order in the maximum diameter of the Voronoi tiles. That is, let $h = \max_j \operatorname{diam}(\Omega_j)$. If $N \leq \mathcal{O}(1/h^2)$, then

$$\sum_{j=1}^{N} \phi(\mathbf{A}_j)(\mathcal{L}c)(\mathbf{A}_j)A(\Omega_j) = \int_{\Omega=\cup_j\Omega_j} \phi(\mathbf{x})\Delta c(\mathbf{x})d\mathbf{x} + \mathcal{O}(h) \tag{2.6}$$

for arbitrary smooth functions ϕ and c with $\hat{\mathbf{n}} \cdot \nabla\phi = 0$. The fact that $\mathcal{L}$ is not pointwise consistent with Δ does not preclude convergence; numerical convergence studies indicate that the method converges linearly in the L^2 norm [2].

Now that we have obtained a discretization of the Laplace operator on the irregular grid, the simplest way to approximate equation (2.1) is by a split–time scheme:

$$c_j^* = c_j^n + (D\Delta t)\mathcal{L}c_j^n \tag{2.7}$$

$$c_j^{n+1} = \mathcal{F}\left(c_j^*, f(\cdot); \Delta t\right). \tag{2.8}$$

The material derivative becomes a simple time derivative since we are solving the equation on a moving grid. In equation (2.8), the operator $\mathcal{F}$ represents any suitable one–step ODE method; in the current simulations a second order Runge-Kutta method is used.

In many situations, stability of the mechanical scheme will require a sufficiently small time step to ensure the stability of the explicit signal diffusion scheme given in (2.7). Otherwise, the diffusion equation (2.7) can be solved semi-implicitly using the Crank-Nicolson method. This improvement has not yet been implemented.

Readers familiar with finite element methods may notice that the discretization given by (2.4) is the same expression for a finite element discretization using piecewise elements on the underlying Delaunay triangulation [2]. Using the Voronoi diagram formulation instead of finite elements is advantageous in the context of modeling mechanochemical coupling because the width of the Voronoi faces will also be used for weighting of forces due to mechanical interactions. These forces will be discussed in Section 3.

2.2. Updating of the Voronoi diagram and Delaunay triangulation. The initial Delaunay triangulation is computed using the code of Tipper [23] or Shewchuk [21] and requires $\mathcal{O}(N \log N)$ steps to compute. Since the nodes move continuously, however, successive updates of both the Delaunay triangulation and the Voronoi diagram can be computed locally for each node and therefore require only $\mathcal{O}(N)$ work in each time step. We detail the updating scheme in this section.

For each node j with position $\mathbf{A}_j$, the Tipper code generates a list of the indices $\mathcal{N}_j \equiv \{i_1, i_2 \ldots i_D\}$ of the Delaunay neighbors of j such that $\mathbf{A}_{i_1}, \ldots \mathbf{A}_{i_D}$ proceed in a clockwise fashion around $\mathbf{A}_j$. At each time step, after the nodes have been moved due to the mechanical scheme, the stored

Delaunay triangulation is assumed to be correct. The Voronoi vertices for Ω_j are simply the circumcenters of the successive Delaunay triangles

$$\triangle \mathbf{A}_j \mathbf{A}_{i_1} \mathbf{A}_{i_2},\ \triangle \mathbf{A}_j \mathbf{A}_{i_2} \mathbf{A}_{i_3},\ \ldots\ \triangle \mathbf{A}_j \mathbf{A}_{i_D} \mathbf{A}_{i_1} \tag{2.9}$$

surrounding $\mathbf{A}_j$. The Delaunay neighbor relationships stored from the previous time step are still correct only if the Voronoi vertices of Ω_j will proceed in a clockwise fashion. If a pair of vertices defining the Voronoi edge between nodes j and i_k appear in the wrong order, the Delaunay neighbor relationship between node j and node i_k is no longer valid. This neighbor relationship is deleted and replaced by a neighbor relationship between the two mutual neighbors of j and i_k, that is, between i_{k-1} and i_{k+1}. Figure 4 illustrates this updating strategy.

Another type of Voronoi diagram degeneracy which must be considered is the "encroachment" of an interior node upon the boundary of the domain Ω. If an interior node j makes an angle greater than $\pi/2$ with two successive neighbors which lie on the boundary of Ω, one of the Voronoi vertices of Ω_j will lie outside of Ω, resulting in incorrect computation of shared edge lengths between node j and its boundary neighbors, as well as incorrect computation of the area of Ω_j. This problem can be corrected by inserting an additional boundary node midway between the two encroached boundary nodes. Figure 5 illustrates the encroachment problem as well as the remedy.

The Voronoi vertices that need to be recomputed upon the insertion of a new node form a connected set [2]. Furthermore, one of the vertices of Ω_j is certain to require re-computation; in other words, no global search is required to determine which vertices need to be recomputed, so that a single encroachment fix involves $\mathcal{O}(1)$ operations. At the time of this writing, the encroachment detection scheme has been implemented but the node insertion algorithm has not; no boundary encroachments occur in the test runs described in Section 3.

2.3. Overview of the computational scheme. There are two alternatives for the initialization of the algorithm. In the first, a randomly perturbed hexagonal grid of nodes filling the domain interior is generated, then the code of Tipper [23] is used to compute the Delaunay neighbors of each node in clockwise order. In the second, the boundary of the region is hand-drawn using a developed MATLAB routine, which then invokes the TRIANGLE code of Shewchuk [21]. This external code generates interior nodes which satisfy maximum area and minimum angle constraints for the Delaunay triangles. A second MATLAB routine converts the output of TRIANGLE to a list of Delaunay neighbors, ordered clockwise, associated with each node. In both cases, the initial data consists of nodes $1, \ldots j \ldots N$ with positions $\mathbf{A}_1, \ldots \mathbf{A}_j \ldots \mathbf{A}_N$. $\mathcal{N}_j$ gives the indices of the neighbors of j arranged clockwise. In addition there is a boundary indicator function ∂ such that $\partial(j) = 1$ if $\mathbf{A}_j$ is in $\partial\Omega$ and $\partial(j) = 0$ otherwise.

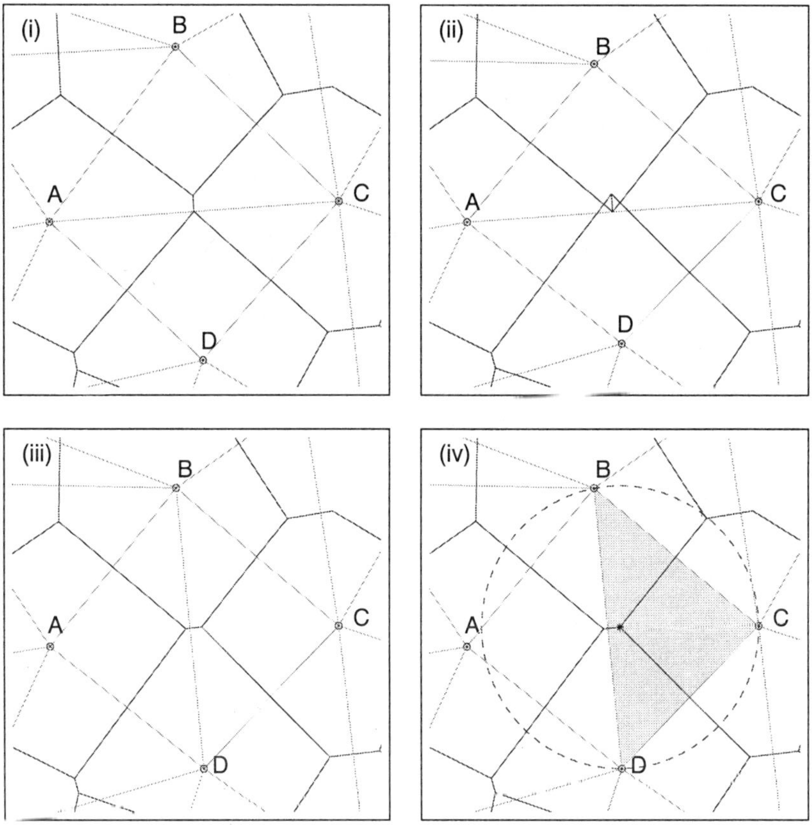

FIG. 4. *Updating of Voronoi diagram and Delaunay triangulation. The correct Delaunay neighbor relationships and Voronoi vertices and edges are shown in (i). After a small displacement of nodes B and D toward each other (ii), the Voronoi polygons for A and C self-intersect, indicating that the mutual Delaunay neighbor relationship between A and C has become "stale" and needs to be replaced with a relationship between B and D, as shown in (iii). The magnitude of the degeneracy is exaggerated for illustrative purposes; in the actual implementation the neighbors are switched at a much lower tolerance, preventing the accumulation of significant errors due to the spurious edges. Notwithstanding these tiny errors, the sudden switching of neighbors does not introduce a discontinuity because at the moment of switching, the Voronoi edge lengths — and therefore the "weights" — attributed to the neighbor relations in question are zero. In (iv) the Voronoi vertex shared by B, C, and D is expressed as the circumcenter of* $\triangle BCD$.

A single time step of the computational algorithm proceeds as follows:

Step 1. **Update Voronoi tesselation and Delaunay triangulation.** For each node the Voronoi polygon Ω_j is computed based on the Delaunay neighbor information saved from the previous time step. If the Voronoi polygon is degenerate (Section 2.2), the Delaunay

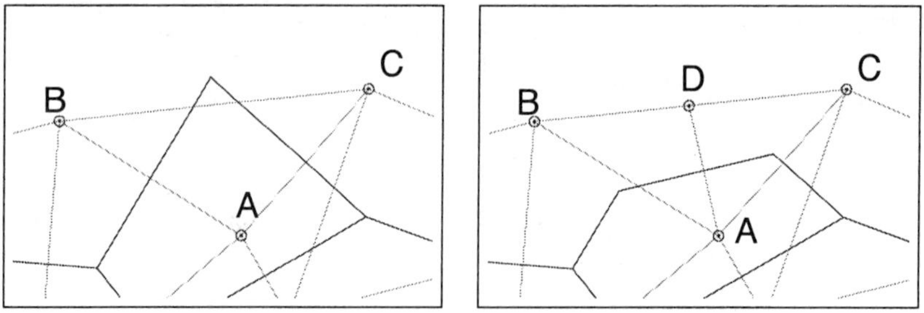

(a) (b)

FIG. 5. *The problem of boundary encroachment. (a): Since $\triangle BAC$ is obtuse, one of the vertices of the Voronoi tile belonging to node A lies outside of the domain. (b): The insertion of node D midway between B and C eliminates the rogue vertex.*

neighbors are updated locally and the Voronoi polygon is recomputed. Switching of Delaunay neighbors corresponds mechanically to the breaking of a link between two previously neighboring nodes and the forming of a new elastic link between two newly neighboring nodes.

Step 2. **Diffusion.** Given the local concentrations c_j^n from the previous time step, we solve the discretized diffusion equation according to (2.7) (forward Euler) to obtain the intermediate values c_j^*.

Step 3. **Reaction.** The reaction dynamics are solved locally by an ODE solver according to equation (2.8) to obtain c_j^{n+1} from c_j^*. In the cell movement application, the reaction dynamics include the production or influx of intracellular signal due to ligand binding to membrane receptors.

Step 4. **Signaling to mechanics.** The mechanical properties of the elastic links joining any two neighboring actin nodes $\mathbf{A}_j$ and $\mathbf{A}_i$ are modified according to local interpolated signal concentration $\bar{c}_{ij} = \frac{1}{2}(c_j + c_i)$. This may affect the resting length ℓ_{ij} or stiffness coefficient κ_{ij} of the elastic element between the two nodes. The value of $\bar{c}_{ij}$ may result directly in protrusive or contractile force generation and therefore the modification of the net force $\mathbf{f}_j$ and $\mathbf{f}_i$ at each node.

Step 5. **Mechanical interactions.** The movement of nodes in the previous time step and the modification of the mechanical properties of the node-node interactions in step Step 4 result in a strain e_{ij} on the link joining each pair of neighboring nodes $\mathbf{A}_j$ and $\mathbf{A}_i$. A penalty force is added to $\mathbf{f}_j$ and $\mathbf{f}_i$ in response to this strain. In the case of a simple Hooke's law approximation, the penalty force added to $\mathbf{f}_j$ is:

$$\sum_{i \in \mathcal{N}_j} \kappa_{ij} e_{ij} \, ||\mathbf{s}_{ij}|| \, \frac{\mathbf{A}_i - \mathbf{A}_j}{||\mathbf{A}_i - \mathbf{A}_j||}. \tag{2.10}$$

Step 6. **Mechanics to signaling.** If we are modeling signal generation in response to mechanical stress (for example, stretch-activated Ca^{2+} channels), we modify c_j and c_i based on the mechanical strain e_{ij} or stress $\kappa_{ij} e_{ij}$ measured at the link between neighboring nodes $\mathbf{A}_j$ and $\mathbf{A}_i$.

Step 7. **Movement of nodes.** In the immersed boundary formulation [19, 5], the net forces $\mathbf{f}_j$ on the nodes are spread to the fluid lattice nodes as the explicit force term $\mathbf{F}(\mathbf{x})$ in the Navier Stokes equations, which are then solved to determine the new fluid velocity $\mathbf{u}(\mathbf{x})$. The nodes are then moved according to:

$$\mathbf{A}_j^{n+1} = \mathbf{A}_j^n + \Delta t \mathbf{u}(\mathbf{A}_j^n). \tag{2.11}$$

3. Preliminary results. We now present test runs of the RDA solver on a deforming domain. The RDA solver has not yet been incorporated into the immersed boundary mechanical model described in [4, 5]; rather, a simplified mechanical scheme is used to accelerate the code for testing purposes.

3.1. Intracellular signaling model. To test the code developed to solve the RDA equations on a deforming domain, we present a simple test problem. A nonconvex, irregularly shaped "blob" boundary and interior nodes were hand-drawn using MATLAB as the input interface. Boundary tension causes the blob to assume a circular shape as the simulation progresses over 30 seconds of simulated time. During the first two seconds of the simulation the component of c corresponding to $[Ca^{2+}]_i$ is increased in the five uppermost boundary nodes in the blob, modeling the influx of Ca^{2+} due to opening of membrane channels. The simulation described below required approximately six minutes of computational time on a Sun Ultra 10 workstation. The parameter values used for this run are summarized in Appendix B, Table 1.

We now summarize the time-stepping scheme for the test run. To emphasize the simplifications and assumptions specific to the test run we follow the outline of the general scheme given in Section 2.3. There are no changes from the original steps of the scheme unless otherwise stated.

Step 1. **Update Voronoi tesselation and Delaunay triangulation.**

Step 2. **Diffusion.** Only the component of c corresponding to $[Ca^{2+}]_i$ diffuses, although the code allows the user to specify nonzero diffusion coefficients for any subset of the components of c. In this run diffusion is treated explicitly according to (2.7).

Step 3. **Reaction.** A temporal increase of $[Ca^{2+}]_i$ is imposed at the uppermost five nodes on the boundary for the first 2 seconds of the

30-second simulation. The Tang-Othmer [17] equations for IP_3-controlled $[Ca^{2+}]_i$ dynamics are used. The second order Runge-Kutta method is used to solve the equations locally. Details are given in Appendix A.

Step 4. **Signaling to mechanics.** There is no feedback from signaling to mechanics.

Step 5. **Mechanical interactions.**

(a) Boundary tension: The resting lengths $\ell_{i,i+1}$ of the links joining two successive boundary nodes are set to $\frac{1}{2}\ell^0$, where ℓ^0 is the initial average distance between boundary nodes. This tension drives the deformation of the blob toward a circular shape. The resulting penalty force $\mathbf{f}_j^t$ is added to the net force $\mathbf{f}_j$ on each node. In contrast to equation (2.10) the tensional forces in this run are not weighted by the shared "area" term $||\mathbf{s}_{ij}||$.

(b) Incompressibility: Due to the simplified mechanics (neglecting hydrodynamics) in the test run, incompressibility of the flow governing the nodes must be enforced locally. Let $A_j^0 \equiv A(\Omega_j^0)$ denote the initial area of Voronoi tile Ω_j. At each time step, we compute the "pressure" on Ω_j by

$$P_j = \frac{A_j^0 - A(\Omega_j)}{A_j^0}. \tag{3.1}$$

The penalty force $\mathbf{f}_i^p$ applied to each neighbor i of node j due to pressure on Ω_j is given by

$$\mathbf{f}_i^p = P_j \, ||\mathbf{s}_{ij}|| \, \frac{\mathbf{A}_i - \mathbf{A}_j}{||\mathbf{A}_i - \mathbf{A}_j||} \tag{3.2}$$

and the balancing force on node j is

$$\mathbf{f}_j^p = - \sum_{i \in \mathcal{N}_j} \mathbf{f}_i^p. \tag{3.3}$$

The pressure penalty forces $\mathbf{f}_j^p$ are then added to the net force $\mathbf{f}_j$ on each node.

Step 6. **Mechanics to signaling.** There is no direct feedback from mechanical strain to local signal concentrations.

Step 7. **Movement of nodes.** Instead of using the immersed boundary formulation we assume that each node j moves at a velocity proportional to the net force on that node. The negligible effects of inertia at these scales (Re $\ll 1$) result in the instantaneous balance between drag force $\mathbf{f}_j^{drag} = -\mu(d\mathbf{A}_j/dt)$ due to sliding along the substrate and the net force $\mathbf{f}_j$ applied to the node by its neighboring nodes:

$$\mathbf{u}(\mathbf{A}_j) \equiv \frac{\mathbf{f}_j}{\mu}. \tag{3.4}$$

The results of the simulation are shown in Figure 6. The neighbor relationships for many of the nodes change throughout the course of the simulation as the region assumes a more regular, circular shape. The $[Ca^{2+}]_i$ wave propagates through the deforming region as expected based on the scaling between the $[Ca^{2+}]_i$ diffusion coefficient and the domain size. This is a very coarse simulation ($N = 100$ nodes), and we see a "breaking up" of the wave front as the simulation progresses. For the code to work on more generalized and refined node configurations we will need to add a routine to insert boundary nodes to prevent the boundary encroachment problem described in Section 2.2. Nevertheless the results on the coarsely discretized deforming region are promising.

3.2. Multicellular mechanochemical interactions. Intercellular signaling and multicellular mechanical interactions play key roles during important events such as the early stages of embryonic development. The test run described in this section, inspired in part by the one-cell-thick multicellular Dictyostelium slug experiments of Bonner [1], demonstrates the flexibility of the Voronoi diagram-based code for modeling intercellular signaling.

In the run, each cell is represented by a single Voronoi tile. Two-dimensional sheets of many cell types are in fact well-approximated by Voronoi diagrams [15]; for this reason various researchers have used Voronoi diagrams to model mechanical interactions among many cells [15, 25, 22]. To our knowledge, Voronoi diagrams have not previously been used to numerically model coupling between the diffusion and relay of intercellular signals and the coordinated movement of the cell mass.

The multicellular mass is confined to a fixed annular region, to temporarily avoid the issue of modeling the properties of the deformable slime sheath. There are $N = 530$ Voronoi tiles, 157 of which lie on the boundary of the region. The boundary nodes are fixed in space. In this run only a single marked cell exerts active locomotory forces; the other cells are mechanically passive, but chemically excitable.

We now describe the details of a time step in the numerical scheme; for each sub-step, we list only the differences between the current scheme and the deforming blob scheme described in Section 3.1. The parameter values used for this run are summarized in Appendix B, Table 2.

Step 1. **Update Voronoi tesselation and Delaunay triangulation.**
Step 2. **Diffusion.**
Step 3. **Reaction.** A temporal increase of $[Ca^{2+}]_i$ is imposed in a single marked cell, that is, to a single node j^*. Since Ca^{2+} is not the intercellular signal for Dictyostelium, the Ca^{2+} equations should be thought of as a generic model of excitable reaction dynamics.

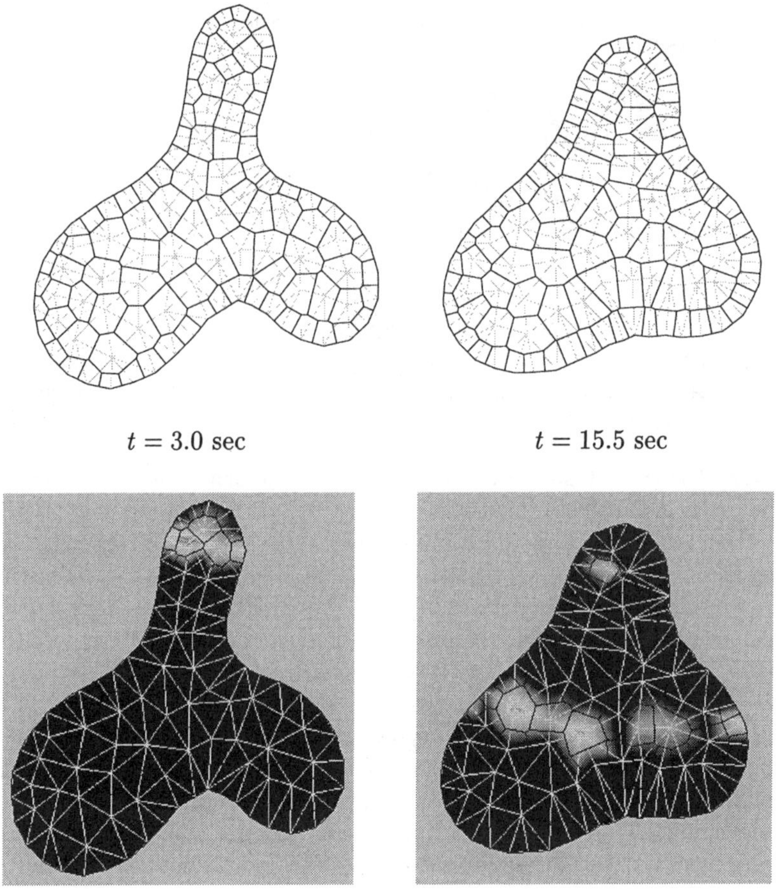

FIG. 6. *Results of the intracellular RDA equations test run. Selected frames at time $t = 3.0$ and $t = 15.5$ seconds. In the top frames the Voronoi diagram is shown to emphasize the changes in neighbor relations that occur throughout the run. In the bottom frames we see that the Ca^{2+} wave has moved from the top portion of the region out to the wider bottom portion.*

Step 4. **Signaling to mechanics.** There is no feedback from signaling to mechanics. The motile cell j^* has a "desired direction" $\vec{d}$ counterclockwise given by

$$\vec{d} = \begin{bmatrix} 0 & -1 \\ 1 & 0 \end{bmatrix} \frac{\mathbf{A}_{j^*}}{||\mathbf{A}_{j^*}||}. \tag{3.5}$$

Step 5. **Mechanical interactions.**

(a) Incompressibility: each cell maintains constant area using the same method as that described in Step 5 in Section 3.1.

(b) Active forces: the motile cell exerts active forces only on its neighboring cells and generates no locomotive force from interactions with the substrate "beneath" it. Where $j = j^*$, we express the Voronoi edge between nodes i and j as $\mathbf{s}_{ij}$ and the projection vector of $\vec{d}$ onto $\mathbf{s}_{ij}$ as $P_{\mathbf{s}_{ij}}\vec{d}$. The active force applied by the motile cell onto its neighbor i is simply $-F_d P_{\mathbf{s}_{ij}}\vec{d}$, where F_d is the locomotive force magnitude constant. The net active force $\mathbf{f}_j^a$ on node $j = j^*$ is therefore

$$\mathbf{f}_j^a = F_d \sum_{i \in \mathcal{N}} P_{\mathbf{s}_{ij}} \vec{d}. \tag{3.6}$$

These forces contribute the net force $\mathbf{f}_j$ on each node.

Step 6. **Mechanics to signaling.** There is no direct feedback from mechanical strain to local signal concentrations.

Step 7. **Movement of nodes.** The boundary nodes are not moved. We again use the Stokes approximation (3.4) to move each interior node proportionally to the net force exerted on it. This movement rule represents an instantaneous balance between the net active and incompressibility forces $\mathbf{f}_j^a + \mathbf{f}_j^p$ and the drag force $\mathbf{f}^{drag}$ due to movement *relative to the substrate* beneath the cells. Cell-cell drag is neglected in this simulation and will need to be incorporated in the future, particularly in three dimensions, where cells in the middle of the cell mass experience drag only from interactions with neighboring cells.

Figure 7 shows two snapshots from the test run. The simulation successfully captures the propagation of excitation waves around the annular region. Since the motile cell moves several cell diameters from its initial position during the run, a significant number of neighbor changes take place. The code's ability to handle these local changes in neighbor topology illustrates the power of the Voronoi diagram formulation.

4. Discussion. We have presented a technique for solving reaction-advection-diffusion (RDA) equations in a deforming irregular domain. This method is well-suited for incorporation into the immersed boundary mechanical cell model [4, 5] because it exploits the pre-existing network of actin nodes in the cell interior. Furthermore, the construction of a Voronoi diagram based on the nodes can enhance the mechanical model by providing a measure of "cross-sectional area" (in two dimensions, the length of the Voronoi edge) shared by two neighboring nodes. For example, the tension forces between two nodes can be scaled proportionally to this shared area term.

While other methods to solve PDEs on an irregular deforming domain have been developed [6], the first order method employed in this paper has the advantage that interactions between neighboring nodes can be understood completely in terms of fluxes across the Voronoi faces. Furthermore,

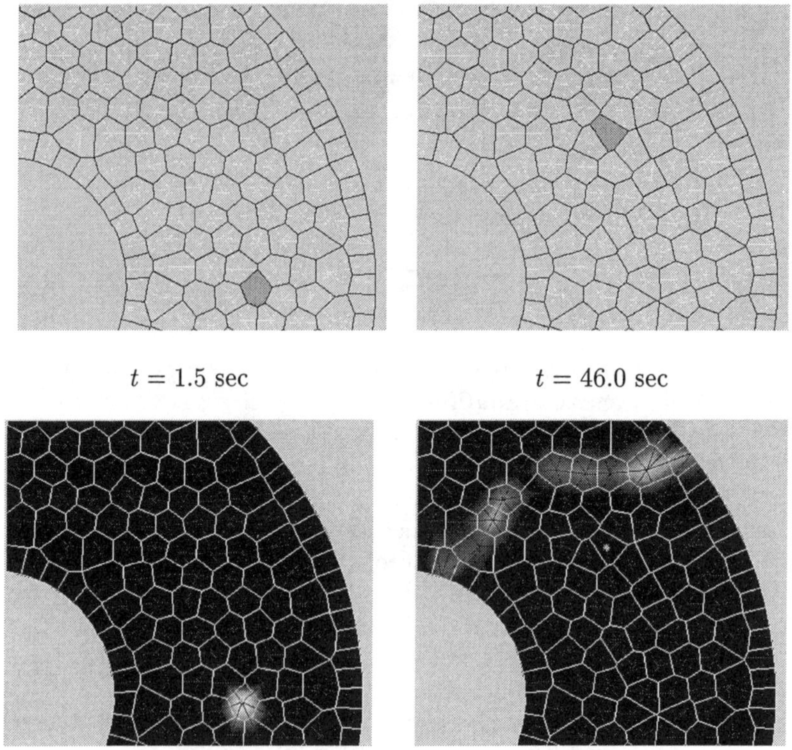

FIG. 7. *Results of the intercellular interaction RDA solver test run. In this run each Voronoi tile represents one cell. The motile cell (shaded, top frames) has an entirely different set of neighboring cells at $t = 46.0$ than it has at $t = 1.5$ sec. At time $t = 1.5$ the marked cell is initiating an excitation wave; at time $t = 46.0$ the wave front has developed and moved ahead of the moving cell. The other wave front has moved downward and off the visible portion of the figure.*

the finite volume approach allows modeling of spatially inhomogeneous cytoplasmic properties by assigning different properties to each Voronoi tile. Such spatially inhomogeneous systems may be difficult to formulate or analyze in terms of PDEs.

Forthcoming improvements to the existing code include the simplification of the user interface, implicit solution of the diffusion equation, and development of the boundary encroachment-fixing algorithm discussed in Section 2.2. More recent implementations of the code include a MATLAB interface which allows the user to hand-draw the region boundary. The external TRIANGLE code [21] is then invoked to triangulate the interior of the region, providing the necessary initialization data for the FORTRAN code described in this paper. The boundary encroachment remedy is required for the code to successfully handle a wide class of deformations; the runs shown in this paper have the property that no interior nodes encroach upon the boundary. In the case of the deforming region test in Section 3.1, boundary encroachments were the rule rather than the exception in preliminary runs. The tendency for the diffusion component of the solver to have instabilities when node spacing is small relative to the time step can be remedied by implementing implicit methods. These improvements will allow the rapid input of complex shapes and the stable solution of the RDA equations on a much larger class of deformations than currently possible.

The next major step toward the goal of an integrated signaling and mechanical model of cell movement is the incorporation of the RDA solver into the immersed boundary mechanical model developed in [4, 5]. This will involve developing models for the production of intracellular signal based upon binding of chemoattractant to membrane bound receptors, the response of the mechanical internode links to local changes in signal concentration, and possibly the coupling of mechanical strain to production and influx of intracellular signals. A simplified model will also be developed which will follow more closely the methods used in the test run on the deforming region in Section 3.1. By replacing the fluid dynamics in the cell interior by simpler movement rules we can greatly reduce the computational expense of simulating mechanical interactions.

Modifications of the same code could possibly be used to model multicellular Dd slug movement, primitive streak formation in avian development, Ca^{2+} wave propagation in glial cells, and pattern formation in two-dimensional cell layers.

4.1. Acknowledgements. I would like to thank Eirikur Palsson, Hans Othmer, Robert Dillon, Micah Dembo, David Eyre, Gary Odell, and John Wagner for their helpful suggestions. The Delaunay triangulation initialization code for the test runs shown in this paper was kindly provided by John C. Tipper. The user interface allowing hand-drawn domains relies on the Delaunay triangulation code TRIANGLE, a nice piece of software freely provided by Jonathan R. Shewchuk. Finally, I thank Sue for her continual patience. This research was supported in part by NSF grant DMS 9805494 and NIH grant GM 29123.

APPENDIX

A. Othmer-Tang IP_3 controlled Ca^{2+} dynamics. The equations used are:

$$\dot{y_1} = -l_1 y_6 y_1 + l_{-1} y_2 \tag{A.1}$$

$$\dot{y_2} = -(l_{-1} + l_2 y_5) y_2 + l_1 y_1 y_6 + l_{-2} y_3 \tag{A.2}$$

$$\dot{y_3} = -(l_{-2} + l_3 y_5) y_3 + l_2 y_5 y_2 + l_{-3}(1 - (y_1 + y_2 + y_3)) \tag{A.3}$$

$$\dot{y_4} = -g_1(y_4 - y_5) + p_{+1} - J_1 \tag{A.4}$$

$$\dot{y_5} = v_r(g_1(y_4 - y_5) - p_{+1} + J_1) + J_j^{in} \tag{A.5}$$

$$\dot{y_6} = m_1(I_0) - m_2 y_6 \tag{A.6}$$

where $y_5 = [\mathrm{Ca}^{2+}]_i$ and

$$J_1 = Ch_1 y_3 (y_4 - y_5) \tag{A.7}$$

$$p_{+1} = \frac{p_1 y_5^2}{y_5^2 + p_2^2} \tag{A.8}$$

$$J_j^{in} = \begin{cases} J^0 & , t < 2 \text{ and } j \in \mathcal{M} \\ 0 & , \text{otherwise.} \end{cases} \tag{A.9}$$

$\mathcal{M}$ is the index set of the five nodes at the top of the test region where Ca^{2+} influx is imposed for the first two seconds of the simulation. The parameter values (see [17] for explanations) used are:

$$\begin{array}{rclrcl} l_1 & = & 120.0 & l_2 & = & 18.0 \\ l_3 & = & 1.0 & p_1 & = & 60.0 \\ q_1 & = & 50.0 & Ch_1 & = & 18.0 \\ m_1 & = & 2.0 & v_r & = & 0.185 \\ I_0 & = & 0.1 & l_{-1} & = & 8.0 \\ l_{-2} & = & 2.45 & l_{-3} & = & 0.16 \\ p_2 & = & 0.04 & m_2 & = & 1.0 \\ g_1 & = & 0.1 & J^0 & = & 0.2. \end{array} \tag{A.10}$$

The initial values used for $y_1, \ldots, y_6$ are

$$\begin{array}{rcl} y_1(0) & = & 0.17797157808109 \\ y_2(0) & = & 0.53391419104710 \\ y_3(0) & = & 0.21467621423065 \\ y_4(0) & = & 9.9203817821662 \\ y_5(0) & = & 0.054729370299254 \\ y_6(0) & = & 0.20000000000000. \end{array} \tag{A.11}$$

Rewriting the system above as $\dot{\mathbf{y}} = \mathbf{G}(t, \mathbf{y})$, at each time step we increment the concentration values using the second order Runge-Kutta method [20]:

$$\mathbf{y}(t + \Delta t) = \mathbf{y}(t) + \Delta t \mathbf{G}\left(t + \frac{1}{2}\Delta t, \mathbf{y}(t) + \frac{1}{2}\Delta t \mathbf{G}\left(\mathbf{y}(t)\right)\right). \tag{A.12}$$

In order to ensure stability the time step Δt must satisfy

$$\left|1 + \lambda \Delta t + \frac{1}{2}(\lambda \Delta t)^2\right| < 1 \tag{A.13}$$

for all eigenvalues λ of the linearization of $\mathbf{G}$. For negative real eigenvalues λ the stability requirement reduces to: $\Delta t < -2/\lambda$.

B. Parameter values for test runs.

TABLE 1

Simulation parameters used for the "intracellular signaling" run described in Section 3.1.

VARIABLE	DESCRIPTION	VALUE
μ	Drag coefficient	100.0 g·cm
$\kappa_{ij} \equiv \kappa$	Link stiffness	0.001 dyne
$D_{calcium}$	Diffusion coefficient for $[Ca^{2+}]_i$	5×10^{-6} cm^2/sec
diam(Ω)	Approximate diameter of region	0.1 cm
N	Number of nodes	100
Δt	Time step	5×10^{-5} sec
t_{end}	Total simulated time	30 sec

TABLE 2

Simulation parameters used for the "multicellular interactions" run described in Section 3.2.

VARIABLE	DESCRIPTION	VALUE
μ	Drag coefficient	100.0 g·cm
$\mathbf{F}_d$	Locomotive force	10.0 dyn
$D_{calcium}$	Diffusion coefficient for $[Ca^{2+}]_i$	5×10^{-6} cm^2/sec
diam(Ω)	Approximate diameter of region	0.2 cm
N	Number of nodes	530
Δt	Time step	5×10^{-5} sec
t_{end}	Total simulated time	60 sec

REFERENCES

[1] JOHN T. BONNER. A way of following individual cells in the migrating slugs of *Dictyostelium discoideum*. *Proc. Nat. Acad. Sci*, **95**:9355–9359, Aug. 1998.

[2] CHRISTOPH BÖRGERS AND CHARLES PESKIN. A Lagrangian fractional step method for the incompressible Navier–Stokes equations on a periodic domain. *J. Comput. Phys.*, **70**:397–438, 1987.

[3] DEAN BOTTINO. *An Immersed Boundary Model of Ameboid Deformation and Locomotion.* PhD thesis, Tulane University, 1996.

[4] DEAN BOTTINO. Modeling viscoelastic networks and cell deformation in the context of the immersed boundary method. *J. Comput. Phys*, **147**(1), 1998.

[5] DEAN BOTTINO AND LISA J. FAUCI. A computational model of ameboid deformation and locomotion. *European Biophysics Journal*, **27**:532–539, 1998.

[6] JEAN BRAUN AND MALCOLM SAMBRIDGE. A numerical method for solving partial differential equations on highly irregular evolving grids. *Nature*, **376**:655–660, Aug. 1995.

[7] DENNIS BRAY. *Cell Movements.* Garland Publishing, New York, 1992.

[8] MICAH DEMBO. Field theories of the cytoplasm. *Comments on Theoretical Biology*, **1**:59–157, 1989.

[9] MICAH DEMBO. On free boundary problems and amoeboid motion. In Nuri Akkas, editor, *Biomechanics of Active Movement and Deformation of Cells*. Springer Verlag, 1994.

[10] MICAH DEMBO AND FRANCIS HARLOW. Cell motion, contractile networks, and the physics of interpenetrating reactive flow. *Biophys. J.*, **50**:109–121, 1986.

[11] ROBERT DILLON AND HANS OTHMER. A mathematical model for outgrowth and spatial patterning of the vertebrate limb bud. *Journal of Theoretical Biology*, **197**(3):295–330, 1999.

[12] K. FANG, E. IONIDES, G. OSTER, R. NUCCITELLI, AND R. ISSEROFF. Receptor redistribution and tyrosine kinase activity of epidermal growth factor receptor regulates directional migration of keratinocytes in DC electric fields. Submitted, 1998.

[13] C. A. J. FLETCHER. *Computational Techniques for Fluid Dynamics: Volume I.* Springer Verlag, second edition, 1991.

[14] M.J. FRITTS, WILLIAM CROWLEY, AND HAROLD TREASE, editors. *The Free-Lagrange method : proceedings of the First International Conference on Free-Lagrange methods.* Springer-Verlag, 1985.

[15] HISAO HONDA. Geometrical models for cells in tissues. *Int. Rev. Cyto.*, **81**:191–248, 1983.

[16] A. OKABE, B. BOOTS, AND K. SUGIHARA. *Spatial Tesselations: Concepts and Applications of Voronoi Diagrams.* Wiley, Chichester, 1992.

[17] HANS OTHMER AND YUANHUA TANG. Oscillations and waves in a model of $InsP_3$–controlled calcium dynamics. In H. Othmer, P. Maini, and J. Murray, editors, *Experimental and Theoretical Advances in Biological Pattern Formation*, pages 277–300. Plenum Press, 1993.

[18] CAROLE PARENT, BRENDA BLACKLOCK, WENDY FROEHLICH, DOUGLAS MURPHY, AND PETER DEVREOTES. G protein signaling events are activated at the leading edge of chemotactic cells. *Cell*, **95**:81–91, 1998.

[19] CHARLES S. PESKIN. Numerical analysis of blood flow in the heart. *Journal of Computational Physics*, **25**(3):220–252, November 1977.

[20] ANTHONY RALSTON AND PHILIP RABINOWITZ. *A First Course in Numerical Analysis.* McGraw Hill, 1978.

[21] JONATHAN RICHARD SHEWCHUK. Triangle: Engineering a 2D Quality Mesh Generator and Delaunay Triangulator. In Ming C. Lin and Dinesh Manocha, editors, *Applied Computational Geometry: Towards Geometric Engineering*, volume 1148 of *Lecture Notes in Computer Science*, pages 203–222. Springer-Verlag, May 1996. From the First ACM Workshop on Applied Computational Geometry.

[22] D. SULSKY, S. CHILDRESS, AND J. K. PERCUS. A model of cell sorting. *J. Theor. Biol.*, **106**:275–301, 1986.

[23] JOHN C. TIPPER. Fortran programs to construct the planar Voronoi diagram. *Computers and Geosciences*, **17**(5):597–632, 1991.

[24] MICHAEL G. VICKER. The regulation of chemotaxis and chemokinesis in *Dictyostelium* amoebae by temporal signals and spatial gradients of cyclic AMP. *J. Cell. Sci.*, **107**:659–667, 1994.

[25] MICHAEL WELIKY AND GEORGE OSTER. The mechanical basis of cell rearrangement: I. epithelial morphogenesis during *Fundulus* epiboly. *Development*, **109**:373–386, 1990.

[26] DEBORAH WESSELS AND DAVID SOLL. Computer–assisted characterization of the behavioral defects of cytoskeletal mutants of *Dictyostelium discoideum.* In David Soll and Deborah Wessels, editors, *Motion Analysis of Living Cells*, pages 101–140. Wiley–Liss, New York, 1998.

[27] SHIGEHIKO YUMURA, KISHIO FURUYA, AND IKUO TAKEUCHI. Intracellular free calcium responses during chemotaxis of *Dictyostelium* cells. *J. Cell. Sci.*, **109**:2673–2678, 1996.

LIST OF WORKSHOP PARTICIPANTS

- Amber Anderson, Department of Mathematics and Statistics, University of New Mexico
- Kevin Anderson, Institute for Mathematics and its Applications, University of Minnesota
- Bruce Ayati, Institute for Mathematics and its Applications, University of Minnesota
- Eshel Ben-Jacob, School of Physics and Astronomy, Tel-Aviv University
- Dean Bottino, Department of Mathematics, University of Utah
- Nicholas Britton, School of Mathematics, University of Bath
- Nicolas Coult, Institute for Mathematics and its Applications, University of Minnesota
- Edmund Crampin, Centre for Mathematical Biology, Oxford University
- Shangbin Cui, Department of Mathematics, Lanzhou University
- John Dallon, Department of Mathematics, Brigham Young University
- Micah Dembo, Department of Biomedical Engineering, Boston University
- Robert Dillon, Department of Mathematics, Washington State University
- Fred Dulles, Institute for Mathematics and its Applications, University of Minnesota
- Pat Fahey, University of Minnesota
- Roseanne M. Ford, Department of Chemical Engineering, University of Virginia
- Avner Friedman, MCIM, University of Minnesota
- Eamonn Gaffney, Mathematical Institute, Centre for Mathematical Biology
- Gabriela Gomes, Department of Biological Sciences, University of Warwick
- Thomas Hillen, Department of Mathematics, University of Utah
- Thomas Hoefer, Theoretical Biolphysics, Institute of Biology, Humboldt University Berlin
- Trachette Jackson, Duke University
- Jaap Kaandorp, Parallel Scientific Computing & Simulation Group, University of Amsterdam
- David Knecht, Department of Biological Sciences, University of Connecticut

- Paul Kulesa, The Sloan Center for Theoretical Neurobiology, California Institute of Technology
- Rene Lefever, Service de Chimie-Physique, CP 231, Universite Libra de Bruxelles
- Herbert Levine, Department of Physics, University of California–San Diego
- Howard Levine, Department of Mathematics, Iowa State University
- Philip Maini, Centre for Mathematical Biology, Mathematical Institute, University of Oxford
- Maia Martcheva, Institute for Mathematics and its Applications, University of Minnesota
- Georgiy Medvedev, Department of Mathematics, Boston University
- Hans Meinhardt, Max-Planck Institut fuer Entwicklungsbiologie, Tübingen
- Alexandra Milik, Institut fuer Angewandte und Numerische Mathematik, Technische Universitaet Wien
- Willard Miller, Institute for Mathematics and its Applications, University of Minnesota
- Atsushi Mochizuki, Department of Biology, Kyushu University
- Alexander Mogilner, Department of Mathematics, University of California-Davis
- Nick Monk, Developmental Genetics Programme University of Sheffield
- Patrick Nelson, Department of Mathematics, Duke University
- Wei-Ming Ni, School of Mathematics, University of Minnesota
- Hans G. Othmer, Department of Mathematics, University of Utah
- Kevin Painter, Department of Mathematics, University of Utah
- Eirikur Palsson, Department of Mathematics, University of Utah
- John Pearson, Applied Theoretical Physics Division, Los Alamos National Laboratory
- Michel Rascle, Laboratoire de Mathematiques, University of Nice
- Kathleen Rogers, Mathematics Department, University of Maryland Baltimore County
- Faustino Sanchez-Garduno, Departamento de Matematicas, Facultad de Ciencias, Cuidad Universitaria
- Ofer Schochet, School of Physics and Astronomy, Tel-Aviv University
- Lee Segel, Department of Applied Mathematics & Computer Science, Weizmann Institute of Science

- David Sharp, Los Alamos National Laboratory
- Todd Shaw, Department of Mathematics, University of Utah
- Jonathan A. Sherratt, Department of Mathematics Heriot-Watt University
- Angela Stevens, Max-Planck-Institute for Mathematics in the Sciences, Leipzig
- Kristin Rae Swanson, Department of Applied Math, University of Washington
- Moxun Tang, School of Math, University of Minnesota
- Robert Tranquilo, Department of Chemical Eng. and Material Sciences, University of Minnesota
- Warren Weckesser, Department of Mathematics, University of Michigan
- Kees Weijer, Department of Anatomy and Physiology, University of Dundee
- Hans Weinberger, Department of Mathematics, University of Minnesota
- Ralf Wittenberg, Department of Mathematics, University of Michigan
- Carla Wofsy, Department of Mathematics University of New Mexico
- David Wollkind, Department of Pure and Applied Mathematics, Washington State University

IMA SUMMER PROGRAMS

1987 Robotics
1988 Signal Processing
1989 Robust Statistics and Diagnostics
1990 Radar and Sonar (June 18–29)
New Directions in Time Series Analysis (July 2–27)
1991 Semiconductors
1992 Environmental Studies: Mathematical, Computational, and Statistical Analysis
1993 Modeling, Mesh Generation, and Adaptive Numerical Methods for Partial Differential Equations
1994 Molecular Biology
1995 Large Scale Optimizations with Applications to Inverse Problems, Optimal Control and Design, and Molecular and Structural Optimization
1996 Emerging Applications of Number Theory (July 15–26)
Theory of Random Sets (August 22–24)
1997 Statistics in the Health Sciences
1998 Coding and Cryptography (July 6–18)
Mathematical Modeling in Industry (July 22–31)
1999 Codes, Systems, and Graphical Models (August 2–13, 1999)
2000 Mathematical Modeling in Industry - A Workshop for Graduate Students (July 19–28)
2001 Geometric Methods in Inverse Problems and PDE Control (July 16–27)

IMA "HOT TOPICS" WORKSHOPS

- Challenges and Opportunities in Genomics: Production, Storage, Mining and Use, April 24–27, 1999
- Decision Making Under Uncertainty: Energy and Environmental Models, July 20–24, 1999,
- Analysis and Modeling of Optical Devices, September 9–10, 1999
- Decision Making under Uncertainty: Assessment of the Reliability of Mathematical Models, September 16–17, 1999
- Scaling Phenomena in Communication Networks, October 22–24, 1999
- Text Mining, April 17–18, 2000
- Mathematical Challenges in Global Positioning Systems (GPS), August 16-18, 2000
- Modeling and Analysis of Noise in Integrated Circuits and Systems, August 29–30, 2000
- Mathematics of the Internet: E-Auction and Markets, December 3–5, 2000
- Analysis and Modeling of Industrial Jetting Processes, January 10–13, 2001

SPRINGER LECTURE NOTES FROM THE IMA

The Mathematics and Physics of Disordered Media
Editors: Barry Hughes and Barry Ninham
(Lecture Notes in Math., Volume 1035, 1983)

Orienting Polymers
Editor: J.L. Ericksen
(Lecture Notes in Math., Volume 1063, 1984)

New Perspectives in Thermodynamics
Editor: James Serrin
(Springer-Verlag, 1986)

Models of Economic Dynamics
Editor: Hugo Sonnenschein
(Lecture Notes in Econ., Volume 264, 1986)

The IMA Volumes in Mathematics and its Applications

Current Volumes:

1 **Homogenization and Effective Moduli of Materials and Media** J. Ericksen, D. Kinderlehrer, R. Kohn, and J.-L. Lions (eds.)
2 **Oscillation Theory, Computation, and Methods of Compensated Compactness** C. Dafermos, J. Ericksen, D. Kinderlehrer, and M. Slemrod (eds.)
3 **Metastability and Incompletely Posed Problems** S. Antman, J. Ericksen, D. Kinderlehrer, and I. Muller (eds.)
4 **Dynamical Problems in Continuum Physics** J. Bona, C. Dafermos, J. Ericksen, and D. Kinderlehrer (eds.)
5 **Theory and Applications of Liquid Crystals** J. Ericksen and D. Kinderlehrer (eds.)
6 **Amorphous Polymers and Non-Newtonian Fluids** C. Dafermos, J. Ericksen, and D. Kinderlehrer (eds.)
7 **Random Media** G. Papanicolaou (ed.)
8 **Percolation Theory and Ergodic Theory of Infinite Particle Systems** H. Kesten (ed.)
9 **Hydrodynamic Behavior and Interacting Particle Systems** G. Papanicolaou (ed.)
10 **Stochastic Differential Systems, Stochastic Control Theory, and Applications** W. Fleming and P.-L. Lions (eds.)
11 **Numerical Simulation in Oil Recovery** M.F. Wheeler (ed.)
12 **Computational Fluid Dynamics and Reacting Gas Flows** B. Engquist, M. Luskin, and A. Majda (eds.)
13 **Numerical Algorithms for Parallel Computer Architectures** M.H. Schultz (ed.)
14 **Mathematical Aspects of Scientific Software** J.R. Rice (ed.)
15 **Mathematical Frontiers in Computational Chemical Physics** D. Truhlar (ed.)
16 **Mathematics in Industrial Problems** A. Friedman
17 **Applications of Combinatorics and Graph Theory to the Biological and Social Sciences** F. Roberts (ed.)
18 ***q*-Series and Partitions** D. Stanton (ed.)
19 **Invariant Theory and Tableaux** D. Stanton (ed.)
20 **Coding Theory and Design Theory Part I: Coding Theory** D. Ray-Chaudhuri (ed.)
21 **Coding Theory and Design Theory Part II: Design Theory** D. Ray-Chaudhuri (ed.)
22 **Signal Processing Part I: Signal Processing Theory** L. Auslander, F.A. Grünbaum, J.W. Helton, T. Kailath, P. Khargonekar, and S. Mitter (eds.)

23 **Signal Processing Part II: Control Theory and Applications of Signal Processing** L. Auslander, F.A. Grünbaum, J.W. Helton, T. Kailath, P. Khargonekar, and S. Mitter (eds.)

24 **Mathematics in Industrial Problems, Part 2** A. Friedman

25 **Solitons in Physics, Mathematics, and Nonlinear Optics**
P.J. Olver and D.H. Sattinger (eds.)

26 **Two Phase Flows and Waves**
D.D. Joseph and D.G. Schaeffer (eds.)

27 **Nonlinear Evolution Equations that Change Type**
B.L. Keyfitz and M. Shearer (eds.)

28 **Computer Aided Proofs in Analysis**
K. Meyer and D. Schmidt (eds.)

29 **Multidimensional Hyperbolic Problems and Computations**
A. Majda and J. Glimm (eds.)

30 **Microlocal Analysis and Nonlinear Waves**
M. Beals, R. Melrose, and J. Rauch (eds.)

31 **Mathematics in Industrial Problems, Part 3** A. Friedman

32 **Radar and Sonar, Part I**
R. Blahut, W. Miller, Jr., and C. Wilcox

33 **Directions in Robust Statistics and Diagnostics: Part I**
W.A. Stahel and S. Weisberg (eds.)

34 **Directions in Robust Statistics and Diagnostics: Part II**
W.A. Stahel and S. Weisberg (eds.)

35 **Dynamical Issues in Combustion Theory**
P. Fife, A. Liñán, and F.A. Williams (eds.)

36 **Computing and Graphics in Statistics**
A. Buja and P. Tukey (eds.)

37 **Patterns and Dynamics in Reactive Media**
H. Swinney, G. Aris, and D. Aronson (eds.)

38 **Mathematics in Industrial Problems, Part 4** A. Friedman

39 **Radar and Sonar, Part II**
F.A. Grünbaum, M. Bernfeld, and R.E. Blahut (eds.)

40 **Nonlinear Phenomena in Atmospheric and Oceanic Sciences**
G.F. Carnevale and R.T. Pierrehumbert (eds.)

41 **Chaotic Processes in the Geological Sciences** D.A. Yuen (ed.)

42 **Partial Differential Equations with Minimal Smoothness and Applications** B. Dahlberg, E. Fabes, R. Fefferman, D. Jerison, C. Kenig, and J. Pipher (eds.)

43 **On the Evolution of Phase Boundaries**
M.E. Gurtin and G.B. McFadden

44 **Twist Mappings and Their Applications**
R. McGehee and K.R. Meyer (eds.)

45 **New Directions in Time Series Analysis, Part I**
D. Brillinger, P. Caines, J. Geweke, E. Parzen, M. Rosenblatt, and M.S. Taqqu (eds.)

46 **New Directions in Time Series Analysis, Part II**
D. Brillinger, P. Caines, J. Geweke, E. Parzen, M. Rosenblatt, and M.S. Taqqu (eds.)
47 **Degenerate Diffusions**
W.-M. Ni, L.A. Peletier, and J.-L. Vazquez (eds.)
48 **Linear Algebra, Markov Chains, and Queueing Models**
C.D. Meyer and R.J. Plemmons (eds.)
49 **Mathematics in Industrial Problems, Part 5** A. Friedman
50 **Combinatorial and Graph-Theoretic Problems in Linear Algebra**
R.A. Brualdi, S. Friedland, and V. Klee (eds.)
51 **Statistical Thermodynamics and Differential Geometry of Microstructured Materials**
H.T. Davis and J.C.C. Nitsche (eds.)
52 **Shock Induced Transitions and Phase Structures in General Media** J.E. Dunn, R. Fosdick, and M. Slemrod (eds.)
53 **Variational and Free Boundary Problems**
A. Friedman and J. Spruck (eds.)
54 **Microstructure and Phase Transitions**
D. Kinderlehrer, R. James, M. Luskin, and J.L. Ericksen (eds.)
55 **Turbulence in Fluid Flows: A Dynamical Systems Approach**
G.R. Sell, C. Foias, and R. Temam (eds.)
56 **Graph Theory and Sparse Matrix Computation**
A. George, J.R. Gilbert, and J.W.H. Liu (eds.)
57 **Mathematics in Industrial Problems, Part 6** A. Friedman
58 **Semiconductors, Part I**
W.M. Coughran, Jr., J. Cole, P. Lloyd, and J. White (eds.)
59 **Semiconductors, Part II**
W.M. Coughran, Jr., J. Cole, P. Lloyd, and J. White (eds.)
60 **Recent Advances in Iterative Methods**
G. Golub, A. Greenbaum, and M. Luskin (eds.)
61 **Free Boundaries in Viscous Flows**
R.A. Brown and S.H. Davis (eds.)
62 **Linear Algebra for Control Theory**
P. Van Dooren and B. Wyman (eds.)
63 **Hamiltonian Dynamical Systems: History, Theory, and Applications**
H.S. Dumas, K.R. Meyer, and D.S. Schmidt (eds.)
64 **Systems and Control Theory for Power Systems**
J.H. Chow, P.V. Kokotovic, R.J. Thomas (eds.)
65 **Mathematical Finance**
M.H.A. Davis, D. Duffie, W.H. Fleming, and S.E. Shreve (eds.)
66 **Robust Control Theory** B.A. Francis and P.P. Khargonekar (eds.)
67 **Mathematics in Industrial Problems, Part 7** A. Friedman
68 **Flow Control** M.D. Gunzburger (ed.)

69 **Linear Algebra for Signal Processing**
A. Bojanczyk and G. Cybenko (eds.)
70 **Control and Optimal Design of Distributed Parameter Systems**
J.E. Lagnese, D.L. Russell, and L.W. White (eds.)
71 **Stochastic Networks** F.P. Kelly and R.J. Williams (eds.)
72 **Discrete Probability and Algorithms**
D. Aldous, P. Diaconis, J. Spencer, and J.M. Steele (eds.)
73 **Discrete Event Systems, Manufacturing Systems, and Communication Networks**
P.R. Kumar and P.P. Varaiya (eds.)
74 **Adaptive Control, Filtering, and Signal Processing**
K.J. Åström, G.C. Goodwin, and P.R. Kumar (eds.)
75 **Modeling, Mesh Generation, and Adaptive Numerical Methods for Partial Differential Equations** I. Babuska, J.E. Flaherty, W.D. Henshaw, J.E. Hopcroft, J.E. Oliger, and T. Tezduyar (eds.)
76 **Random Discrete Structures** D. Aldous and R. Pemantle (eds.)
77 **Nonlinear Stochastic PDEs: Hydrodynamic Limit and Burgers' Turbulence** T. Funaki and W.A. Woyczynski (eds.)
78 **Nonsmooth Analysis and Geometric Methods in Deterministic Optimal Control** B.S. Mordukhovich and H.J. Sussmann (eds.)
79 **Environmental Studies: Mathematical, Computational, and Statistical Analysis** M.F. Wheeler (ed.)
80 **Image Models (and their Speech Model Cousins)**
S.E. Levinson and L. Shepp (eds.)
81 **Genetic Mapping and DNA Sequencing**
T. Speed and M.S. Waterman (eds.)
82 **Mathematical Approaches to Biomolecular Structure and Dynamics**
J.P. Mesirov, K. Schulten, and D. Sumners (eds.)
83 **Mathematics in Industrial Problems, Part 8** A. Friedman
84 **Classical and Modern Branching Processes**
K.B. Athreya and P. Jagers (eds.)
85 **Stochastic Models in Geosystems**
S.A. Molchanov and W.A. Woyczynski (eds.)
86 **Computational Wave Propagation**
B. Engquist and G.A. Kriegsmann (eds.)
87 **Progress in Population Genetics and Human Evolution**
P. Donnelly and S. Tavaré (eds.)
88 **Mathematics in Industrial Problems, Part 9** A. Friedman
89 **Multiparticle Quantum Scattering With Applications to Nuclear, Atomic and Molecular Physics** D.G. Truhlar and B. Simon (eds.)
90 **Inverse Problems in Wave Propagation** G. Chavent, G. Papanicolau, P. Sacks, and W.W. Symes (eds.)
91 **Singularities and Oscillations** J. Rauch and M. Taylor (eds.)

92 **Large-Scale Optimization with Applications, Part I: Optimization in Inverse Problems and Design**
L.T. Biegler, T.F. Coleman, A.R. Conn, and F. Santosa (eds.)

93 **Large-Scale Optimization with Applications, Part II: Optimal Design and Control**
L.T. Biegler, T.F. Coleman, A.R. Conn, and F. Santosa (eds.)

94 **Large-Scale Optimization with Applications, Part III: Molecular Structure and Optimization**
L.T. Biegler, T.F. Coleman, A.R. Conn, and F. Santosa (eds.)

95 **Quasiclassical Methods**
J. Rauch and B. Simon (eds.)

96 **Wave Propagation in Complex Media**
G. Papanicolaou (ed.)

97 **Random Sets: Theory and Applications**
J. Goutsias, R.P.S. Mahler, and H.T. Nguyen (eds.)

98 **Particulate Flows: Processing and Rheology**
D.A. Drew, D.D. Joseph, and S.L. Passman (eds.)

99 **Mathematics of Multiscale Materials** K.M. Golden, G.R. Grimmett, R.D. James, G.W. Milton, and P.N. Sen (eds.)

100 **Mathematics in Industrial Problems, Part 10** A. Friedman

101 **Nonlinear Optical Materials** J.V. Moloney (ed.)

102 **Numerical Methods for Polymeric Systems** S.G. Whittington (ed.)

103 **Topology and Geometry in Polymer Science** S.G. Whittington, D. Sumners, and T. Lodge (eds.)

104 **Essays on Mathematical Robotics** J. Baillieul, S.S. Sastry, and H.J. Sussmann (eds.)

105 **Algorithms For Parallel Processing** M.T. Heath, A. Ranade, and R.S. Schreiber (eds.)

106 **Parallel Processing of Discrete Problems** P.M. Pardalos (ed.)

107 **The Mathematics of Information Coding, Extraction, and Distribution** G. Cybenko, D.P. O'Leary, and J. Rissanen (eds.)

108 **Rational Drug Design** D.G. Truhlar, W. Howe, A.J. Hopfinger, J. Blaney, and R.A. Dammkoehler (eds.)

109 **Emerging Applications of Number Theory** D.A. Hejhal, J. Friedman, M.C. Gutzwiller, and A.M. Odlyzko (eds.)

110 **Computational Radiology and Imaging: Therapy and Diagnostics**
C. Börgers and F. Natterer (eds.)

111 **Evolutionary Algorithms** L.D. Davis, K. De Jong, M.D. Vose, and L.D. Whitley (eds.)

112 **Statistics in Genetics** M.E. Halloran and S. Geisser (eds.)

113 **Grid Generation and Adaptive Algorithms** M.W. Bern, J.E. Flaherty, and M. Luskin (eds.)

114 **Diagnosis and Prediction** S. Geisser (ed.)

115 **Pattern Formation in Continuous and Coupled Systems: A Survey Volume**
M. Golubitsky, D. Luss, and S.H. Strogatz (eds.)

116 **Statistical Models in Epidemiology, the Environment, and Clinical Trials**
M.E. Halloran and D. Berry (eds.)

117 **Structured Adaptive Mesh Refinement (SAMR) Grid Methods**
S.B. Baden, N.P. Chrisochoides, D.B. Gannon, and M.L. Norman (eds.)

118 **Dynamics of Algorithms**
R. de la Llave, L.R. Petzold, and J. Lorenz (eds.)

119 **Numerical Methods for Bifurcation Problems and Large-Scale Dynamical Systems**
E. Doedel and L.S. Tuckerman (eds.)

120 **Parallel Solution of Partial Differential Equations**
P. Bjørstad and M. Luskin (eds.)

121 **Mathematical Models for Biological Pattern Formation**
P.K. Maini and H.G. Othmer (eds.)

Forthcoming Volumes:

1998–1999: Mathematics in Biology

Computational Modeling in Biological Fluid Dynamics

Membrane Transport and Renal Physiology

Mathematical Approaches for Emerging and Reemerging Infectious Diseases

Decision Making under Uncertainty: Energy and Environmental Models

1999 Summer Program: Codes, Systems, and Graphical Models

1999–2000: Reactive Flow and Transport Phenomena

Fire

Confinement and Remediation of Environmental Hazards and Resource Recovery

Dispersive Corrections to Transport Equations, Simulation of Transport in Transition Regimes, Multiscale Models for Surface Evolution and Reacting Flows